电子封装技术丛书
Series of Electronic Packaging Technology

Through-Silicon Vias for 3D Integration

三维电子封装的硅通孔技术

【美】 刘汉诚 （John H. Lau）

中国电子学会电子制造与封装技术分会
《电子封装技术丛书》编辑委员会 组织译审

秦飞 曹立强 译
朱文辉 审

化学工业出版社
·北京·

内 容 提 要

本书系统讨论了用于电子、光电子和微机电系统（MEMS）器件的三维集成硅通孔（TSV）技术的最新进展和可能的演变趋势，详尽讨论了三维集成关键技术中存在的主要工艺问题和潜在解决方案。首先介绍了半导体工业中的纳米技术和三维集成技术的起源和演变历史，然后重点讨论 TSV 制程技术、晶圆减薄与薄晶圆在封装组装过程中的拿持技术、三维堆叠的微凸点制作与组装技术、芯片与芯片键合技术、芯片与晶圆键合技术、晶圆与晶圆键合技术、三维器件集成的热管理技术以及三维集成中的可靠性问题等，最后讨论了具备量产潜力的三维封装技术以及 TSV 技术的未来发展趋势。

本书适合从事电子、光电子、MEMS 等器件三维集成的工程师、科研人员和技术管理人员阅读，也可以作为相关专业大学高年级本科生和研究生教材和参考书。

图书在版编目（CIP）数据

三维电子封装的硅通孔技术/[美]刘汉诚著；秦飞，曹立强译.—北京：化学工业出版社，2014.5
（电子封装技术丛书）
书名原文：Through-Silicon Vias for 3D Integration
ISBN 978-7-122-19897-6

Ⅰ.①三… Ⅱ.①刘…②秦…③曹… Ⅲ.①电子器件-封装工艺 Ⅳ.①TN605

中国版本图书馆 CIP 数据核字（2014）第 036735 号

Through-Silicon Vias for 3D Integration/by John Lau
ISBN 9780071785143

本书中文简体字版由原著作者 John H. Lau 博士授权化学工业出版社独家出版发行。未经许可，不得以任何方式复制或抄袭本书的任何部分，违者必究。
北京市版权局著作权合同登记号：01-2013-8575

责任编辑：吴　刚　　　　　　　　　　文字编辑：孙　科
责任校对：宋　玮　　　　　　　　　　装帧设计：韩　飞

出版发行：化学工业出版社（北京市东城区青年湖南街 13 号　邮政编码 100011）
印　　装：北京虎彩文化传播有限公司
710mm×1000mm　1/16　印张 25¾　字数 533 千字　2014 年 7 月北京第 1 版第 1 次印刷

购书咨询：010-64518888　　　　　　　售后服务：010-64518899
网　　址：http://www.cip.com.cn
凡购买本书，如有缺损质量问题，本社销售中心负责调换。

定　　价：148.00 元　　　　　　　　　　　　　　　版权所有　违者必究

《电子封装技术丛书》编辑委员会

顾　问：俞忠钰　王阳元　邹世昌　杨玉良　许居衍
　　　　于寿文　龚　克　段宝岩　王　曦　汪　敏
　　　　徐晓兰

主　任：毕克允

副主任：汤小川　秦　飞　张蜀平　武　祥

编　委：（以姓氏笔画为序）
　　　　丁冬雁　万里分　马莒生　王　红　王春青
　　　　王新潮　孔令文　石　磊　石明达　田艳红
　　　　史训清　冯小龙　毕克允　朱文辉　朱颂春
　　　　刘　胜　刘兴军　任爱光　李　明　李可为
　　　　李维平　汤小川　杨士勇　杨银堂　杨崇峰
　　　　杨道国　肖　斐　肖胜利　张　弓　张　宏
　　　　张小建　张国旗　张建华　张蜀平　陈长生
　　　　武　祥　尚金堂　罗　乐　郑宏宇　郑学军
　　　　郑晓光　赵　宁　赵元富　贵大勇　禹胜林
　　　　秦　飞　柴广跃　柴志强　恩云飞　徐忠华
　　　　郭永兴　陶建中　曹立强　韩江龙　程　凯
　　　　赖志明　蔡　坚　樊学军

译　序

集成电路产业作为国民经济和社会发展的战略性、基础性和先导性产业，具有极强的创新能力和融合力，已经渗透到人民生活、生产以及社会安全的方方面面。拥有强大的集成电路技术和产业，已成为迈向创新性国家的重要标志。特别是当前云计算、物联网、移动互联网等成为各界关注和投资的热点，没有强大的集成电路产业作为支撑和基础，这些战略性新兴产业无疑是建立在流沙基础之上，产业发展可能面临"空芯化"的局面。迄今为止集成电路技术一直沿着摩尔定律卓有成效地发展，但随着32nm以下线宽技术在单一芯片集成、高密度和多功能等方面进展越来越困难，成本也越来越高，于是出现了新的解决方案，即超越摩尔定律。

三维（3D）集成技术是目前被认定为超越摩尔定律可持续实现小型化、高密度、多功能化的首选解决方案，而硅通孔（TSV）技术，则被认为是三维（3D）集成的核心，即是三维硅（3D Si）集成技术和三维芯片（3D IC）集成技术的关键。TSV 技术具有六大关键工序，可实现芯片与芯片间距离最短、间距最小的互连，与引线键合互连相比有六大优点。

为了适应我国日新月异的电子封装业的发展，满足广大电子封装工程技术人员的迫切需求，中国电子学会电子制造与封装技术分会成立了《电子封装技术丛书》编辑委员会，组织丛书的编译工作。

近年来，丛书编辑委员会已先后组织编写、翻译出版了《集成电路试验手册》（1998年电子工业出版社出版）、《微电子封装手册》（2001年电子工业出版社出版）、《微电子封装技术》（2003年中国科学技术大学出版社出版）、《电子封装材料与工艺》（2006年化学工业出版社出版）、《MEMS/MOEMS封装技术》（2008年化学工业出版社出版）、《电子封装工艺设备》（2012年化学工业出版社出版）、《电子封装与可靠性》（2012年化学工业出版社出版）、《系统级封装导论》（2014年化学工业出版社出版）共八本书籍。《三维电子封装的硅通孔技术》一书是这一系列丛书中第九本。正在编写中的系列丛书之五《光电子封装》，也将于近期出版，以飨读者。

本书译自 John H. Lau（刘汉诚）博士编写的 "Through-Silicon Vias for 3D Integration"，该书的内容涉及半导体技术产业中的纳米技术和三维集成技术的起源和演变历史，并重点讨论 TSV 制作技术、晶圆减薄与薄晶圆在封装过程中的拿持技术、晶圆与晶圆键合技术、三维器件集成的热管理技术及三维集成中的可靠性问题等。最后讨论了量产三维封装技术及 TSV 未来发展趋势。该书对从事电子封装及相关行业的科研、生产、应用工作者都会有较高的使用价值。对高等院校、相关师生也具有一定的参考价值。

我相信本书中译本的出版发行将对我国电子封装产业及系统级集成技术的发展起到积极的推动作用。在本书的翻译出版过程中，长期从事电子封装技术研究的秦飞教授、朱文辉博士、曹立强博士做了许多工作，在此表示由衷的谢意。同时，我也向北京工业大学参与组织该书翻译的全体师生及出版社工作人员，表示衷心的感谢！

译者的话

摩尔定律自 1965 年被提出以来，一直卓有成效地指引电子器件技术的发展方向。但随着 32nm 以下线宽技术的出现，使人们日益认识到在单一芯片集成更高密度的电路和实现更多的功能越来越困难，成本也越来越高，于是出现了"超越摩尔（More than Moore）"的呼声。三维（3D）集成技术目前被认定为是超越摩尔定律，持续实现器件小型化、高密度、多功能化的首选解决方案，而硅通孔（TSV）技术则被认为是 3D 集成的核心。近年来越来越多的企业和研发机构投入大量人力和物力从事 TSV 及其相关技术的研究，以期在未来的 3D 时代取得竞争优势。这些企业的工程师、研发人员和技术管理人员以及研发机构的科学家都迫切需要深入了解 3D 集成 TSV 及其相关技术，如 TSV 制程、晶圆减薄与拿持技术、3D 堆叠的微凸点技术、芯片到芯片键合技术、芯片到晶圆键合技术、晶圆到晶圆键合技术以及 3D 集成中遇到的可靠性问题等。

John H. Lau 博士 2013 年出版的英文版专著"Through-Silicon Vias for 3D Integration"（《三维电子封装的硅通孔技术》）基于最新的研究成果和发展趋势，系统详尽地讨论了 3D 集成 TSV 技术的关键工艺问题及其潜在的解决方案，内容涵盖了 3D 集成技术领域的几乎所有方面。对于希望掌握 3D 集成 TSV 相关技术的人员来说，该书不可不读。

为满足国内企业技术人员和科研人员的需要，中国电子学会电子制造与封装技术分会、《电子封装技术丛书》编辑委员会组织了英文版专著的翻译工作。全书由秦飞、曹立强翻译，由秦飞教授统稿，朱文辉博士对全书内容进行了认真审阅和校正。《电子封装技术丛书》编辑委员会主任毕克允教授对翻译工作给予了大力支持和精心指导。北京工业大学先进电子封装技术与可靠性实验室的安彤、夏国峰、别晓锐、武伟、陈思等参与了图片、参考文献资料的整理和翻译工作。

翻译过程中，力求准确再现英文版的技术细节，并在认真核对的基础上，对英文版中的个别地方进行了补漏。对于英文版中采用的英制单位，中文版中未做处理，但以附录的形式给出了英制和国际单位制之间的换算表，以方便读者。考虑到书中大量使用缩略语，中文版以附录形式列出了缩略语表。

书中不妥之处，敬请读者批评指正。

译　者

英文版序

　　三维（3D）集成技术特别是3D IC集成技术正席卷半导体工业，主要表现在：（1）3D集成技术影响芯片制造、集成电路设计、晶圆制造、器件集成制造、封装与测试、材料与设备制造等众多企业，同时也影响大学和研究机构。（2）3D集成技术吸引了来自全球的研发人员和工程师参加相关的会议、讲座、讨论会和论坛，分享他们的成果、交流信息、学习最新的技术和寻求解决方案，并规划他们的未来。（3）3D集成技术推动半导体产业建立新的标准、新的产业体系和基础设施。这样的事情是前所未有的。

　　这是一个完美风暴。业界认为摩尔定律正在谢幕，3D集成即将登场。为了在将来的竞争中处于优势，众多企业和机构在3D IC集成技术研发方面纷纷投入大量人力和物力。3D IC集成定义为：薄晶圆或转接板通过硅通孔（TSV）和微凸点实现堆叠互连。因此，TSV制作、薄晶圆或芯片的拿持、微凸点制作、键合技术以及热管理成为3D IC集成的关键技术。

　　然而，大多数工程师、技术管理人员、研发人员以及科学家对于TSV制作、薄晶圆的强度测量和拿持、微凸点制作、芯片与芯片（C2C）键合、芯片与晶圆（C2W）键合、晶圆与晶圆（W2W）键合以及3D集成相关的可靠性等问题并没有太深入的了解。因此，工业界和研究机构迫切需要一本能全面介绍这些关键技术领域现状的书籍。本书不仅可以使读者尽快了解3D集成相关技术的最新发展和趋势，而且可以帮助企业界技术领导人做出关于3D集成技术的正确决策。

　　为达到上述目标，来自电子与光电子研究实验室的John H. Lau（刘汉诚）博士搜集了大量最新的技术文献，并撰写了《三维电子封装的硅通孔技术》一书。对于半导体工业界、研究机构以及大学来说，这是一本优秀专著。除此以外，对于刚刚进入该领域的人来说，本书提供了3D IC集成、3D Si集成以及3D IC封装的入门知识；对于已涉足3D集成互连设计与工艺的人员来说，本书可作为了解最新技术发展的参考书。

　　本书共11章，涵盖了3D集成技术从基础到最新发展的全部内容。第1章简要讨论半导体工业中的纳米技术和3D集成技术。第2章讲述TSV制程的6个关键工艺步骤：孔制作、介电层沉积、阻挡层和种子层沉积、孔填充、化学机械抛光和铜外露。第3章讨论TSV的力学行为、热行为和电学行为。第4章和第5章分别讨论薄晶圆的强度测量和封装组装中的拿持问题。第6章讨论微凸点制作、组装以及组装中微凸点的可靠性问题。第7章讨论微凸点的电迁移问题。第8章讨论C2C、C2W和W2W的瞬态液相键合技术。第9章讨论3D IC SiP集成技术中的热管理问题。最后，第10章讨论与3D IC集成和3D Si集成相比成本较低并已接近量产的竞争性技术，如 3D 封装技术。

本书提供了 3D Si 集成、3D IC 集成、3D IC 封装以及它们在高密度、高性能、低功耗、宽带宽、轻薄以及绿色产品中应用的最新信息。本书对于希望掌握 TSV 技术、薄晶圆强度测量与拿持技术、微凸点技术、封装与组装技术、热管理技术、成本效益设计技术以及高良率制造工艺技术的专业人员来说，是不可或缺的。本书涵盖了 3D 集成这一快速发展技术领域的所有方面。

3D 集成技术能不能像摩尔定律那样成为未来世界所能依赖的基础呢？本书也许不能回答这个问题，但可以帮助电子与光电子设计与制造人员更好地理解我们当前需要做什么、如何回答这个问题、如何规划未来以及如何促成其变为现实。

Ian Yi-Jen Chan（詹益仁）博士
电子与光电子研究实验室　主任
工业技术研究院　副主席
中国台湾

英文版前言

硅通孔（TSV）技术是三维硅（3D Si）集成技术和三维芯片（3D IC）集成技术的核心和关键。TSV 技术可以实现芯片与芯片间距离最短、间距最小的互连。与引线键合等互连技术相比，TSV 技术的优势包括：（1）更好的电性能；（2）更低的功耗；（3）更宽的带宽；（4）更高的密度；（5）更小的外形尺寸；（6）更轻的质量。

TSV 是一项颠覆性技术。所有的颠覆性技术都要回答的问题是：它将取代什么技术？成本如何？不幸的是，TSV 正试图取代目前高良率、低成本而且最成熟的引线键合技术。制作 TSV 需要 6 个关键工艺：深反应离子刻蚀（DRIE）制作 TSV 孔，等离子增强化学气相沉积（PECVD）制作介电层，物理气相沉积（PVD）制作阻挡层和种子层，电镀铜（Cu）填孔，化学机械抛光（CMP）去除多余 Cu，TSV Cu 外露。可见，与引线键合相比，TSV 技术十分昂贵！然而，正如昂贵的倒装芯片技术一样，由于其独特优势，仍然在高性能、高密度、低功耗以及宽带宽产品中得到应用。

3D Si 集成与 3D IC 集成的另外一些关键技术包括薄晶圆的强度测量和拿持，以及热管理等。3D IC 集成中的关键技术包括晶圆微凸点制作、组装技术以及电迁移问题的处理等。本书将对所有这些方面进行讨论。

本书内容主要有 7 个部分：（1）半导体工业中的纳米技术和 3D 集成技术（第 1 章）；（2）TSV 技术与 TSV 的力学、热学与电学行为（第 2 章和第 3 章）；（3）薄晶圆的强度测量与拿持技术（第 4 章和第 5 章）；（4）晶圆微凸点制作、组装可靠性和电迁移问题（第 6 章和第 7 章）；（5）瞬态液相键合技术（第 8 章）；（6）热管理技术（第 9 章）；（7）3D IC 封装技术（第 10 章）。本书最后在第 11 章给出了到 2020 年该领域发展趋势的几点想法。

第 1 章简要介绍半导体产业中的纳米技术及其展望，并给出 3D Si 集成与 3D IC 集成近期的进展、挑战和发展趋势。此外，还介绍了一些嵌入式 3D IC 集成的例子。

第 2 章详细讨论 TSV 制程的 6 个关键工艺，简要讨论了 3D Si 集成与 3D IC 集成中的 TSV 工艺。TSV 的力学、热学与电学行为在第 3 章中进行了讨论。

第 4 章介绍用于薄晶圆强度测量的应力传感器的设计、制作和校准，讨论晶圆背面减薄（磨削）对 Cu-low-k 芯片力学行为的影响。用于薄晶圆拿持的临时键合与解键合过程中遇到的问题及其解决方法在第 5 章讨论。对无载体的薄晶圆拿持技术也进行了介绍和讨论。

第 6 章讨论晶圆微凸点制作与组装工艺、细节距无铅焊锡接点的可靠性评估以及超细节距微凸点遇到的问题。第 7 章讨论微凸点的电迁移问题及其失效机制。

第 8 章讨论芯片与芯片（C2C）、芯片与晶圆（C2W）以及晶圆与晶圆（W2W）的低温瞬态多相键合方法。还将讨论用于金属间化合物（IMC）观察的扫描电子显微镜（SEM）、离子束聚焦（FIB）、透射电子显微镜（TEM）、X 射线衍射（XDR）、差示扫描量热法（DSC）以及 C 模态扫描超声显微镜（C-SAM）等仪器。

第 9 章讨论 TSV 芯片或转接板对 3D IC SiP 器件热性能的影响和 3D 堆叠存储芯片的热性能。此外，还讨论了 TSV 芯片的厚度对热阻的影响。最后，介绍了用于 3D SiP 热管理的 TSV 和流体微通道技术。

第 10 章给出了阻碍 TSV 技术量产应用的 3D IC 封装技术的最新进展，如 Cu-low-k 芯片堆叠的引线键合技术、采用焊锡凸点的 C2C 技术以及采用焊锡凸点的扇出埋入晶圆级封装（WLP）到芯片互连技术。还讨论了引线键合的可靠性问题。

本书的读者设定为：(1) 正在或即将从事 TSV 技术、薄晶圆拿持技术、晶圆焊锡微凸点制作与组装技术、电迁移控制技术、热管理以及低温 C2C、C2W、W2W 键合技术的研发人员；(2) 在研发过程中已经遇到了 TSV 相关关键技术问题，急于深入了解和解决问题的技术人员；(3) 在产品中不得不使用高性能、高密度、低功耗、宽带宽 3D 集成技术的人。本书还可以作为那些有志于成为未来电子/光电子领域领导者、科学家和工程师的大学高年级本科生和研究生的教材。

应用于 3D Si 集成和 3D IC 集成的 TSV 技术出现了不断增长的挑战性问题，对于面临这些挑战的人来说，我希望本书能成为他们有价值的参考书。我同样希望本书有助于推动 TSV 及其相关技术的研发进程，有助于推动 TSV 技术在 3D 集成产品中的应用。

掌握了 3D 集成 TSV 技术的组织和机构有潜力在电子/光电子产业取得重要进展，并从所研发产品的性能、功能、集成密度、功耗、带宽、品质、尺寸和质量等方面获得利益。希望本书的内容有助于扫清 TSV 相关技术研发中的路障，避免走弯路，并加速在设计、材料和工艺等方面的研发进程。

John H. Lau（刘汉诚）博士

致　谢

　　本书得以出版是许多具有奉献精神的人们共同努力的结果，谢谢他们！特别感谢 McGraw-Hill 出版社的 Bridget Thoreson，Pamela Pelton 和 Stephen Smith，Cenveo 出版社的 Sapna Rastogi 以及 Jame K. Madru，感谢他们坚定的支持和鼓励。特别感谢 Michael Penn 和 Steve Chapman，他们帮我实现了出版此书的梦想。他们不仅批准和资助了这个项目，而且耐心聆听我一再推迟计划的解释，帮助我解决书稿准备过程中出现的各种问题。与他们一起工作，最终把我混乱的手稿变成精美的印刷品，这实在是令人愉悦和富有成就感的经历。

　　本书的素材有多种来源，包括个人、公司和组织。我尝试在书中适当的地方通过引用来体现我得到的这些协助，但显然不可能一一列出对本书出版提供帮助的每一个人。虽然如此，我内心充满对他们的真挚谢意。感谢美国机械工程师协会（ASME）允许本书使用其会议（如 International Intersociety Electronic Packaging Conference）论文集和会刊（如 Journal of Electronic Packaging）中的部分内容；感谢国际电气电子工程师协会（IEEE）允许本书使用其会议（如 Electronic Components and Technology Conference）论文集和会刊（如 Advanced Packaging, Components and Packaging Technologies, and Manufacturing Technology）的部分内容；感谢国际微电子与封装协会（IMAPS）允许本书使用其会议（如 International Symposium on Microelectronics）论文集和会刊（如 International Journal of Microcircuits and Electronic Packaging）的部分内容。

　　感谢我的前雇主，香港科技大学、新加坡微电子研究所（IME）、安捷伦以及 HPL 公司。他们为我提供了良好的、人性化的工作环境，满足了我对工作的渴望，同时还提升了我的职业声誉。感谢 HPL 公司的 Don Rice 博士、安捷伦公司的 Steve Erasmus 博士、新加坡微电子研究所的 Dim-Lee Kwong 教授、香港科技大学的 Ricky Lee 教授，感谢他们的好意和友谊。感谢台湾电子与光电子研究实验室（EOL）主任兼副主席 Ian Yi-Jen 博士，感谢他对我的信任、尊重以及对我在台湾工业技术研究院（ITRI）工作的支持。最后，感谢我的同事们，与他们富有灵感的讨论为本书增色颇多。他们是：Zhang Xiaowu 博士、C. S. Premachandran 先生、Vincent Lee 先生、V. N. Sekhar 博士、D. Pinjala 先生、Tang Gongyue 博士、Ricky Lee 博士、M. S. Zhang 博士、Y. S. Chan 博士、Sharon Lim Pei-Siang 女士、Vempati Srinivasa Rao 先生、Vaidyanathan Kripesh 博士、Juan Milla 先生、Andy Fenner 先生、于大全博士、Aibin Yu 博士、Navas Khan 先生、Li Ling Yan 博士、Won Kyoung Chou 博士、Seung Wock Yoon 博士、Cheryl Selvanayagam 女士、Chai Tai Chong 先生、Shiguo Liu 先生、Charles Vath 先生、

John Doricko 先生、Germaine Yen 女士、朱文辉博士、Jui-Chin Chen 博士、Ching-Kuan Lee 博士、Tao-Chih Chang 博士、Yu-Min Lin 博士、Chau-Jie Zhan 博士、Pei-Jer Tzeng 博士、Cha-Hsin Lin 博士、Shin-Yi Huang 博士、Chun-Hsien Chien 先生、Chien-Ying Wu 先生、Yu-Chen Hsin 女士、Shang-Chun Chen 先生、Chien-Chou Chen 先生、Hsiang-Hung Chang 先生、Jing-Ye Juang 先生、Wen-Li Tsai 先生、Chia-Wen Chiang 女士、Cheng-Ta Ko 先生、Ra-Min Tain 博士、Heng-Chieh Chien 博士、Sheng-Tsai Wu 先生、Ming-Ji Dai 先生、Yu-Lin Chao 先生、Shyh-Shyuan Sheu 先生、Zhe-Hui Lin 先生、Jui-Feng Hung 先生、Shih-Hsien Wu 先生、Shinn-Juh Lai 先生、Peng-Shu Chen 先生、Li Li 博士、Yu-Hua Chen 博士、Tai-Hung Chen 先生、Chih-Sheng Lin 先生、Tzu-Kun Ku 博士、Wei-Chung Lo 博士和 Ming-Jer Kao 博士。当然，还要感谢我在台湾工业技术研究院、香港科技大学、新加坡微电子研究所、安捷伦、EPS 公司、HPL 公司以及美国圣蒂亚国家实验室的杰出同事们，感谢他们的帮助、坚强支持和富有灵感的讨论。和他们工作和交往是我一生的荣幸和奇遇，从他们那里我学到很多关于生活的智慧，也学到很多先进 IC 封装和三维 IC 集成的技术。

　　最后，感谢我的女儿 Judy 和我的妻子 Teresa，她们的爱、关心和耐心使我可以静心投入本书的编写工作。她们相信我可以为电子工业做出点贡献，她们的这种信心成为我强大的动力。一想到 Judy、Teresa 和我都很健康，我情不自禁地要感谢上帝的慷慨赐福。

John H. Lau（刘汉诚）博士

作者简介

John H. Lau（刘汉诚）博士，于 2010 年 1 月当选台湾工业技术研究院院士。之前，刘博士曾作为访问教授在香港科技大学工作 1 年，作为新加坡微电子研究所（IME）所属微系统、模组与元器件实验室主任工作 2 年，作为资深科学家在位于加利福尼亚的 HPL、安捷伦公司工作超过 25 年。

刘汉诚博士是电子器件、光电子器件、发光二极管（LED）和微机电系统（MEMS）等领域的著名专家，多年从事器件、基板、封装和印制电路板（PCB）的设计、分析、材料表征、工艺制造、品质与可靠性测试以及热管理等方面工作，尤其专注于表面贴装技术（SMT）、晶圆级倒装芯片封装技术、硅通孔（TSV）技术、三维（3D）IC 集成技术以及 SiP 封装技术。

在超过 36 年的研究、研发与制造业经历中，刘汉诚博士发表了 310 多篇技术论文，编写和出版书籍 120 多章，申请和授权专利 30 多项，并在世界范围内做了 270 多场学术报告。独自或与他人合作编写和出版了 17 部关于 TSV、3D MEMS 封装、3D IC 集成可靠性、先进封装技术、球栅陈列（BGA）封装、芯片尺寸封装（CSP）、载带键合（TAB）、晶圆级倒装芯片封装、高密度互连、板上芯片（COB）、SMT、无铅焊料、钎焊与可靠性等方面的教材。

刘汉诚博士在伊利诺伊大学（香槟校区）获得理论与应用力学博士学位，在不列颠哥伦比亚大学获得第一个硕士学位（结构工程），在威斯康辛大学（麦迪逊）获得第二个硕士学位（工程物理），在费尔莱迪金森大学获得第三个硕士学位（管理科学），在台湾大学获得土木工程专业学士学位。

刘汉诚博士曾担任多家学术期刊编委。这些期刊包括美国机械工程师协会（ASME）会刊 Journal of Electronic Packaging；美国电气电子工程师协会（IEEE）会刊 Components, Packaging, and Manufacturing Technology；Circuit World；Soldering and Surface Mount Technology 等。1990 年至 1995 年，担任 IEEE 电子元件与技术会议（ECTC）主席和技术委员会主席；1987 年至 1992 年，担任 International Electronic Manufacturing Technology Symposium 会议主席和技术委员会主席；1987 年至 2002 年，为 ASME 冬季年会 Solder Mechanics Symposium 会议组织者；为 ASME IMECE 2010 3D IC Integration Symposium 会议组织者；1995 年至 2006 年，担任 IEEE ECTC 会议论文集出版主席。刘汉诚博士曾服务于 IEEE 元件封装与制造技术（CPMT）理事会，并在过去的 11 年里每年都是理事会最杰出的讲师之一。

刘汉诚博士获得 ASME、IEEE、美国制造工程师协会（SME）等协会颁发的多个奖项。刘汉诚博士为 ASME 院士和 IEEE 院士（1994）。

目 录

第1章 半导体工业中的纳米技术和3D集成技术 ························ 1
1.1 引言 ·· 1
1.2 纳米技术 ·· 1
1.2.1 纳米技术的起源 ······································ 1
1.2.2 纳米技术的重要里程碑 ································ 1
1.2.3 石墨烯与电子工业 ···································· 3
1.2.4 纳米技术展望 ·· 3
1.2.5 摩尔定律：电子工业中的纳米技术 ······················ 4
1.3 3D集成技术 ··· 5
1.3.1 TSV技术 ·· 5
1.3.2 3D集成技术的起源 ··································· 7
1.4 3D Si集成技术展望与挑战 ································· 8
1.4.1 3D Si集成技术 ······································· 8
1.4.2 3D Si集成键合组装技术 ······························· 9
1.4.3 3D Si集成技术面临的挑战 ····························· 9
1.4.4 3D Si集成技术展望 ··································· 9
1.5 3D IC集成技术的潜在应用与挑战 ·························· 10
1.5.1 3D IC集成技术的定义 ································ 10
1.5.2 移动电子产品的未来需求 ····························· 10
1.5.3 带宽和宽I/O的定义 ·································· 11
1.5.4 存储带宽 ··· 11
1.5.5 存储芯片堆叠 ······································· 12
1.5.6 宽I/O存储器 ·· 13
1.5.7 宽I/O动态随机存储器（DRAM） ······················· 13
1.5.8 宽I/O接口 ·· 17
1.5.9 2.5D与3D IC集成（无源与有源转接板）技术 ············ 17
1.6 2.5D IC集成（转接板）技术的最新进展 ···················· 18
1.6.1 用作中间基板的转接板 ······························· 18
1.6.2 用于释放应力的转接板 ······························· 20
1.6.3 用作载板的转接板 ··································· 22
1.6.4 用于热管理的转接板 ································· 23
1.7 3D IC集成无源TSV转接板技术的新趋势 ···················· 23
1.7.1 双面贴装空腔式转接板技术 ··························· 24

1.7.2 有机基板开孔式转接板技术 …………………………………… 25
   1.7.3 设计举例 …………………………………………………………… 25
   1.7.4 带散热块的有机基板开孔式转接板技术 …………………… 27
   1.7.5 超低成本转接板 ………………………………………………… 27
   1.7.6 用于热管理的转接板技术 ……………………………………… 28
   1.7.7 用于LED和SiP封装的带埋入式微流体通道的转接板技术 …… 29
 1.8 埋入式3D IC集成技术 …………………………………………………… 32
   1.8.1 带应力释放间隙的半埋入式转接板 ………………………… 33
   1.8.2 用于光电子互连的埋入式3D混合IC集成技术 ……………… 33
 1.9 总结与建议 ………………………………………………………………… 34
 1.10 参考文献 …………………………………………………………………… 35

第2章 TSV 技术 …………………………………………………………………… 39
 2.1 引言 ………………………………………………………………………… 39
 2.2 TSV的发明 ………………………………………………………………… 39
 2.3 采用TSV技术的量产产品 ………………………………………………… 40
 2.4 TSV孔的制作 ……………………………………………………………… 41
   2.4.1 DRIE与激光打孔 ………………………………………………… 41
   2.4.2 制作锥形孔的DRIE工艺 ………………………………………… 44
   2.4.3 制作直孔的DRIE工艺 …………………………………………… 46
 2.5 绝缘层制作 ………………………………………………………………… 56
   2.5.1 热氧化法制作锥形孔绝缘层 …………………………………… 56
   2.5.2 PECVD法制作锥形孔绝缘层 …………………………………… 58
   2.5.3 PECVD法制作直孔绝缘层的实验设计 ……………………… 58
   2.5.4 实验设计结果 ……………………………………………………… 60
   2.5.5 总结与建议 ………………………………………………………… 61
 2.6 阻挡层与种子层制作 ……………………………………………………… 62
   2.6.1 锥形TSV孔的Ti阻挡层与Cu种子层 ………………………… 63
   2.6.2 直TSV孔的Ta阻挡层与Cu种子层 …………………………… 64
   2.6.3 直TSV孔的Ta阻挡层沉积实验与结果 ……………………… 65
   2.6.4 直TSV孔的Cu种子层沉积实验与结果 ……………………… 67
   2.6.5 总结与建议 ………………………………………………………… 67
 2.7 TSV电镀Cu填充 …………………………………………………………… 69
   2.7.1 电镀Cu填充锥形TSV孔 ………………………………………… 69
   2.7.2 电镀Cu填充直TSV孔 …………………………………………… 70
   2.7.3 直TSV盲孔的漏电测试 ………………………………………… 72
   2.7.4 总结与建议 ………………………………………………………… 73
 2.8 残留电镀Cu的化学机械抛光（CMP） ………………………………… 73
   2.8.1 锥形TSV的化学机械抛光 ……………………………………… 73

2.8.2　直TSV的化学机械抛光 ································· 74
　　2.8.3　总结与建议 ··· 82
2.9　TSV Cu外露 ··· 83
　　2.9.1　CMP湿法工艺 ··· 83
　　2.9.2　干法刻蚀工艺 ··· 86
　　2.9.3　总结与建议 ··· 89
2.10　FEOL与BEOL ··· 90
2.11　TSV工艺 ··· 90
　　2.11.1　键合前制孔工艺 ··· 91
　　2.11.2　键合后制孔工艺 ··· 91
　　2.11.3　先孔工艺 ··· 91
　　2.11.4　中孔工艺 ··· 91
　　2.11.5　正面后孔工艺 ··· 91
　　2.11.6　背面后孔工艺 ··· 92
　　2.11.7　无源转接板 ··· 93
　　2.11.8　总结与建议 ··· 93
2.12　参考文献 ··· 94

第3章　TSV的力学、热学与电学行为 ····························· 97
3.1　引言 ··· 97
3.2　SiP封装中TSV的力学行为 ····································· 97
　　3.2.1　有源/无源转接板中TSV的力学行为 ······················· 97
　　3.2.2　可靠性设计（DFR）结果 ································ 100
　　3.2.3　含RDL层的TSV ·· 102
　　3.2.4　总结与建议 ·· 105
3.3　存储芯片堆叠中TSV的力学行为 ······························· 105
　　3.3.1　模型与方法 ·· 105
　　3.3.2　TSV的非线性热应力分析 ································ 106
　　3.3.3　修正的虚拟裂纹闭合技术 ································ 108
　　3.3.4　TSV界面裂纹的能量释放率 ······························ 110
　　3.3.5　TSV界面裂纹能量释放率的参数研究 ···················· 110
　　3.3.6　总结与建议 ·· 115
3.4　TSV的热学行为 ·· 116
　　3.4.1　TSV芯片/转接板的等效热导率 ·························· 116
　　3.4.2　TSV节距对TSV芯片/转接板等效热导率的影响 ········· 119
　　3.4.3　TSV填充材料对TSV芯片/转接板等效热导率的影响 ····· 120
　　3.4.4　TSV Cu填充率对TSV芯片/转接板等效热导率的影响 ···· 120
　　3.4.5　更精确的计算模型 ······································ 123
　　3.4.6　总结与建议 ·· 125

3.5 TSV 的电学性能 ··· 125
　3.5.1 电学结构 ··· 125
　3.5.2 模型与方程 ·· 126
　3.5.3 总结与建议 ·· 127
3.6 盲孔 TSV 的电测试 ·· 128
　3.6.1 测试目的 ··· 128
　3.6.2 测试原理与仪器 ··· 128
　3.6.3 测试方法与结果 ··· 131
　3.6.4 盲孔 TSV 电测试指引 ··· 133
　3.6.5 总结与建议 ·· 136
3.7 参考文献 ·· 136

第 4 章　薄晶圆的强度测量 ·· 140
4.1 引言 ··· 140
4.2 用于薄晶圆强度测量的压阻应力传感器 ···························· 140
　4.2.1 压阻应力传感器及其应用 ·· 140
　4.2.2 压阻应力传感器的设计与制作 ··································· 140
　4.2.3 压阻应力传感器的校准 ··· 142
　4.2.4 背面磨削后晶圆的应力 ··· 144
　4.2.5 切割胶带上晶圆的应力 ··· 149
　4.2.6 总结与建议 ·· 150
4.3 晶圆背面磨削对 Cu-low-k 芯片力学行为的影响 ·············· 151
　4.3.1 实验方法 ··· 151
　4.3.2 实验过程 ··· 152
　4.3.3 结果与讨论 ·· 154
　4.3.4 总结与建议 ·· 160
4.4 参考文献 ·· 161

第 5 章　薄晶圆拿持技术 ·· 163
5.1 引言 ··· 163
5.2 晶圆减薄与薄晶圆拿持 ·· 163
5.3 黏合是关键 ··· 163
5.4 薄晶圆拿持问题与可能的解决方案 ··································· 164
　5.4.1 200mm 薄晶圆的拿持 ·· 165
　5.4.2 300mm 薄晶圆的拿持 ·· 172
5.5 切割胶带对含 Cu/Au 焊盘薄晶圆拿持的影响 ···················· 176
5.6 切割胶带对含有 Cu-Ni-Au 凸点下金属（UBM）薄晶圆拿持的影响 ········ 177
5.7 切割胶带对含 RDL 和焊锡凸点 TSV 转接板薄晶圆拿持的影响 ·········· 178
5.8 薄晶圆拿持的材料与设备 ·· 180
5.9 薄晶圆拿持的黏合剂和工艺指引 ······································ 181

  5.9.1 黏合剂的选择 ………………………………………………………… 181
  5.9.2 薄晶圆拿持的工艺指引 ……………………………………………… 182
 5.10 总结与建议 …………………………………………………………………… 182
 5.11 3M 公司的晶圆支撑系统 …………………………………………………… 183
 5.12 EVG 公司的临时键合与解键合系统 ……………………………………… 186
  5.12.1 临时键合 ……………………………………………………………… 186
  5.12.2 解键合 ………………………………………………………………… 186
 5.13 无载体的薄晶圆拿持技术 …………………………………………………… 187
  5.13.1 基本思路 ……………………………………………………………… 187
  5.13.2 设计与工艺 …………………………………………………………… 187
  5.13.3 总结与建议 …………………………………………………………… 189
 5.14 参考文献 ……………………………………………………………………… 189

第 6 章 微凸点制作、组装与可靠性 …………………………………………… 192

 6.1 引言 ……………………………………………………………………………… 192
 A 部分：晶圆微凸点制作工艺 ……………………………………………………… 193
 6.2 内容概述 ………………………………………………………………………… 193
 6.3 普通焊锡凸点制作的电镀方法 ………………………………………………… 193
 6.4 3D IC 集成 SiP 的组装工艺 …………………………………………………… 194
 6.5 晶圆微凸点制作的电镀方法 …………………………………………………… 194
  6.5.1 测试模型 ………………………………………………………………… 194
  6.5.2 采用共形 Cu 电镀和 Sn 电镀制作晶圆微凸点 ……………………… 195
  6.5.3 采用非共形 Cu 电镀和 Sn 电镀制作晶圆微凸点 …………………… 200
 6.6 制作晶圆微凸点的电镀工艺参数 ……………………………………………… 202
 6.7 总结与建议 ……………………………………………………………………… 203
 B 部分：超细节距晶圆微凸点的制作、组装与可靠性评估 …………………… 203
 6.8 细节距无铅焊锡微凸点 ………………………………………………………… 204
  6.8.1 测试模型 ………………………………………………………………… 204
  6.8.2 微凸点制作 ……………………………………………………………… 204
  6.8.3 微凸点表征 ……………………………………………………………… 205
 6.9 C2C 互连细节距无铅焊锡微凸点的组装 …………………………………… 210
  6.9.1 组装方法、表征方法与可靠性评估方法 …………………………… 210
  6.9.2 C2C 自然回流焊组装工艺 …………………………………………… 211
  6.9.3 C2C 自然回流焊组装工艺效果的表征 ……………………………… 211
  6.9.4 C2C 热压键合（TCB）组装工艺 …………………………………… 212
  6.9.5 C2C 热压键合（TCB）组装工艺效果的表征 ……………………… 214
  6.9.6 组装可靠性评估 ………………………………………………………… 214
 6.10 超细节距晶圆无铅焊锡微凸点的制作 ……………………………………… 219
  6.10.1 测试模型 ……………………………………………………………… 219

  6.10.2 微凸点制作 ·········· 219
  6.10.3 超细节距微凸点的表征 ·········· 219
 6.11 总结与建议 ·········· 221
 6.12 参考文献 ·········· 221

第7章 微凸点的电迁移 224

 7.1 引言 ·········· 224
 7.2 大节距大体积微焊锡接点 ·········· 224
  7.2.1 测试模型与测试方法 ·········· 224
  7.2.2 测试步骤 ·········· 226
  7.2.3 测试前试样的微结构 ·········· 226
  7.2.4 140℃、低电流密度条件下测试后的试样 ·········· 227
  7.2.5 140℃、高电流密度条件下测试后的试样 ·········· 229
  7.2.6 焊锡接点的失效机理 ·········· 231
  7.2.7 总结与建议 ·········· 232
 7.3 小节距小体积微焊锡接点 ·········· 233
  7.3.1 测试模型与方法 ·········· 233
  7.3.2 结果与讨论 ·········· 235
  7.3.3 总结与建议 ·········· 241
 7.4 参考文献 ·········· 241

第8章 芯片到芯片、芯片到晶圆、晶圆到晶圆键合 245

 8.1 引言 ·········· 245
 8.2 低温焊料键合基本原理 ·········· 245
 8.3 低温 C2C 键合 [($SiO_2/Si_3N_4/Ti/Cu$) 到
   ($SiO_2/Si_3N_4/Ti/Cu/In/Sn/Au$)] ·········· 246
  8.3.1 测试模型 ·········· 246
  8.3.2 拉力测试结果 ·········· 248
  8.3.3 X射线衍射与透射电镜观察结果 ·········· 250
 8.4 低温 C2C 键合 [($SiO_2/Ti/Cu/Au/Sn/In/Sn/Au$) 到
   ($SiO_2/Ti/Cu/Sn/In/Sn/Au$)] ·········· 252
  8.4.1 测试模型 ·········· 252
  8.4.2 测试结果评估 ·········· 253
 8.5 低温 C2W 键合 [($SiO_2/Ti/Au/Sn/In/Au$) 到 ($SiO_2/Ti/Au$)] ·········· 254
  8.5.1 焊料设计 ·········· 255
  8.5.2 测试模型 ·········· 255
  8.5.3 用于3D IC芯片堆叠的 InSnAu 低温键合 ·········· 257
  8.5.4 InSnAu IMC 层的 SEM、TEM、XDR、DSC 分析 ·········· 258
  8.5.5 InSnAu IMC 层的弹性模量和硬度 ·········· 259
  8.5.6 三次回流后的 InSnAu IMC 层 ·········· 259

- 8.5.7 InSnAu IMC 层的剪切强度 ……260
- 8.5.8 InSnAu IMC 层的电阻 ……262
- 8.5.9 InSnAu IMC 层的热稳定性 ……263
- 8.5.10 总结与建议 ……264
- 8.6 低温 W2W 键合 [TiCuTiAu 到 TiCuTiAuSnInSnInAu] ……264
  - 8.6.1 测试模型 ……265
  - 8.6.2 测试模型制作 ……265
  - 8.6.3 低温 W2W 键合 ……265
  - 8.6.4 C-SAM 检测 ……267
  - 8.6.5 微结构的 SEM/EDX/FIB/TEM 分析 ……268
  - 8.6.6 氦泄漏率测试与结果 ……271
  - 8.6.7 可靠性测试与结果 ……272
  - 8.6.8 总结与建议 ……273
- 8.7 参考文献 ……275

第 9 章 3D IC 集成的热管理 ……278

- 9.1 引言 ……278
- 9.2 TSV 转接板对 3D SiP 封装热性能的影响 ……279
  - 9.2.1 封装的几何参数与材料的热性能参数 ……279
  - 9.2.2 TSV 转接板对封装热阻的影响 ……280
  - 9.2.3 芯片功率的影响 ……280
  - 9.2.4 TSV 转接板尺寸的影响 ……281
  - 9.2.5 TSV 转接板厚度的影响 ……281
  - 9.2.6 芯片尺寸的影响 ……282
- 9.3 3D 存储芯片堆叠封装的热性能 ……282
  - 9.3.1 均匀热源 3D 堆叠 TSV 芯片的热性能 ……282
  - 9.3.2 非均匀热源 3D 堆叠 TSV 芯片的热性能 ……282
  - 9.3.3 各带一个热源的两个 TSV 芯片 ……283
  - 9.3.4 各带两个热源的两个 TSV 芯片 ……284
  - 9.3.5 交错热源作用下的两个 TSV 芯片 ……285
- 9.4 TSV 芯片厚度对热点温度的影响 ……287
- 9.5 总结与建议 ……287
- 9.6 3D SiP 封装的 TSV 和微通道热管理系统 ……288
  - 9.6.1 测试模型 ……288
  - 9.6.2 测试模型制作 ……289
  - 9.6.3 晶圆到晶圆键合 ……291
  - 9.6.4 热性能与电性能 ……292
  - 9.6.5 品质与可靠性 ……293
  - 9.6.6 总结与建议 ……295

9.7 参考文献 ·· 296

第10章 3D IC封装 299
10.1 引言 299
10.2 TSV 技术与引线键合技术的成本比较 300
10.3 Cu-low-k 芯片堆叠的引线键合 301
10.3.1 测试模型 301
10.3.2 Cu-low-k 焊盘上的应力 301
10.3.3 组装与工艺 304
10.3.4 总结与建议 312
10.4 芯片到芯片的面对面堆叠 313
10.4.1 用于3D IC 封装的 AuSn 互连 313
10.4.2 测试模型 313
10.4.3 C2W 组装 316
10.4.4 C2W 实验设计 319
10.4.5 可靠性测试与结果 322
10.4.6 用于3D IC 封装的 SnAg 互连 323
10.4.7 总结与建议 325
10.5 用于低成本、高性能与高密度 SiP 封装的面对面互连 326
10.5.1 用于超细节距 Cu-low-k 芯片的 Cu 柱互连技术 326
10.5.2 可靠性评估 327
10.5.3 一些新的设计 328
10.6 埋入式晶圆级封装（eWLP）到芯片的互连 328
10.6.1 2D eWLP 与再布线芯片封装（RCP）互连 328
10.6.2 3D eWLP 与再布线芯片封装（RCP）互连 329
10.6.3 总结与建议 329
10.7 引线键合可靠性 330
10.7.1 常用芯片级互连技术 330
10.7.2 力学模型 330
10.7.3 数值结果 332
10.7.4 实验结果 333
10.7.5 关于 Cu 引线的更多结果 334
10.7.6 关于 Au 引线的结果 334
10.7.7 Cu 引线与 Au 引线的应力应变关系 335
10.7.8 总结与建议 336
10.8 参考文献 338

第11章 3D集成的发展趋势 344
11.1 引言 344
11.2 3D Si 集成发展趋势 344

11.3　3D IC 集成发展趋势 …………………………………………………… 345
11.4　参考文献 ……………………………………………………………… 346
附录 A　量度单位换算表 …………………………………………………… 347
附录 B　缩略语表 …………………………………………………………… 351
附录 C　TSV 专利 …………………………………………………………… 355
附录 D　推荐阅读材料 ……………………………………………………… 366
D.1　TSV、3D 集成与可靠性 ……………………………………………… 366
D.2　3D MEMS 与 IC 集成 ………………………………………………… 380
D.3　半导体 IC 封装 ………………………………………………………… 384

第1章 半导体工业中的纳米技术和3D集成技术

1.1 引言

本章简略介绍纳米技术发展过程中重要的里程碑事件，重点关注纳米技术在电子工业中的应用前景。1985年8月9日，诺贝尔物理学奖获得者费曼（Richard Feynman）在东京学习院大学（Gakushuin University）做了一篇题为"未来的计算机（Computing Machines in the Future）"的纪念演讲。在这个演讲中，费曼不仅告诉大家要朝着三维（3D）集成的方向发展，而且还告诉人们如何做到这一点。本书的核心是3D集成技术，重点为3D集成电路（IC）集成及其最近的一些进展和未来的发展趋势。同时，简单介绍3D硅（Si）集成。

1.2 纳米技术

1.2.1 纳米技术的起源

1959年12月29日晚，费曼在加州理工学院为美国物理学会做了一篇题为"There's Plenty of Room at the Bottom"的演讲。虽然后来东京大学的谷口纪男（Norio Taniguchi）教授在1974年的国际生产工程会议论文集（Proceedings of the International Conference on Production Engineering）发表了题为"On the Basic Concept of Nano-Technology"的论文，并在文中首次使用"纳米技术（nanotechnology）"这个词，但费曼的那次演讲仍被公认为是关于纳米技术最早的论述。在谷口纪男教授的论文中，将纳米技术定义为$1\mu m$以下尺度的制造方法。例如，半导体制程工艺中的薄膜沉积和离子束铣，这些工艺均需要在纳米量级进行尺度控制。然而，今天工业界对纳米技术的定义为不超过$0.1\mu m$或100nm的制造技术。

1.2.2 纳米技术的重要里程碑

费曼是一位令人惊讶的预言家，他的见解富有远见。在1959年的演讲中，他告诉大家要把东西做得"越来越小"，正如摩尔（Gordon Moore）在1965年提出的摩尔定律中指明的那样[1]。值得一提，他们两个都出自加州理工学院，费曼曾是那里的教授，而摩尔在那里获得了博士学位。在众多有远见的想法中，费曼在他1959年的演讲中说道："如果我们能够将原子一个一个地排列成我们所需要的形式，那将会发生什么？"自此以后，世界范围内许多科学家开始了相关的研究。这里将一些标志性的成果列举如下：①1974年，第一个分子电子器件专利诞生。②1981年，IBM发明了可以在纳米尺度对结构进行测量和识别的扫描探针显微镜（SPM）。SPM可以实现在一个平面内对单独的分子和原子进行操控，如图1.1(a)所示。③1985年，柯尔（Robert F. Curl）、克鲁托（H. W. Kroto）和斯莫利（R. E. Smalley）发现了巴基球，即包含50~500个碳原子的球形稳定分子团。他们采用激光气化碳技术制得的巴基球如图1.1(b)所示。他们3人于1996年获得诺贝尔化学奖。④1989年，IBM阿尔马登研究中心利用35个氙原子组成了IBM图案。⑤1991年，NEC研究实验室的饭岛澄男（Sumio Ijima）发现了碳纳米管，如图1.1(c)所示。⑥2004年，英特尔（Intel）公司推出了基于90nm工艺技术制作的奔腾四PRESCOTT处理器。虽然没有排布任何原子，但这是电子工业采用纳米技术第一次实现量产，是基于纳米技术延续摩尔定律的重要里程碑。⑦2004年，英国曼彻斯特大学的盖姆（Andre Geim）和诺沃肖洛夫（Konstantin Novoselov）（于2010年获得诺贝尔物理学奖）发现了一种简单的方法用于分离石墨的单原子层，即现在熟知的石墨烯，如图1.1(d)所示。

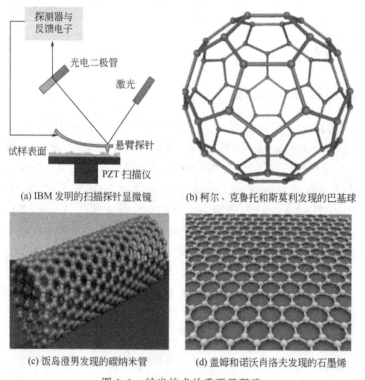

图1.1 纳米技术的重要里程碑

1.2.3 石墨烯与电子工业

盖姆和诺沃肖洛夫发现了如何制作稳定的石墨烯,并制作了只有一个原子厚、由碳原子构成的蜂窝板,这个结果发表在 2004 年 10 月的《科学》杂志上(Science,2004,306:666-669)。自那时起,超过 4000 篇的研究论文引用了他们的文献,并且有超过 2000 篇同石墨烯相关的论文发表在世界最著名的期刊《物理学评论快报》上,而此前只有 20 篇。6 年后,盖姆和诺沃肖洛夫获得了诺贝尔物理学奖(2010 年 10 月)。石墨烯具有比钢材更高的强度、比铜更优越的电性能,同时能够以惊人的速度传递电信号,有可能成为制作更快、功能更强大电子器件的备选材料。但可以肯定地说,将其投入到大规模生产中还面临着许多挑战和难题。不过幸运的是,IBM、三星(Samsung)以及其他电子设备制造商正在积极开展基于石墨烯的电子器件的研究,在《应用物理学快报》上已经有超过 500 篇有关基于石墨烯电子器件的论文。

1.2.4 纳米技术展望

2006 年,贝鲁博(David Berube)在美国国家科学基金会(NSF)的报告"Nano Hype:The Truth Behind the Nanotechnology"中告诫人们要停止所有关于纳米技术的炒作,否则"纳米泡沫"便不远了。需要研发能够量产、具有竞争力的纳米技术产品。对于电子工业来说,基于石墨烯的基板就有可能是其中一个。例如,三星电子正计划利用它来制造柔性显示器。

为了制造能够量产的纳米技术产品,必须研发用于纳米技术器件封装的高性价比、高可靠性的下一代互连技术。目前,从事纳米技术研究的多数科学家(尤其在学术机构和研究机构的科学家)往往忽略了纳米技术的这些特性,致使他们的研究成果与量产应用的距离较大。

纳米技术器件并不是一个孤岛,它必须与具有输入/输出(I/O)互连系统中的其他 IC 或光电元件进行数据交换。除此之外,纳米技术器件必然也是需要通电的,而且其中的埋入式电路和器件单元都很脆弱,这就需要适当的封装为其提供支撑和保护。因此,纳米技术封装的主要功能有:①为器件提供供电电路;②对器件的输入和输出信号进行分配;③对电路产生的热量进行散热;④在不利的环境中为器件提供支撑和保护。所有的这些都解释了为什么半导体行业细分出一个被称作封装测试厂商或外包半导体封装测试(OSAT)厂的关键子行业,封装测试对基于纳米技术的半导体产品的大规模生产起到了支撑作用。

表 1.1 列出了半导体工业中五个主要子行业的最新世界排名。表 1.2 列出了半导体买家,即原始设备制造商(OEM)或代工厂的世界排名。全世界无晶圆集成电路设计公司的排名是:①高通(Qualcomm);②博通(Broadcom);③英伟达(Nvidia);④闪迪(SanDisk);⑤迈威尔(Marvell);⑥罗技(LSI Logic);⑦赛灵思(Xilinx);⑧联发科技(Media Tek);⑨阿尔特拉(Altera);⑩科胜讯(Conexant)。

表 1.1 半导体工业五个主要子行业排名

半导体厂商	晶圆代工厂商	封装测试厂商	基板/PCB厂商	专业电子制造服务商
Intel	TSMC	ASE	Unimicron	Foxconn
Samsung	UMC	Amkor	Ibiden	Flextronics
Texas Instruments	Globalfoundries	SPIL	Samsung	Jabil
Toshiba	SMIC	STATSChipPAC	Nippon Mektron	Celestica
Renesas	Dongbu HiTek	Powertech	CMK	Sanmina-SCI
Qualcomm	TowerJazz	UTAC	Nanya	Cal-Comp
STMicroelectronics	Vanguard	ChipMOS	Shiko	Benchmark
Hynix	IBM	KYEC	KB Group	Elcoteq
Micron	MagnaChip	CARSEM	Compeq	Venture
Broadcom	Samsung	Unisem	Multek	Plexus

注：1. 半导体厂商（不含晶圆代工厂）包含了整合元件制造商（IDM）和无晶圆集成电路设计公司。
2. 整合元件制造商（IDM）设计、制造、并且销售它们的芯片，例如 Intel 和 Samsung。
3. 无晶圆集成电路设计公司设计并且销售它们的芯片，但是把制造外包给晶圆代工厂，例如 Qualcomm、Broadcom 和 Nvidia。
4. 晶圆代工厂制造设计好的芯片并且由其客户（尤其是来自无晶圆集成电路设计公司）负责销售，例如 TSMC 和 UMC。
5. 封装测试厂商或外包半导体封装测试（OSAT）厂，例如 ASE、Amkor 和 SPIL。
6. 专业电子制造服务商（EMS），例如 Foxconn（HH）和 Flextronics。

表 1.2 2010、2011 年半导体买家前十名

2010年排名	2011年排名	公司	2010年[①]	2011年[①]	增长率/%	市场份额/%
3	1	Apple	12819	17257	34.6	5.7
2	2	Samsung Electronics	15272	16681	9.2	5.5
1	3	HP	17585	16618	−5.5	5.5
5	4	Dell	10497	9792	−6.7	3.2
4	5	Nokia	11318	9042	−20.1	3.0
6	6	Sony	9020	8210	−9.0	2.7
7	7	Toshiba	7768	7589	−2.3	2.5
10	8	Lenovo	6091	7537	23.7	2.5
8	9	LG Electronics	6738	6645	−1.4	2.2
9	10	Panasonic	6704	6267	−6.5	2.1
		其他公司	195552	196413	0.4	65.0
		合计	299364	302051	0.9	100.0

① 数据单位为百万美元。源自：Gartner（2012 年 1 月）。

1.2.5 摩尔定律：电子工业中的纳米技术

摩尔定律如图 1.2 所示[1]。今天，32nm 制程技术已经用于大规模生产。台积

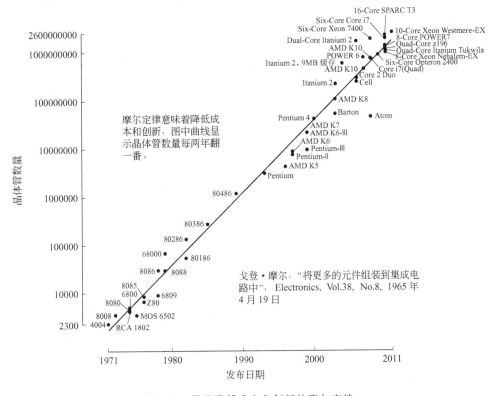

图 1.2 促进降低成本和创新的摩尔定律

电（TSMC）2011 年 9 月宣布已经验证了他们的 28nm 制程技术。三星电子也在 2011 年 9 月 23 日宣布在他们已出货的部分 2GB 动态随机存储器（DRAM）中使用了 20nm 技术。东芝（Toshiba）和闪迪闪存联盟将会很快把 19nm 制程技术应用到他们的 NAND 闪存芯片中。2011 年 2 月 18 日，英特尔宣布将投资 50 亿美元在亚利桑那州的钱德勒附近建造一个 14nm 制程的工厂，同时 10nm 制程已经列入计划。台积电的总裁 Morris Chang 在 2011 年 9 月 2 日宣称摩尔定律仍将有 10 年以上的辉煌期。在过去的 46 年中，摩尔定律可以说是半导体行业降低成本、促进创新的主要推动力。摩尔定律强调通过光刻技术缩小电路线宽或者通过片上系统（SoC）将所有的功能集成到一个芯片的二维平面上。

1.3 3D 集成技术

1.3.1 TSV 技术

本书中，为了将基于硅通孔（Through-Silicon Vias，TSV）技术的晶圆/芯片与传统的遵循摩尔定律的晶圆/芯片加以区分，简称前者为 TSV 晶圆/芯片，后者为摩尔晶圆/芯片。3D 集成定义为将摩尔晶圆/芯片在垂直于晶圆/芯片平面的方向上进行堆叠。如图 1.3 所示，3D 集成包括 3D IC 封装、3D IC 集成和 3D Si 集成[2,3]，它们是互不相同的。一般而言，TSV 可将 3D IC 封装、3D IC 集成和 3D Si 集成区分开来，这是因为后两者都使用了 TSV，而 3D IC 封装却没有。

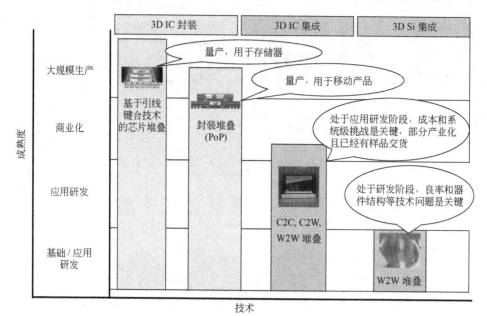

图 1.3　3D 集成技术

TSV 技术是 3D IC 或 3D Si 集成的核心，其带来的新概念是：每个芯片或转接板的正反两面都有可能制作电路。TSV 缩短了芯片和芯片之间的互连距离，同时与摩尔定律也并不冲突。事实上，通过 TSV 将摩尔晶圆/芯片集成到 3D 器件中，可以使产品具备更加出色的电学性能、更小的外观尺寸、更轻的质量，同时也意味着更低的成本。

TSV 是一项颠覆性的技术。与所有的颠覆性技术一样，所面临的问题是"什么将被取代"以及"成本如何"。不幸的是，TSV 技术正试图取代目前最成熟、高良率以及低成本的引线键合技术[4]。TSV 的制程包括 6 个关键的工艺步[2,3]，即采用深反应离子刻蚀（DRIE）进行刻孔，采用等离子体增强化学气相沉积（PECVD）制作介电层，采用物理气相沉积（PVD）制作阻挡层和种子层，采用电镀工艺对孔进行铜（Cu）填充，采用化学机械抛光（CMP）去除覆盖的 Cu，最后是 TSV Cu 外露。因此，与引线键合技术相比，TSV 技术是十分昂贵的。然而，就像采用焊锡微凸点的倒装芯片封装技术一样[5,6]，由于它们与众不同的优势，TSV 将会在很长一段时间内用于高性能和高密度封装。

本章介绍 3D 集成技术的起源，同时也会对 3D IC 和 3D Si 集成技术的发展、挑战、展望进行讨论；最后讨论一些基于无源 TSV 转接板的通用、低成本、散热性能强的 3D IC 集成系统级封装（SiP）。

1.3.2 3D集成技术的起源

由于本书重点关注的是TSV技术，因此，3D IC封装仅在本书的最后（第10章）进行介绍。3D集成是一个非常老的想法[7]，它将两层或者更多层的有源器件通过TSV（过去被称作垂直互连）在垂直方向上集成到一起。绝缘硅（SOI）技术的进步触发了人们3D集成的想法。30多年前，半导体从业者认为摩尔定律到20世纪90年代将会走到尽头（后来的事实证明并非如此），因此，Gat和他的同事们提出了SOI技术[8]。

在20世纪80年代早期，存在两种观点[7]：一种是利用TSV和倒装微凸点技术将芯片堆叠起来，即3D IC集成，如图1.4中右侧所示；另一种则是只利用TSV将晶圆/芯片进行堆叠，这就是无凸点工艺的3D Si集成，如图1.4中左侧所示。3D Si集成与3D IC集成相比其优势在于：①电性能更好；②功耗更低；③外形更薄；④质量更轻；⑤I/O更多。一般而言，工业界更偏爱3D Si集成。

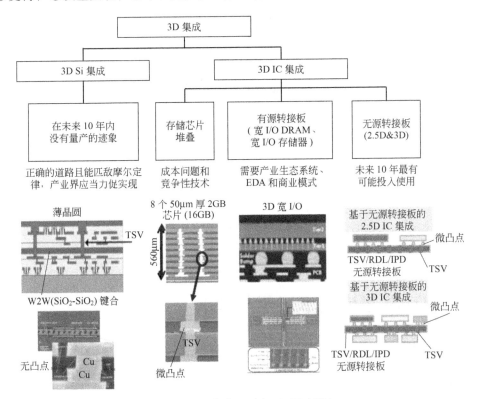

图1.4　3D集成（不含3D IC封装）

对3D集成最强有力的推动者当属1965年诺贝尔物理学奖获得者费曼。1985年8月9日，费曼在东京的学习院大学的演讲中说："提升计算能力的另外一个方法是制作三维的物理器件，代替（现在）将所有东西都集中在芯片表面。开始可以先做几层，随着时间推移，可以加更多的层。"可见在29年前，费曼不仅告诉人们要朝3D方向发展,而且还告诉了人们如何做。费曼接着说:"另一个重要的器件是这样一个装置:它可以自动探测芯片上的失效单元,并使芯片自动重新接线,从而避开这些失效单元而不影响整个芯片的功能。"这再次让我们感到惊讶:费曼在29年前就告诉人们去做内置自测试(BIST)和内置自修复(BISR)器件了。他完全了解3D集成中已知合格芯片(KGD)的重要性,并且他认为这是第一位的事情。图1.5展示了费曼对电子工业中的纳米技术和3D集成技术方面富有远见的贡献。

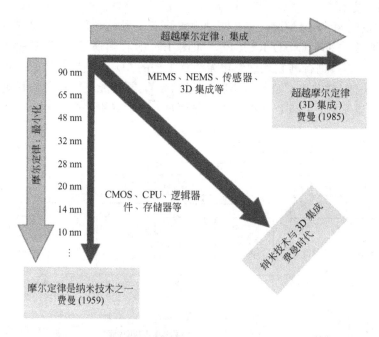

图1.5 费曼对电子工业的纳米技术和3D集成的贡献

1.4 3D Si集成技术展望与挑战

1.4.1 3D Si集成技术

如图1.4左侧所示,3D Si集成是不使用任何焊锡微凸点而将TSV晶圆/芯片堆叠起来。从根本上说,采用无凸点方法进行晶圆到晶圆(W2W)的键合是一种合乎逻辑的方法,但良率是一个大问题。例如,可能将坏芯片强制性键合到好芯片上。除此之外,晶圆/芯片之间无间隙也使得热管理成为问题。而且,对键合的环境条件如表面清洁度、表面平整度以及3D Si集成所需净化间的等级等要求非常高[9~12]。

1.4.2　3D Si 集成键合组装技术

对于 3D Si 集成来说，至少有两种不同的 W2W 无凸点键合方法，分别是 Cu 到 Cu 键合和 SiO_2 到 SiO_2 键合，如图 1.4 中左侧所示。对于 3D Si 集成值得注意和强调的是[9~12]：①它是无凸点的；②对于 Cu 到 Cu 键合，TSV 在键合之前制作；③对于 SiO_2 到 SiO_2 键合，TSV 在键合之后制作；④采用的 TSV 直径非常小（通常不超过 $1\mu m$），因此它的深宽比很大；⑤通常利用 PVD/CVD 工艺用钨（W）或铜（Cu）填充 TSV。除此之外，对于异种器件系统，由于芯片尺寸和引脚的不同，W2W 键合是十分困难的。在这种情况下，芯片到晶圆键合（C2W）是一种可行的方法，同时可以提升良率。此外，由于对中问题，还需要混合键合，如 Cu 到 SiO_2 的键合。

1.4.3　3D Si 集成技术面临的挑战

将 3D Si 集成技术应用于实际生产之前，还有大量工作要做[9~12]。除了热管理、TSV 孔的制作和填充、超薄晶圆的拿持之外，更多的研发工作应致力于降低成本、优化设计和工艺参数、改进键合条件、W2W 键合对中、晶圆变形、检测和测试、键合性能与可靠性、制造良率等问题。另外一个巨大的挑战来自如何以高性价比、高可靠性的系统技术对 3D Si 集成模块进行封装，以实现与下一级结构的互连。

除了技术上的问题，3D Si 集成的核心——电子设计自动化（EDA）还不是很完善，针对 3D Si 集成的产业生态系统（如标准和基础设施等）也需要建立。在这些工作都完成后，EDA 才能将以下指引写入设计、仿真、分析、验证、制造准备以及测试软件[9~12]：

① 从高级描述到布局生成/优化的设计自动化；
② 所有专用 3D 集成的验证；
③ 强调在第三维度方向上不同于封装凸点，即应当是无凸点的；
④ 解决真正第三维度的布局，包括分区、元件平面布置、自动放置及布线；
⑤ 实现第三维度方向的全提取，完整的 3D 设计规则检查（DRC），同一个数据库中所有叠层的 3D 电路布局验证（LVS）；
⑥ 3D Si 集成应当视为分布在不同层中的一个整体系统，而不仅仅是预定义芯片的堆叠。

1.4.4　3D Si 集成技术展望

在接下来的几年中，除了利基（niche）明显的应用外，产业界不太愿意制造基于 3D Si 集成技术的产品。但应该强调的是，3D Si 集成是超越摩尔定律的一条正确道路，产业界应力促其实现[9~12]。

1.5　3D IC 集成技术的潜在应用与挑战

1.5.1　3D IC 集成技术的定义

与 3D Si 集成不同，3D IC 集成（图 1.4 右侧）是将摩尔芯片在垂直方向（z 向）借助 TSV、薄芯片/转接板和微凸点[1,2,9~12]进行堆叠，以实现高性能、低功耗、宽带宽、小尺寸和低成本的目标。与 29 年前不同，现在大部分人倾向于 3D IC 集成，其最有可能的应用如图 1.6 所示。

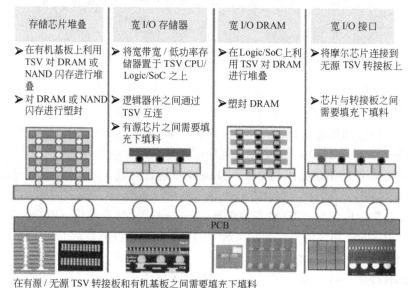

图 1.6　3D IC 集成的潜在应用

1.5.2　移动电子产品的未来需求

现在的智能手机和平板电脑等移动电子产品都是个性化产品，注重使用者的个人体验。因此，设计人员和制造商都急于推出外观更好、功能更多更强的产品，而这需要更丰富的想象力和先进的技术。

移动电子产品的这些特性驱动了性能和封装密度的不断提升。手机和平板电脑都希望做到多任务处理，即拥有 800 万～1200 万像素的摄像头、全高清（HD）视频、3D 游戏，同时还要满足高速、高密度、低功耗和高带宽的要求。例如，平板电脑要具备 2GB 的移动动态随机存储器（DRAM），再加上 128GB 的嵌入式多媒体卡（eMMC）以及 160Mb/s 的运行速度。

智能手机中移动 DRAM 的用量增长是 PC DRAM 的好几倍，2011 年智能手机对 DRAM 的需求量超过了 PC。未来移动 DRAM 的基本要求是：①性能不断提升且带宽更宽，同时功耗更低；②在更薄的封装中实现更高的密度。闪存的基本要求是：①采用 eMMC 实现更快的存储速度和更高的存储密度；②用于轻薄器件的独特固态驱动器（SSD）。今天，有超过 75% 的移动内存由韩国的三星和海力士（Hynix）提供。他们甚至将移动 DRAM 与 NAND 闪存整合在一个多芯片模块上并封装在一起，称之为多芯片封装（MCP）。

对带宽的需求正以每年翻番的速度增长，2012 年的带宽为 25GB/s。在这种情况下宽输入/输出（I/O）是一个解决方案，因此需要 TSV 技术。对于芯片堆叠，TSV 使得内存/处理器/逻辑器件的整体封装成为可能。所以 2013 年是 TSV 的起步之年，2014 年是提升之年，2015 则可能是大规模应用之年。

1.5.3 带宽和宽 I/O 的定义

每秒传递的数据总量被定义为带宽。典型的动态随机存储器（DRAM）具备 4、8、16 或 32bit 的数据位宽（Data width），用于同 CPU/logic/SoC 和（或）外部进行通信。相应地，称这些为×4、×8、×16 或×32bit I/O。宽 I/O 是指×512 比特 I/O 或 512 比特数据位宽，甚至更高。

1.5.4 存储带宽

存储带宽与存储 I/O 数据位宽是成比例的。例如，DDR3（双倍速率同步动态随机存储器，型号 3）-1600 芯片每个 I/O 具备 1600Mb/s 的传输速率。如果这个 DDR3-1600 芯片拥有×32bit I/O，芯片能够达到的总存储带宽为 $32 \times 1600 = 51200Mb/s = 51.2Gb/s$ 或 6.4GB/s[1B(byte)=8b(bit)]。数据位宽越大，存储带宽也就越大。

此外，对于具有宽 I/O 数据位宽的存储芯片，其存储带宽也会随功耗的增加而增加，而且增加的幅度相同。例如，ST-Ericsson 公司的杜马斯（Sophie Dumas，JEDEC JC42.6 的主席）在移动存储研讨会上曾展示，2011 年 6 月，单倍数据倍率（SDR）为 200MHz 的 512bit 宽 I/O LPDDR3 存储器在相同功耗下，其带宽为 2 个单倍数据倍率为 400MHz 的 LPDDR2（低功耗双倍速率同步动态随机存储器，型号 2）的 2 倍（见图 1.7）。

数据位宽受到 IC 封装技术的限制。TSV 技术可以提供非常小的孔径（5~10μm）和节距（20~40μm），有希望实现更宽的 I/O 数据位宽，如 512bit 的数据位宽。另一方面，如果采用引线键合技术（其焊盘尺寸与节距要比 TSV 大许多倍），要达到 512bit 的数据位宽的话，芯片尺寸将会大幅增加，且成本也大幅增加。这就是为什么 TSV 技术在存储带宽方面会如此受关注的原因。如果在一个×512bit 数据位宽的 4-DRAM 堆叠中使用 TSV，那么就可以达到同 DDR3-1600 芯片相当的效果，总的存储带宽为 $512 \times 1600 = 819.2Gb/s = 102.4GB/s$。当然，这种 DRAM 芯片堆叠需要与 Logic/SoC 等器件进行互连来实现这样的

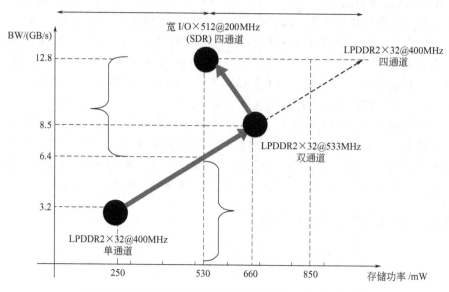

图 1.7　在相同功耗下宽 I/O 数据带宽是 LPDDR2 的两倍
（Sophie Dumas，ST-Ericsson）

带宽。

1.5.5　存储芯片堆叠

图 1.6 左侧为三星公司在 2006 年公布的一款存储芯片堆叠（图 1.8）的简单示意图，这些芯片可能是 I/O 数小于 100（准确地说是 78 个）的 DRAM 或 NAND 闪存。即便是 8 个芯片堆叠在一起，其总高度 560μm 仍然要比一个普通的芯片要薄，这一点值得注意。但是，由于成本以及引线键合技术的竞争，基于 TSV 技术的芯片堆叠现在还没有在消费类产品领域实现量产。例如，苹果的 iPhone 4（http://www.infoneedle.com/posting/99670?snc=20641）采用的就是引线键合技术。当前，三星正在瞄准下一代服务器产品，TSV 最有可能应用于 DDR4（双倍速率同步动态随机存储器，型号 4）。

已知合格芯片（KGD）对 3D 集成来说是重要问题之一，因为它关系到 3D 集成的良率，最终关系到成本问题。对于同种结构，如存储芯片堆叠，良率 y 可由公式 $y=Y^N$ 给出，其中 Y 是芯片良率，N 为堆叠芯片的个数。例如，如果将 8 个同类芯片进行堆叠（$N=8$），同时芯片良率 Y 为 80%，则 3D 芯片堆叠的良率为 17%。但是，如果芯片良率从 80% 提升到 99%，3D 芯片堆叠良率将显著提高到 92%。进一步，如果 $Y=100\%$，即已知堆叠的芯片全部是好芯片，则 3D 芯片堆叠良率是 100%。当然这是在假设组装良率为 100%，且所有可能的缺陷都被检出的情况下。由此可见，"好"或者"高"芯片良率对于 3D 集成来说是十分重要的。

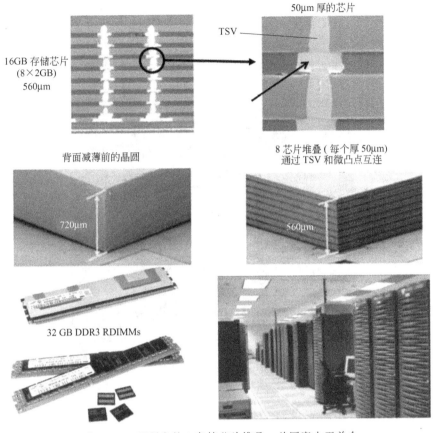

图 1.8 三星研发的 8 存储芯片堆叠，总厚度小于单个普通芯片，可能应用于下一代服务器产品

1.5.6 宽 I/O 存储器

图 1.6 中左边第二列所示为宽 I/O 存储器，包含一个低功耗、宽带宽存储芯片以及几千个接口引脚，这种存储器由 TSV CPU/logic 或 SoC 支撑[13]。由于移动电子产品对这种存储器有需求，现在已经发布了一些样品。如图 1.9 所示为三星生产的一款类似产品。但是，由于相关的产业标准、供应链、商业模式等产业基础条件还不具备，成本也没有竞争性，从样品到量产应用还需要经历一段时间。同时，一些低成本解决方案，如 PoP 技术，也阻碍了 TSV 宽 I/O 存储器在智能手机和平板电脑等消费类产品中的应用。

1.5.7 宽 I/O 动态随机存储器（DRAM）

图 1.6 中右边第二列所示为一类用于移动电子产品的宽 I/O DRAM。三星在过去的三年中已经在这方面发表了许多论文，如文献 [14]，并且于 2011 年在洛杉

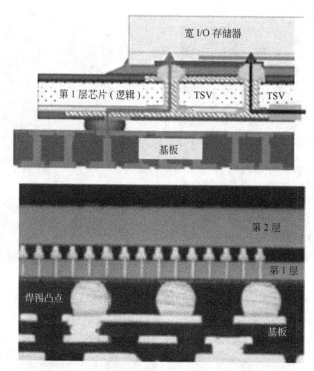

图1.9 三星用于移动电话的宽I/O存储器
(3D半导体集成与封装，Burlingame, CA, 2010年12月)

矶举办的IEEE国际固态电路会议（ISSCC）上展示了一个利用TSV将两层DRAM集成在一个主逻辑芯片上的例子（见图1.10）。这款DRAM的接口引脚数刚好超过了1000个，国际电子器件工程联合会（JEDEC）标准提出将1200个引脚分布在四个通道中（http://www.jedec.org/）。然而，同样由于基础设施欠缺、价格不具竞争性以及低成本嵌入式的PoP等技术的竞争，目前尚未在大规模生产的移动消费类产品中得到应用。

异种器件3D集成的良率y由公式$y=(X^R)(Y^S)(Z^T)$给出，其中X、Y和Z为不同类型芯片的良率，R、S和T为不同类型芯片的数目。例如，如果对宽I/O DRAM进行堆叠，其中包含1个逻辑芯片和4个DRAM，它们芯片良率分别为66%和68%。那么，这个宽I/O DRAM堆叠的良率$y=(0.66)^1(0.68)^4=21\%$。当逻辑芯片和DRAM的良率分别为90%和95%时，堆叠的良率则为$y=(0.90)^1(0.95)^4=73\%$。因此内置自检测和内置自修复功能对于3D Si和3D IC集成的大规模应用来说是极其需要的，电子工业界仍在为此努力。

最近，包括美光科技（Micron）、英特尔、阿尔特拉、三星、Open Silicon、赛灵思以及IBM等公司在内的混合内存立方（Hybrid Memory Cube, HMC）联盟，创造出一整套新的技术（http://www.micron.com/innovations/hmc.html）。最终的结果可能是一个更宽带宽（15倍于DDR3）、更低功耗（每比特功耗比

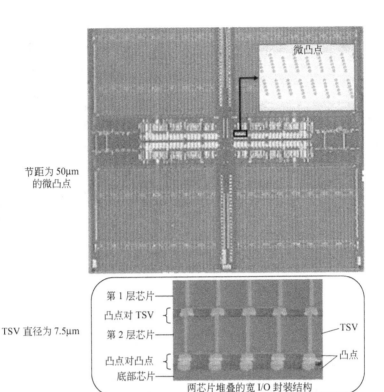

图 1.10 三星用于移动电话的宽 I/O 存储器
(IEEE/ISSCC，2011 年 2 月)

DDR3 低 70%）以及更小外形（比 RDIMMS 小 90%）的器件，如图 1.11(a) 所示。同时，这种产品计划采用 IBM 的一种中孔（via-middle）TSV 技术制造，如图 1.11(b) 所示[15]。图 1.12(a) 为截面示意图，逻辑内存接口（LMI）符合 JEDEC 宽 I/O SDR（JESD229）标准（http://www.jedec.org/），如图 1.12(b) 所示。

图 1.12(b) 显示了宽 I/O 在 LMI 上定义的 4 个存储通道，每个通道的数据带宽为 128bit，因此总共为 512bit。每个通道有 300 个连接器（6 行 50 列），因此共有 1200 个连接器。LMI 的整体面积为 $5.25mm \times 0.52mm$，并位于中心位置。内存立方的底部芯片和逻辑芯片（SoC）采用面对面互连，引脚/焊盘的位置在通道之间是对称分布的，焊盘间距为 $50\mu m$（在 5.25mm 长度方向上）和 $40\mu m$（在 0.52mm 长度方向上）。每个通道包含所有针对通道的控制、电源以及接地，通道间的电源连接是共享的，但是每个通道对时钟、数据的控制都是独立的。DRAM 拥有 1.2V 的互补金属氧化物半导体（CMOS）信号水平，而且不会终止。对于数据传输率为 266Mb/s 的 SDR，总的存储带宽为 17GB/s 或每个通道 4.26GB/s。

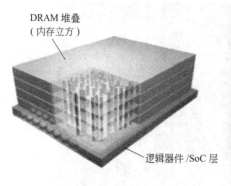
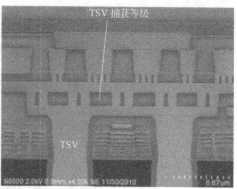

(a)   (b)

图 1.11 (a) 美光科技的混合内存（存储）立方（HMC）；
     (b) IBM 的中孔 TSV 技术

(a)

(b)

图 1.12 (a) HMC 的截面示意图（Dan Skinner，美光科技）；
     (b) JEDEC 标准（JESD229）

1.5.8 宽 I/O 接口

图 1.6 中最右边一列表示的是宽 I/O 接口在 CPU/路由器/通信等领域的应用。其中的摩尔芯片包括内存、专用集成电路（ASIC）以及 CPU 等，采用带有 TSV 和再分布层（RDL）的硅转接板可使其 I/O 数达到几百或上千。图 1.13 所示为来自赛灵思[16~18]的样品，其中的现场可编程门阵列（FPGA）器件采用台积电的 28nm 制程技术制造，转接板是基于 65nm 制程技术。在转接板顶部一共有 4 层 RDL，可以让 4 个 FPGA 在很短的距离内实现通信，转接板中的 TSV 用于连接电源、接地以及传递信号。该样品的带宽提高了 100 倍[16]，并于 2012 年开始交付给赛灵思的客户。

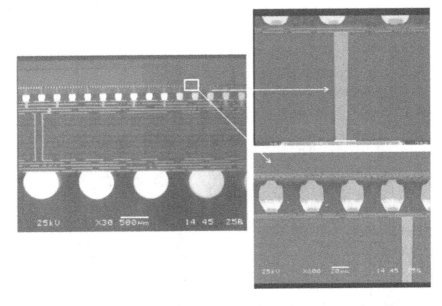

图 1.13 赛灵思/台积电的宽 I/O 接口（IEEE/ECTC，2011 年 5 月）

1.5.9 2.5D 与 3D IC 集成（无源与有源转接板）技术

在图 1.6 中，存储芯片堆叠、宽 I/O 内存和宽 I/O DRAM 都是 3D IC 集成的例子，这是由于有源芯片在 3D 方向被堆叠起来，其中至少一个有源芯片使用了 TSV。宽 I/O 接口是 2.5D 集成的一个例子，这是因为 TSV 制作在一片哑硅片中（即 TSV 无源转接板或称为简单转接板），而并不是制作在其上方的有源摩尔芯片中。由于带有 TSV 的有源 CPU/Logic/SoC 芯片为宽 I/O 内存提供了支撑，带有 TSV 的主逻辑有源芯片为内存立方（宽 I/O DRAM）提供了支撑，因此它们被称为 TSV 有源转接板或称为简单有源转接板[9~12]。

1.6 2.5D IC 集成（转接板）技术的最新进展

对于 2.5D IC 集成，TSV 无源转接板或简单转接板可以用作[9~12]：①中间基板；②应力释放（可靠性）缓冲区；③载板；④热管理工具。关于 TSV 转接板的一些最新进展将在下面讨论。

1.6.1 用作中间基板的转接板

现在最著名的转接板莫过于赛灵思的 FPGA 宽 I/O 接口[16~18]，本书的 1.5.8 节已经提到过。图 1.14 给出了台湾工业技术研究院（ITRI）3D IC 集成 SiP 测试器件的截面图[19]，其中包括一个支撑四个内存芯片的转接板，一个温度传感芯片和一个应力传感芯片。整个器件被塑封。在转接板的顶部和底部都布置有 RDL 层，应力传感器被放置在转接板顶部，集成无源器件（IPD）被制作在转接板体内，转接板平面尺寸为 12.3mm×12.2mm，厚度为 100μm。图 1.15 给出了样品，图 1.16 所示为截面的扫描电子显微镜（SEM）图像以及 SiP 的 X 射线图像[20]。尽管这个器件还不够完美，但作为第一次尝试还是可以接受的。

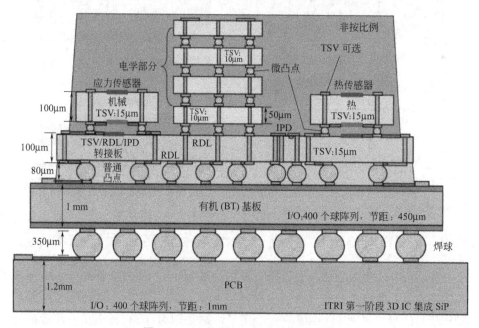

图 1.14 ITRI 的 3D IC 集成测试器件

该测试器件可以蜕变为几种不同的器件：①宽 I/O 存储器——如果封装体中既没有内存芯片堆叠，应力和热传感器芯片也不用 TSV，并且转接板是逻辑电路模块、微处理器或 SoC 之一。②宽 I/O DRAM——如果封装体中没有应力和热传

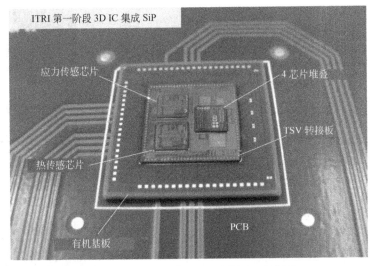

图 1.15 ITRI 的 3D IC 集成测试工具样片

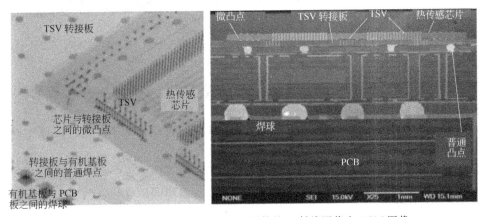

图 1.16 ITRI 的 3D IC 集成测试器件的 X 射线图像和 SEM 图像

感芯片，同时转接板是一个逻辑电路模块。③宽 I/O 接口——如果封装体中没有内存芯片堆叠，应力和热传感器芯片也不用 TSV。

可见，针对该测试器件研发的核心支撑技术（如刻孔、介电层制作、阻挡层和种子层沉积、填孔、CMP、薄晶圆拿持、电路和热管理设计、TSV 测试、超细节距无铅微凸点的制作、C2W 键合、微凸点的电迁移以及微凸点组装的可靠性等）[19~36]是十分有用的，而且有广阔的应用前景。

图 1.17 所示为目前最大的转接板，由新加坡微电子研究所（IME）于 2007 年研发[37,38]，用于支撑特许半导体公司（Charter Semiconductor，现在改名为 Globalfoundries）2006 年基于 65nm Cu-low-k 技术制造的具备 11000 个 I/O 的超大有源芯片（21mm×21mm）。仿真结果[37]显示，由于 Si 和 Cu 的热膨胀系数（CTE）差异较大（Si：2.5×10^{-6}/℃，Cu：17.5×10^{-6}/℃），在竖直方向上，

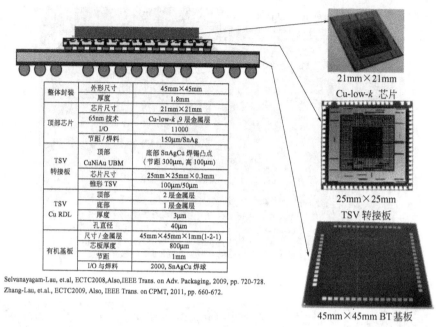

图 1.17　IME 用于支撑 65nm Cu-low-k 芯片的 25mm×25mm 转接板
（2007 年），包含 11000 个 I/O

TSV Cu 会从周围的 Si 中胀出（Cu pumping），如图 1.18 所示。如果 TSV Cu 被 SiO_2 部分覆盖，那么在加热过程中，Cu 胀出可能导致 SiO_2 开裂，如图 1.19 所示。文献 [37] 表明，一般来说，当深宽比（厚度/直径）大于 5 时，深宽比对应力和应变几乎没有影响。

图 1.6 中的宽 I/O 存储器可以重新设计，即把 SoC 放置在 TSV/RDL/IPD 转接板的顶部，把宽带宽和低功耗存储芯片放在 SoC 两侧，这样就没有必要在有源芯片上制作 TSV 了。

1.6.2　用于释放应力的转接板

图 1.20 所示为一个摩尔芯片放置在：①Cu 填充的 TSV 转接板的顶部，转接板放置在有机基板（如双马来酰亚胺三嗪，BT 基板）上；②直接放置在有机基板的顶部。文献 [2]、[37] 和 [38] 已经表明，在摩尔芯片的 Cu-low-k 金属垫处，Cu 填充的 TSV 转接板可以作为应力释放缓冲区，能够把应力水平从 250MPa 降到 125MPa。如果在芯片和转接板中间添加一层专用下填料（填充尺寸非常小），摩尔芯片 Cu-low-k 金属垫处的应力会进一步减小到 42MPa。其原因是由于 BT 基板的热膨胀系数为 $15×10^{-6}/℃$，而 Cu 填充的 TSV 转接板的等效热膨胀系数（与 TSV 孔的数目有关）为 $8\sim10×10^{-6}/℃$。因此，由于 Cu 填充的 TSV 转接板的存在，使得摩尔芯片和转接板之间的热失配程度比芯片和 BT 基板之间的热失配程度要小。

第 1 章 半导体工业中的纳米技术和 3D 集成技术

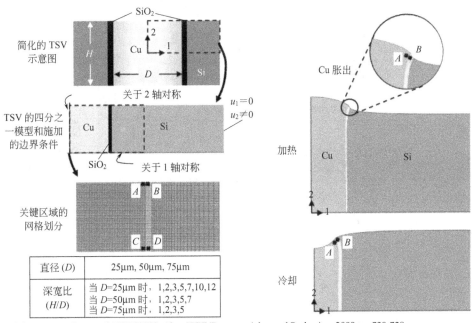

图 1.18 由于局部热失配导致的 TSV Cu 胀出

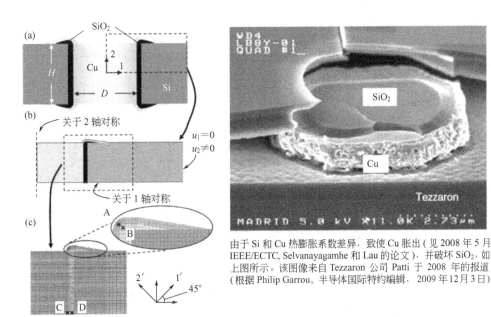

由于 Si 和 Cu 热膨胀系数差异，致使 Cu 胀出（见 2008 年 5 月 IEEE/ECTC, Selvanayagamhe 和 Lau 的论文），并破坏 SiO_2，如上图所示。该图像来自 Tezzaron 公司 Patti 于 2008 年的报道（根据 Philip Garrou，半导体国际特约编辑，2009 年 12 月 3 日）

图 1.19 由于 TSV Cu 胀出导致 SiO_2 破裂

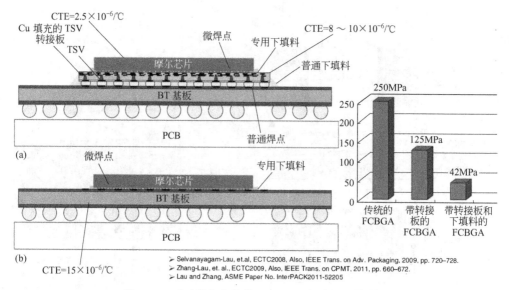

图1.20 对于摩尔芯片的Cu-low-k金属垫，转接板可以作为应力释放（可靠性）缓冲区

1.6.3 用作载板的转接板

图1.21所示为一个3D模组，它包含3个芯片，其中两个芯片堆叠在一起，然后将三个芯片组装在一起[39]。模组的尺寸为12mm×12mm，厚度为1.3mm。硅载板的大小为12mm×12mm×0.2mm，在周围布置了168个孔。底部的硅载板1（carrier1）同一个5mm×5mm的倒装芯片组装在一起，顶部的硅载板2（carrier2）同一个5mm×5mm的倒装芯片以及两个3mm×6mm的引线键合芯片堆叠组装在一起。为了保护引线键合芯片，载板2被再次塑封（利用传递模塑工艺）。

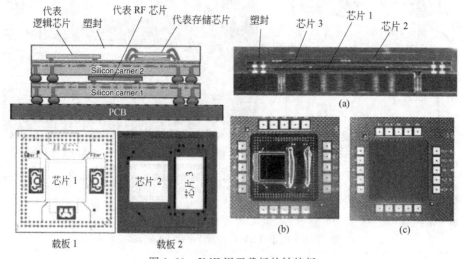

图1.21 IME用于载板的转接板

所有的硅载板都采用 SiO_2 作介电层/钝化层，并且制作了两层金属层。通过载板上的 TSV 来实现电信号的互连。采用直径为 $250\mu m$ 的 SAC305 焊球将载板 1 安装在 FR4 PCB 板上，并填充下填料（在 165℃下固化 3h）。更多相关的技术信息见文献 [39]。

1.6.4 用于热管理的转接板

图 1.22 为两个同样的硅转接板（载板），其中包含用于热管理的微通道以及电信号传导的 TSV[40]。整个系统的大小相当于拳头的四分之一。每个转接板支撑着一个 10mm×10mm 的芯片，同时具备 100W 的散热能力。转接板采用晶圆到晶圆（W2W）键合技术制作，同时在转接板中间埋入经过优化的液体冷却通道结构（采用 DRIE 制作）。利用一个微型泵将冷却水从热交换器抽送到转接板的入口处，经由微通道流至转接板的出口处，然后返回到热交换器，如图 1.22 所示。硅之所以被选作转接板的材料是由于在同一个基板中，硅比较适合利用 DRIE 来制作 TSV 和流体微通道。这两个转接板的不同之处就是底部的硅转接板没有任何出口。TSV 可以设计在转接板的外围处，完成 W2W 键合后，穿过载板的电信号互连通过壁面金属化的 TSV 实现。这种情况下，对 TSV 孔填充 Cu 的工艺就显得不是很有必要。更多的技术细节可以参考文献 [40]。

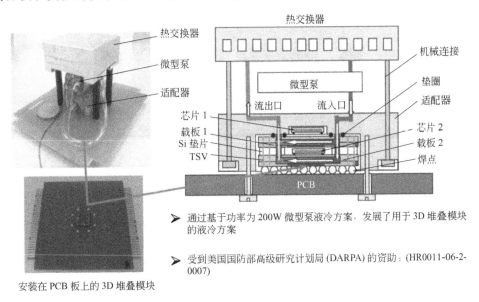

图 1.22　IME 用于热管理的转接板（载板）

1.7　3D IC 集成无源 TSV 转接板技术的新趋势

迄今为止报道的大部分 TSV 无源转接板样品都是将摩尔芯片放置在转接板的

顶部（例如 2.5D IC 集成）。其实可以像两边表面贴装技术（SMT）那样，把摩尔芯片堆叠在 TSV 无源转接板的两侧[9~12,41,42]，以实现更小的转接板外形尺寸、更好的电性能、更轻的质量、更低的功耗、更宽的带宽以及更低的成本。

1.7.1 双面贴装空腔式转接板技术

图 1.6 所示的 3D IC 集成 SiP（宽 I/O 存储器）可以被重新设计为图 1.23 所示的样子，主体为底部开了空腔（可选）的 TSV/RDL/IPD 转接板[9~12]。所有的大功率芯片，如 SoC，微处理器单元（MPU）、图像处理器单元（GPU）、专用 IC（ASIC）、数字信号处理器（DSP）、微控制器单元（MCU）以及射频（RF）芯片都可以按照倒装芯片形式安置在转接板的上表面[4,5]。这些芯片的焊盘节距、大小、尺寸和位置可以各不相同，芯片的背部可以用热界面材料（TIM）与散热器相连，从而将这些大功率芯片产生的大部分热量通过散热器消散掉（如果必要的话还可以使用热沉）。而所有的低功率芯片，如 MEMS、OMEMS、CMOS 图像传感器以及低功耗存储器等，则可以安装在转接板的底部。

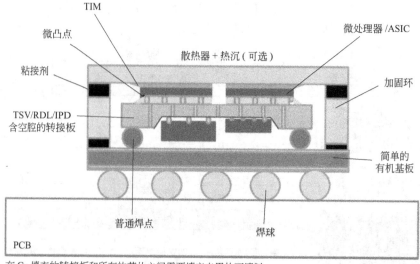

在 Cu 填充的转接板和所有的芯片之间需要填充专用的下填料。
在转接板和有机基板之间只需要普通的下填料。

图 1.23 ITRI 的正面有大功率芯片、背面有低功率
芯片的转接板（含空腔）

图 1.23 中，用于 PCB 组装的、具有标准尺寸和节距焊球的简单有机基板就可以支撑起无源转接板。与有机基板和散热器相连接的加固环为无源转接板 3D IC 集成提供了机械支撑，同时也支撑起安装或不安装热沉的散热器。在 TSV 转接板与倒装芯片之间以及 TSV 转接板和有机基板之间，需要使用下填料进行填充。但是，在有机基板和 PCB 之间是不需要使用下填料的。这种 3D IC 集成 SiP 器件具有以下优点：①不需要新的 EDA 工具；②有源芯片上没有 TSV；③大功率芯片产生的热量可以从背面散去；④由于采用的是标准封装，焊锡接点的可靠性不存在问题。

推荐的这种 3D IC 集成 SiP 器件对整合元件制造商（IDM）、原始设备制造商（OEM）、封测厂（OSAT）、专业电子制造服务商（EMS）极具吸引力，这是因为它采用的是一个标准的塑料球栅阵列（PBGA），这项技术在电子工业已经使用了超过 17 年[43]。

如果存储芯片太厚不能放进转接板和有机基板之间，而且对于有源芯片来说背面磨削减薄的成本又太高，这时可以在转接板的底部制作一个空腔，空腔的制作可以采用激光或湿法各向异性刻蚀的方法，如采用氢氧化钾（KOH）刻蚀方法。

1.7.2 有机基板开孔式转接板技术

图 1.24 所示是一个新的设计，与图 1.23 相比，除了在有机基板的中心开孔外其他几乎一样。这种设计的优点在于：①不需要对转接板进行刻蚀制作空腔；②不需要对摩尔芯片进行减薄；③可以实现更多数量存储芯片的堆叠[9~12]。

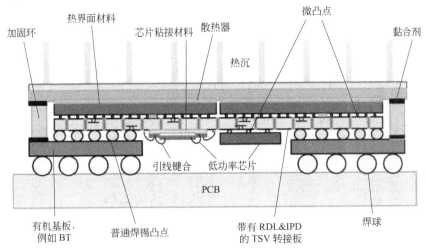

在 TSV 转接板和高、低功率倒装芯片之间需要专门的下填料，在 TSV 转接板和有机基板之间则是普通的下填料；需要对引线键合存储芯片堆叠进行封装。

图 1.24 顶面贴装高功率芯片、底面贴装低功率
芯片的转接板，有机基板中心开孔

1.7.3 设计举例

图 1.25 所示为一个 3D IC 集成 SiP 器件的设计图，它包含 4 个相同的大功率芯片（例如微处理器），并且均配置在 TSV/RDL/IPD 转接板的顶部。同时，还有 16 个相同的低功耗芯片（例如存储芯片）配置在转接板的底部。大功率芯片的尺寸为 10mm×10mm×200μm，低功耗芯片的尺寸为 5mm×5mm×200μm，转接板的尺寸为 35mm×35mm×200μm，转接板中共有 1600 个相同的 TSV 孔，其直径和节距分别为 20μm 和 850μm。这些大功率芯片的背面都通过 100μm 厚的热界面材料与一个散热器相连，散热器背面粘接一个带有 12 个散热片的铝制热沉。

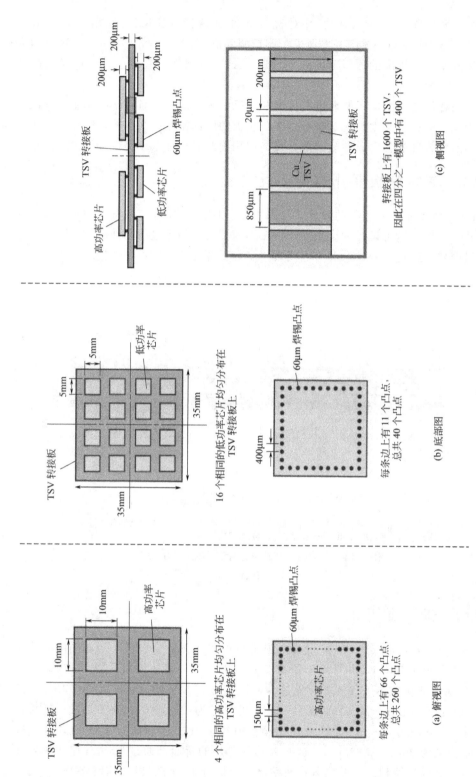

图 1.25 转接板（作为一个性价比高的 3D IC 集成）顶部有四个 CPU，底部有 16 个存储器

采用普通焊球将 3D IC 集成 SiP 安装到 BT 基板上,基板尺寸为 44mm×44mm×0.8mm,基板中心开孔尺寸为 33mm×33mm×0.8mm。基板通过一个铝制加固环连接到带有热沉的散热器上,然后将基板用无铅焊球(SnAgCu)连接到 FR-4 PCB 板上,PCB 尺寸为 50mm×50mm×2.5mm。基于非线性有限元建模和分析,对每个温度循环下的温度分布以及蠕变应变能密度已有相关研究。结果表明,在大多数工作条件下,选取合适的热沉、下填料、微凸点和焊球都能够满足可靠性要求[44]。

1.7.4 带散热块的有机基板开孔式转接板技术

图 1.26 所示的设计与图 1.24 十分相似,不同的是安装在转接板底部的低功率芯片其回流焊温度超过了允许温度。一个解决方案是通过在低功率芯片的背部连接一个散热基座将热量传递到 PCB 板底部的散热器上[45]。

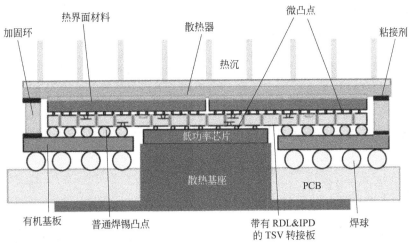

在 TSV 转接板和高、低功率倒装芯片之间需要专用的下填料,在 TSV 转接板和有机基板之间采用普通下填料。

图 1.26 ITRI 基于转接板的 3D IC 集成热管理系统

1.7.5 超低成本转接板

图 1.27 所示为用于 3D IC 集成且成本很低的无源转接板。它其实是一片带有通孔的硅片,通孔采用刻蚀或激光加工,不需要金属化处理,因此被称作硅穿孔(Through-Silicon Hole,TSH)转接板。TSH 转接板可用于支撑顶部和底部的摩尔芯片,同时允许顶部摩尔芯片的信号通过 Cu/Au 线或柱传递给底部摩尔芯片,或者反过来。TSH 转接板的再分布层(RDL)可以使 TSH 转接板顶部和底部的摩尔芯片之间实现通信[46]。前面已提到过,制作 TSV 需要 6 个关键工艺步,而制作 TSH 只需要 DRIE 刻蚀或激光钻孔。

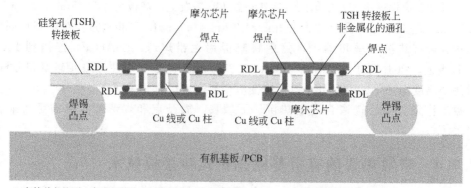

1. 在热载条件下,在摩尔芯片和转接板之间的下填料是可选的。但是,在振动和冲击载荷下,根据芯片的尺寸,也许需要下填料。在 TSH 转接板和有机基板 /PCB 之间,下填料是必要的。
2. 制作 TSV 需要 6 个工艺步:制孔(DRIE)、沉积介电层(PECVD)、沉积阻挡 / 种子层(PVD)、孔填充(电镀 Cu)、CMP、Cu 退火。而制作 TSH 只需要 DRIE(刻蚀)或激光钻孔。

图 1.27　ITRI 用于 3D IC 集成的超低成本硅穿孔(TSH)转接板

1.7.6　用于热管理的转接板技术

由于需要配置散热器和热沉,图 1.23~图 1.27 所示的 3D IC 集成 SiP 器件不能在 z 方向进行 3D 堆叠,而且将转接板顶部的大功率芯片与底部的低功率芯片分离也不是件轻松的事情。图 1.28 和图 1.29 所示的一种新的 3D IC 集成 SiP 设计则可以将任何芯片随意地在转接板的顶部或底部进行堆叠。

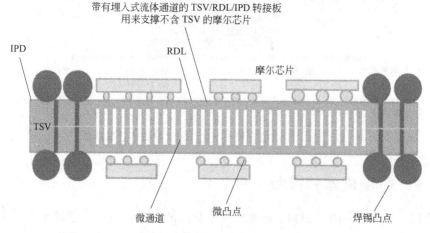

图 1.28　ITRI 带埋入式微通道的转接板(TSV/RDL/IPD),
在转接板的顶部和底部可随意配置摩尔芯片

结构的基本单元如图 1.28 所示,其中包含 TSV/RDL/IPD 转接板,转接板中有埋入的流体通道。这种转接板通过 W2W 键合方式将两个硅板键合在一起,经过优化的液体冷却通道被埋入两个硅板之间,如图 1.30 和图 1.31 所示。在第 1.6.4 节曾提到过,硅之所以被选作转接板材料,是因为在硅基体上适合采用精密微加工

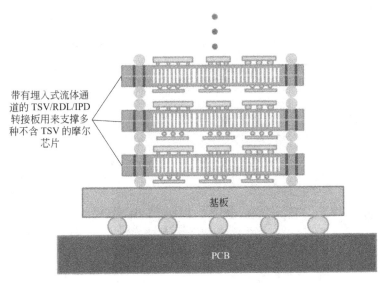

图 1.29 ITRI 带埋入式微通道的多层转接板 3D IC 集成 SiP，可以支撑多种不含 TSV 的摩尔芯片

工艺制作电路和流体通道。上下两个键合硅板的区别是下面的硅板没有流体出入口。TSV 可以制作在转接板的四周，完成 W2W 键合后，转接板电信号的互连通过孔壁金属化的 TSV 来实现。这种情况下，将 TSV 孔填充 Cu 是没有必要的。流体通道通过入口和出口同外界相连，流体通道与个别 TSV 周围制作焊锡（焊料为 Au20Sn，凸点下金属为 TiCuNiAu）密封环，以隔离流体与互连电路，如图 1.30 和图 1.31 所示[40]。

1.7.7 用于 LED 和 SiP 封装的带埋入式微流体通道的转接板技术

图 1.32 所示为带埋入式微通道的 TSV 3D 转接板在 3D IC 集成 SiP 器件中的应用。图 1.33 为 TSV 转接板的详细结构图，TSV 转接板顶部均匀布置 100 个相同的发光二极管（LED），底部均匀布置 4 个相同的逻辑芯片，如专用集成电路（ASIC）等。LED 的尺寸为 1mm×1mm×300μm，逻辑芯片的尺寸为 6mm×6mm×300μm，TSV 转接板的尺寸为 25mm×25mm×1.4mm，微通道高度为 700μm，引脚的宽度和间距分别为 0.1mm、1.25mm，流体的入口/出口尺寸为 20mm×1.5mm。

3D IC 和 LED SiP 器件中各材料的参数为：①水：密度 $\rho=989kg/m^3$，比热容 $C_p=4177J/(kg\cdot K)$，热导率 $k=0.6367W/(m\cdot K)$，黏度为 $5.77\times10^{-4}kg/(m\cdot s)$；②硅 $\rho=2330kg/m^3$，$C_p=660J/(kg\cdot K)$，$k=148W/(m\cdot K)$。边界条件为：①所有暴露的表面都是绝热的，这对于散热来说是最坏的情况；②单个 LED 的功率分别为 1W、1.5W 和 2W；③单个 ASIC 芯片的功率分别为 5W 和 10W；④流量分别为 0.18、0.36、0.54、0.72、0.9、1.08 和 1.26(L/min)。采用有限元法进行分析，所用软件为 ANSYS 和 ICEPAK12.1.6。

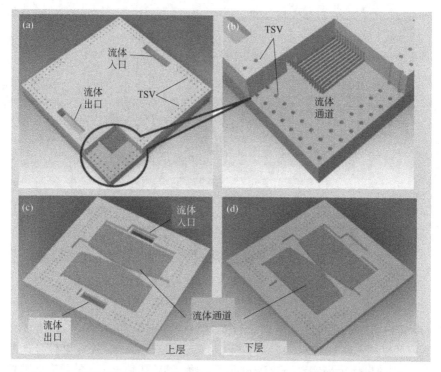

图 1.30 转接板中包含用于电互连的 TSV 和用于热管理的埋入式流体通道

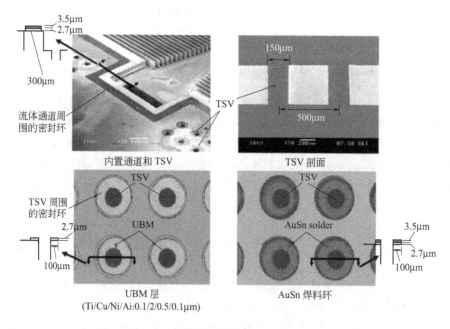

图 1.31 TSV 和微通道转接板（载板）的详细结构

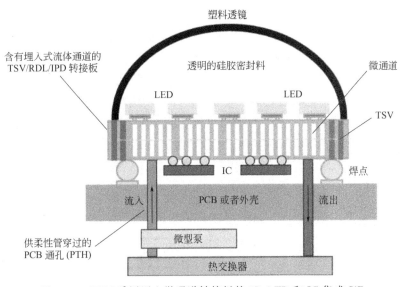

图 1.32 ITRI 采用埋入微通道转接板的 3D LED 和 IC 集成 SiP

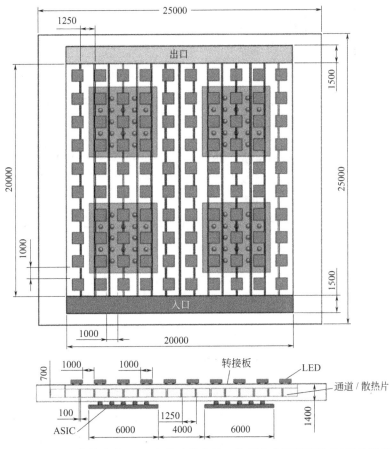

图 1.33 用于支撑多个 LED 和驱动芯片的、带埋入式微通道的 TSV 转接板

图 1.34 所示为仿真结果,可以看到:①靠近流体入口处的 LED 和芯片的温度最低,靠近出口处温度最高,因为微通道内的冷却水把热量带到出口处;②靠近转接板中部的 LED 和芯片的温度要比靠近转接板两边的温度高;③LED 的温度一般比 ASIC 的温度要高。文献 [47] 表明:①LED 的功率越大,其温度越高;②ASIC 的功率越大,其温度越高;③SiP 器件的功率越大,LED、ASIC 芯片以及流出的水的温度越高;④LED 的平均温度变化率要高于 ASIC 芯片和水。此外值得注意的是,当流量为 0.54L/min、入水温度为 20℃时,可从转接板中带走 240W 的热量,LED 与 ASIC 芯片的最高温度分别为 88.7℃和 59.2℃,表明该设计具备较强的冷却能力。

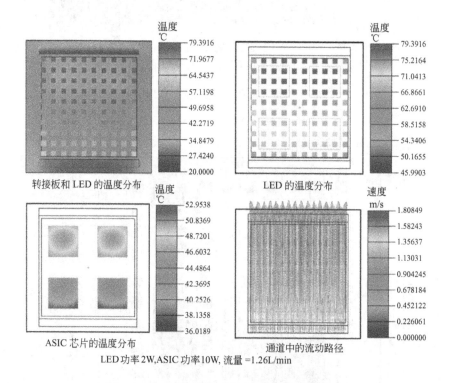

图 1.34 3D LED 和 IC 集成 SiP 中的温度分布

1.8 埋入式 3D IC 集成技术

把器件埋入基板/PCB 板是减小封装尺寸的最好方法。对埋入式结构的基本要求是:①可拆装性好,利于维修;②具有好的热性能;③由于埋入器件需要破坏基板/PCB 板,因此要求破坏性最小;④满足可靠性要求。下面给出两个例子。

1.8.1 带应力释放间隙的半埋入式转接板

图 1.35 所示为带应力释放间隙的半埋入式转接板，其顶部布置有摩尔芯片。这种设计的优点是：①较小的尺寸；②可使用任何无 TSV 的摩尔芯片；③较短的设计周期；④较低的制造成本；⑤借助于 RDL 层可实现芯片间短距离通信；⑥TSV 可用于电源、接地和信号连接；⑦制作下填料前，可拆装性好，利于芯片/转接板模组测试；⑧产生的热量很容易从摩尔芯片背部通过散热片/热沉传出去，或者采用散热块从焊球位置传导到基板/PCB 板底部；⑨由于应力释放间隙的存在，使得转接板 [热膨胀系数为 $(6\sim8)\times10^{-6}/℃$] 与有机基板/PCB 板 [热膨胀系数为 $(15\sim18.5)\times10^{-6}/℃$] 之间的整体热失配大大缓解，从而提高了可靠性；⑩较低的系统成本[45]。

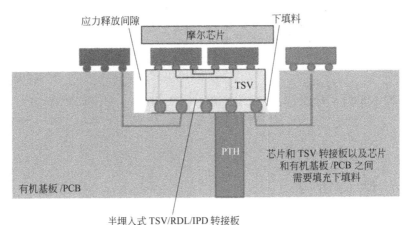

图 1.35　含有应力释放间隙的半埋入式 TSV 转接板

1.8.2 用于光电子互连的埋入式 3D 混合 IC 集成技术

图 1.36 所示为埋入 PCB 板或有机层压基板的低成本、高性能光电系统，包括刚性 PCB 板（或基板）、埋入式光聚合物波导、垂直腔面发射激光器（VCSEL）、驱动芯片、串行器、光电二极管检测器、跨阻放大器（TIA）以及解串行器。将裸 VCSEL、驱动芯片和串行器芯片进行 3D 堆叠，然后安装到 PCB 中埋入的光聚合物波导的一端。类似地，将裸光电二极管探测器、TIA 芯片和解串行器芯片进行 3D 堆叠，然后安装到 PCB 中埋入的光聚合物波导的另一端。驱动芯片、串行器芯片、TIA 芯片和解串行器芯片的背面与散热块粘接，如果需要也可以采用散热器。这种新型结构为芯片到芯片的光互连提供了薄型光电封装。文献 [48] 通过基于光学理论、热传导理论、连续介质力学理论的模拟仿真论证了这种结构的光学性能、热性能和力学性能。

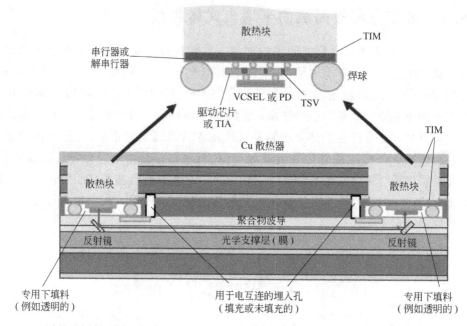

VCSEL=垂直腔面发射激光器（透明的）；PD=光电二极管探测器（透明的）；TIA=跨阻放大器；TIM=热界面材料

图 1.36　用于光电互连的埋入式 3D 混合 IC 集成

1.9　总结与建议

本章简要介绍了半导体工业中纳米技术的一些里程碑事件及其展望，讨论了 3D 集成技术的演变、挑战、研究进展和发展趋势。列举了几种通用的、热增强型的、具有潜在低成本以及包含不同种类无源转接板的 3D IC 集成 SiP 架构，这些架构可用于设计小外形、高性能、高密度、低功耗以及宽带宽器件。下面总结一下得到的重要结果，并给出一些建议。在这本书出版时，IBM 宣布了两个突破性进展：一个是由 12 个原子制作的新型存储器件，另一个是优于 Si 的 9nm 碳纳米管 (CNT) 晶体管。听到这个消息，费曼在天上一定会得意地微笑。

① 关于纳米技术，科学家们已经听从了费曼的建议开始对原子进行排列，最显著的进展是通过排列碳原子，得到了零维的巴基球（1996 年诺贝尔奖）、一维的碳纳米管以及二维的石墨烯（2010 年诺贝尔奖）。也许人们已经到了排列其他原子开发新应用的时代。

② 比钢更硬、比铜的导电性更好、电信号传输速率更惊人的二维石墨烯可能是制作更快、更强大电子器件的替代材料。

③ 对于纳米技术器件，研发系统的、高性价比的以及可靠的封装技术对于量产是必不可少的。否则不论在电镜下看起来有多漂亮，它仍然只是无益的炒作。

④ 2004年奔腾（Pentium）Ⅳ处理器问世，自那时起摩尔定律开始驱动半导体工业纳米技术产品的量产进程，相信摩尔定律仍将驱动至少又一个10年。

⑤ 1985年费曼就告诉人们要朝三维方向发展，并且指明了如何将晶圆/芯片进行堆叠。他还指出了已知合格芯片（KGD）的重要性，并告诉人们可以借助内置自检测（BIST）和内置自修复（BISR）装置解决坏芯片带来的问题。

⑥ TSV是一项颠覆性技术。为了在工业界获得广泛应用，这种颠覆性应当越小越好，至少在最开始的阶段应当如此。因此，目前的无源转接板（2.5D或3D）是最佳方案。等到相关的标准、基础设施以及商业模式都具备之后，就可以在有源芯片上制作TSV来集成异种器件。

⑦ 无源转接板是性价比最高的3D IC集成。它不仅可以作为中间基板、应力释放缓冲区以及载板，同时还可以用作热管理工具。无源转接板应当作为3D IC集成SiP器件的主力。

⑧ 无源转接板提供了与摩尔芯片等器件的灵活集成方式，增强了集成器件的功能，还提供了灵活的组装工艺路线。

⑨ 借助无源转接板，几乎所有3D IC集成SiP器件中的热管理问题都能得到解决。本章给出了采用散热器、散热块、热沉以及流体微通道等进行散热的几个例子。

⑩ 埋入式3D IC集成可以减小产品的尺寸，并且特别适用于移动产品。本章提出了几种带有TSV的新设计。

⑪ 给出了将低成本的硅穿孔（TSH）转接板用于3D IC集成的例子。

⑫ 无微凸点的3D Si集成也是正确的发展方向，可以与摩尔定律匹敌，但仍有很长的路要走。本章提出了需要研发的技术，为建立自动设计工具（EDA）推荐了一些设计准则。工业界应该立即建立TSV产业合作标准以及基础设施，以便EDA开发商开发用于3D Si集成设计、仿真、分析、验证、制造准备以及测试的相关软件，并力促其实现。应当注意的是，虽然焊锡是上帝赐给电子产业的礼物[49]，但是它太"脏"了，半导体厂不得不对其进行处理。

1.10 参考文献

[1] Moore, G., "Cramming More Components Onto Integrated Circuits," *Electronics*, Vol. 38, No. 8, April 19, 1965.

[2] Lau, J. H., *Reliability of RoHS-Compliant 2D and 3D IC Interconnects*, McGraw-Hill, New York, 2011.

[3] Lau, J. H., C. K. Lee, C. S. Premachandran, and A. Yu, *Advanced MEMS Packaging*, McGraw-Hill, New York, 2010.

[4] Lau, J. H., S. Wu, and J. M. Lau, "A Note on the Reliability of Wirebonding," *ASME Transactions, Journal of Electronic Packaging* (in press).

[5] Lau, J. H., *Flip Chip Technology*, McGraw-Hill, New York, 1995.

[6] Lau, J. H., *Low-Cost Flip Chip Technologies for WLCSP*, McGraw-Hill, New York, 2000.

[7] Akasaka, Y., "Three-dimensional IC Trends," *Proceedings of the IEEE*, Vol. 74,

No. 12, December 1986, pp. 1703–1714.
[8] Gat, A., L. Gerzberg, J. Gibbons, T. Mages, J. Peng, and J. Hong, "CW Laser of Polyerystalline Silicon: Crystalline Structure and Electrical Properties," *Applied Physics Letter*, Vol. 33, No. 8, October 1978, pp. 775–778.
[9] Lau, J. H., "Overview and Outlook of TSV and 3D Integrations," *Journal of Microelectronics International*, Vol. 28, No. 2, 2011, pp. 8–22.
[10] Lau, J. H., "3D Integrations," Plenary Keynote at IEEE/EDAPS, Singapore, December 2010.
[11] Lau, J. H., "Evolution, Challenges, and Outlook of 3D IC/Si Integration," Plenary Keynote at IEEE ICEP, Nara, Japan, April 2011, pp. 1–16.
[12] Lau, J. H., "3D IC Integration and 3D Si Integration," Plenary Keynote at IWLPC, San Jose, CA, October 2011, pp. 1–18.
[13] Yu, A., J. H. Lau, Ho, S., Kumar, A., Yin, H., Ching, J., Kripesh, V., Pinjala, D., Chen, S., Chan, C., Chao, C., Chiu, C., Huang, M., and Chen, C., "Three-Dimensional Interconnects with High Aspect Ratio TSVs and Fine Pitch Solder Microbumps," *IEEE/ECTC Proceedings*, San Diego, CA, May 2009, pp. 350–354. Also, *IEEE Transactions in Advanced Packaging* (in press).
[14] Kang, U., H. Chung, S. Heo, D. Park, H. Lee, J. Kim, S. Ahn, S. Cha, J. Ahn, D. Kwon, J. Lee, H. Joo, W. Kim, D. Jang, N. Kim, J. Choi, T. Chung, J. Yoo, J. Choi, C. Kim, and Y. Jun, "8 Gb 3-D DDR3 DRAM Using Through-Silicon-Via Technology," *IEEE Journal of Solid-State Circuits*, Vol. 45, No. 1, January 2010, pp. 111–119.
[15] Farooq, M. G, T. L. Graves-Abe, W. F. Landers, C. Kothandaraman, B. A. Himmel, P. S. Andry, C. K. Tsang, E. Sprogis, R. P. Volant, K. S. Petrarca, K. R. Winstel, J. M. Safran, T. D. Sullivan, F. Chen, M. J. Shapiro, R. Hannon, R. Liptak, D. Berger, and S. S. Iyer, "3D Copper TSV Integration, Testing and Reliability," *Proceedings of IEEE/IEDM*, Washington, DC, December 2011, pp. 7.1.1–7.1.4.
[16] Dorsey, P., "Xilinx Stacked Silicon Interconnect Technology Delivers Breakthrough FPGA Capacity, Bandwidth, and Power Efficiency," Xilinx White Paper: Virtex-7 FPGAs, WP380, October 27, 2010, pp. 1–10.
[17] Banijamali, B., S. Ramalingam, K. Nagarajan, and R. Chaware, "Advanced Reliability Study of TSV Interposers and Interconnects for the 28-nm Technology FPGA," *IEEE/ECTC Proceedings*, Orlando, FL, June 2011, pp. 285–290.
[18] Kim, N., D. Wu, D. Kim, A. Rahman, and P. Wu, "Interposer Design Optimization for High Frequency Signal Transmission in Passive and Active Interposer Using Through Silicon Via (TSV)," *IEEE/ECTC Proceedings*, Orlando, FL, June 2011, pp. 1160–1167.
[19] Lau, J. H., M. Dai, Y. Chao, W. Li, S. Wu, J. Hung, M. Hsieh, J. Chien, R. Tain, C. Tzeng, K. Lin, E. Hsin, C. Chen, M. Chen, C. Wu, J. Chen, J. Chien, C. Chiang, Z. Lin, L. Wu, H. Chang, W. Tsai, C. Lee, T. Chang, C. Ko, T. Chen, S. Sheu, S. Wu, Y. Chen, R. Lo, T. Ku, M. Kao, F. Hsieh, and D. Hu, "Feasibility Study of a 3D IC Integration System-in-Packaging (SiP)," *IEEE/ICEP Proceedings*, Nara, Japan, April 13, 2011, pp. 210–216.
[20] Lau, J. H., C-J Zhan, P.-J. Tzeng, C.-K. Lee, M.-J. Dai, H.-C. Chien, Y.-L. Chao, W. Li, S.-T. Wu, J.-F. Hung, R.-M. Tain, C.-H. Lin, Y.-C. Hsin, C.-C. Chen, S.-C. Chen, C.-Y. Wu, J.-C. Chen, C.-H. Chien, C.-W. Chiang, H. Chang, W.-L. Tsai, R.-S. Cheng, S.-Y. Huang, Y.-M. Lin, T.-C. Chang, C.-D. Ko, T.-H. Chen, S.-S. Sheu, S.-H. Wu, Y.-H. Chen, W.-C. Lo, T.-K. Ku, M.-J. Kao, and D.-Q. Hu, "Feasibility Study of a 3D IC Integration System-in-Packaging (SiP) from a 300-mm Multi-Project Wafer (MPW)," *Proceedings of IMAPS International Conference*, Long Beach, CA, October 2011, pp. 446–454.
[21] Zhan, C.-J., J. H. Lau, P.-J. Tzeng, C.-K. Lee, M.-J. Dai, H.-C. Chien, Y.-L. Chao, W. Li, S.-T. Wu, J.-F. Hung, R.-M. Tain, C.-H. Lin, Y.-C. Hsin, C.-C. Chen, S.-C. Chen, C.-Y. Wu, J.-C. Chen, C.-H. Chien, C.-W. Chiang, H. Chang, W.-L. Tsai, R.-S. Cheng, S.-Y. Huang, Y.-M. Lin, T.-C. Chang, C.-D. Ko, T.-H. Chen, S.-S. Sheu, S.-H. Wu, Y.-H. Chen, W.-C. Lo, T.-K. Ku, M.-J. Kao, and D.-Q. Hu, "Assembly Process and Reliability Assessment of TSV/RDL/IPD Interposer with Multi-Chip-Stacking for 3D IC Integration SiP," *IEEE/ECTC Proceedings* (in press).
[22] Sheu, S., Z. Lin, J. Hung, J. H. Lau, P. Chen, S. Wu, K. Su, C. Lin, S. Lai, T. Ku, W. Lo, M. Kao, "An Electrical Testing Method for Blind Through Silicon Vias

(TSVs) for 3D IC Integration," *Proceedings of IMAPS International Conference*, Long Beach, CA, October 2011, pp. 208–214.
[23] Wu, C., S. Chen, P. Tzeng, J. H. Lau, Y. Hsu, J. Chen, Y. Hsin, C. Chen, S. Shen, C. Lin, T. Ku, and M. Kao, "Oxide Liner, Barrier and Seed Layers, and Cu-Plating of Blind Through Silicon Vias (TSVs) on 300-mm Wafers for 3D IC Integration," *Proceedings of IMAPS International Conference*, Long Beach, CA, October 2011, pp. 1–7.
[24] Chien, H., J. H. Lau, Y. Chao, R. Tain, M. Dai, S. Wu, W. Lo, and M. Kao, "Thermal Performance of 3D IC Integration with Through-Silicon Via (TSV)," *Proceedings of IMAPS International Conference*, Long Beach, CA, October 2011, pp. 25–32.
[25] Chang, H. H., J. H. Lau, W. L. Tsai, C. H. Chien, P. J. Tzeng, C. J. Zhan, C. K. Lee, M. J. Dai, H. C. Fu, C. W. Chiang, T. Y. Kuo, Y. H. Chen, W. C. Lo, T. K. Ku, and M. J. Kao, "Thin Wafer Handling of 300-mm Wafer for 3D IC Integration," *Proceedings of IMAPS International Conference*, Long Beach, CA, October 2011, pp. 202–207.
[26] Lau, J. H., P.-J. Tzeng, C.-K. Lee, C.-J. Zhan, M.-J. Dai, Li Li, C.-T. Ko, S.-W. Chen, H. Fu, Y. Lee, Z. Hsiao, J. Huang, W. Tsai, P. Chang, S. Chung, Y. Hsu, S.-C. Chen, Y.-H. Chen, T.-H. Chen, W.-C. Lo, T.-K. Ku, M.-J. Kao, J. Xue, and M. Brillhart, "Wafer Bumping and Characterizations of Fine-Pitch Lead-Free Solder Microbumps on 12-inch (300-mm) wafer for 3D IC Integration," *Proceedings of IMAPS International Conference*, Long Beach, CA, October 2011, pp. 650–656.
[27] Hsin, Y. C., C. Chen, J. H. Lau, P. Tzeng, S. Shen, Y. Hsu, S. Chen, C. Wn, J. Chen, T. Ku, and M. Kao, "Effects of Etch Rate on Scallop of Through-Silicon Vias (TSVs) in 200-mm and 300-mm Wafers," *IEEE/ECTC Proceedings*, Orlando, FL, June 2011, pp. 1130–1135.
[28] Chen, J. C., J. H. Lau, P. J. Tzeng, S. C. Chen, C. Y. Wu, C. C. Chen, C. H. Lin, Y. C. Hsin, T. K. Ku, and M. J. Kao, "Impact of Slurry in Cu CMP (Chemical Mechanical Polishing) on Cu Topography of Through Silicon Vias (TSVs), Re-distributed Layers, and Cu Exposure," *IEEE/ECTC Proceedings*, Orlando, FL, June 2011, pp. 1389–1394. Also, *IEEE Transactions on CPMT* (in press).
[29] Chien, J., Y. Chao, J. H. Lau, M. Dai, R. Tain, M. Dai, P. Tzeng, C. Lin, Y. Hsin, S. Chen, J. Chen, C. Chen, C. Ho, R. Lo, T. Ku, and M. Kao, "A Thermal Performance Measurement Method for Blind Through Silicon Vias (TSVs) in a 300-mm Wafer," *IEEE/ECTC Proceedings*, Orlando, FL, June 2011, pp. 1204–1210.
[30] Tsai, W., H. H. Chang, C. H. Chien, J. H. Lau, H. C. Fu, C. W. Chiang, T. Y. Kuo, Y. H. Chen, R. Lo, and M. Kao, "How to Select Adhesive Materials for Temporary Bonding and De-Bonding of Thin-Wafer Handling in 3D IC Integration?" *IEEE/ECTC Proceedings*, Orlando, FL, June 2011, pp. 989–998.
[31] Chang, H., J. Huang, C. Chiang, Z. Hsiao, H. Fu, C. Chien, Y. Chen, W. Lo, and K. Chiang, "Process Integration and Reliability Test for 3D Chip Stacking with Thin Wafer Handling Technology," *IEEE/ECTC Proceedings*, Orlando, FL, June 2011, pp. 304–311.
[32] Lee, C. K., T. C. Chang, Y. Huang, H. Fu, J. H. Huang, Z. Hsiao, J. H. Lau, C. T. Ko, R. Cheng, K. Kao, Y. Lu, R. Lo, and M. J. Kao, "Characterization and Reliability Assessment of Solder Microbumps and Assembly for 3D IC Integration," *IEEE/ECTC Proceedings*, Orlando, FL, June 2011, pp. 1468–1474.
[33] Zhan, C., J. Juang, Y. Lin, Y. Huang, K. Kao, T. Yang, S. Lu, J. H. Lau, T. Chen, R. Lo, and M. J. Kao, "Development of Fluxless Chip-on-Wafer Bonding Process for 3D chip Stacking with 30μm Pitch Lead-Free Solder Micro Bump Interconnection and Reliability Characterization," *IEEE/ECTC Proceedings*, Orlando, FL, June 2011, pp. 14–21.
[34] Cheng, R., K. Kao, J. Chang, Y. Hung, T. Yang, Y. Huang, S. Chen, T. Chang, Q. Hunag, R. Guino, G. Hoang, J. Bai, and K. Becker, "Achievement of low Temperature Chip Stacking by a Pre-Applied Underfill Material," *IEEE/ECTC Proceedings*, Orlando, FL, June 2011, pp. 1858–1863.
[35] Huang, S., T. Chang, R. Cheng, J. Chang, C. Fan, C. Zhan, J. H. Lau, T. Chen, R. Lo, and M. Kao, "Failure Mechanism of 20-μm Pitch Microjoint Within a Chip Stacking Architecture," *IEEE/ECTC Proceedings*, Orlando, FL, June 2011, pp. 886–892.
[36] Lin, Y., C. Zhan, J. Juang, J. H. Lau, T. Chen, R. Lo, M. Kao, T. Tian, and K. N. Tu, "Electromigration in Ni/Sn Intermetallic Micro Bump Joint for 3D IC Chip

Stacking," *IEEE/ECTC Proceedings*, Orlando, FL, June 2011, pp. 351–357.

[37] Selvanayagam, C., J. H. Lau, X. Zhang, S. Seah, K. Vaidyanathan, and T. Chai, "Nonlinear Thermal Stress/Strain Analysis of Copper Filled TSV (Through Silicon Via) and Their Flip-Chip Microbumps," *IEEE/ECTC Proceedings*, Orlando, FL, May 27–30, 2008, pp. 1073–1081. Also, *IEEE Transactions on Advanced Packaging*, Vol. 32, No. 4, November 2009, pp. 720–728.

[38] Zhang, X., T. Chai, J. H. Lau, C. Selvanayagam, K. Biswas, S. Liu, D. Pinjala, G. Tang, Y. Ong, S. Vempati, E. Wai, H. Li, B. Liao, N. Ranganathan, V. Kripesh, J. Sun, J. Doricko, and C. Vath, "Development of Through Silicon Via (TSV) Interposer Technology for Large Die (21 × 21 mm) Fine-Pitch Cu/Low-k FCBGA Package," *IEEE/ECTC Proceedings*, May 2009, pp. 305–312. Also, *IEEE Transactions on CPMT*, 2011, pp. 660–672.

[39] Khan, N., V. Rao, S. Lim, S. Ho, V. Lee, X. Zhang, R. Yang, E. Liao, Ranganathan, T. Chai, V. Kripesh, and J. H. Lau, "Development of 3D Silicon Module with TSV for System in Packaging," *IEEE/ECTC Proceedings*, May 2008, pp. 550–555. Also, *IEEE Transactions on Components, Packaging and Manufacturing Technology*, Vol. 33, No. 1, March 2010, pp. 3–9.

[40] Yu, A., N. Khan, G. Archit, D. Pinjalal, K. Toh, V. Kripesh, S. Yoon, and J. H. Lau, "Development of silicon carriers with embedded thermal solutions for high power 3D package," *IEEE/ECTC Proceedings*, May 2008, pp. 24–28. Also, *IEEE Transactions on Components and Packaging Technology*, Vol. 32, No. 3, September 2009, pp. 566–571.

[41] Li, L., P. Su, J. Xue, M. Brillhart, J. H. Lau, P. Tzeng, C. Lee, C. Zhan, M. Dai, H. Chien, and S. Wu, "Addressing the Bandwidth Challenges in Next-Generation High-Performance Network Systems with 3D IC Integration," *IEEE/ECTC Proceedings* , 2012.

[42] Chien, H. C., J. H. Lau, Y. Chao, M. Dai, R. Tain, L. Li, P. Su, J. Xue, M. Brillhart, et al., "Thermal Evaluation and Analyses of 3D IC Integration SiP for Network System Applications," *IEEE/ECTC Proceedings*, 2012.

[43] Lau, J. H., *Ball Grid Array Technology*, McGraw-Hill, New York, 1995.

[44] Lau, J. H., Y. Chan, and R. Lee, "3D IC Integration with TSV Interposers for High-Performance Applications," *Chip Scale Review*, September–October, 2010, pp. 26–29.

[45] Lau, J. H., "TSV Interposers: The Most Cost-Effective Integrator for 3D IC Integration," *Chip Scale Review*, September–October, 2011, pp. 23–27.

[46] Wu, S., J. H. Lau, H. Chien, R. Tain, et al., "Ultra Low-Cost Through Silicon Holes (TSHs) Interposers for 3D IC Integration," *IEEE/ECTC Proceedings*, 2012.

[47] Lau, J. H., H. C. Chien, and R. Tain, "TSV Interposers with Embedded Microchannels for 3D IC and Multiple High-Power LEDs Integration SiP," ASME Paper no. InterPACK2011-52204.

[48] Lau, J. H., M. S. Zhang, and S. W. R. Lee, "Embedded 3D Hybrid IC Integration System-in-Package (SiP) for Opto-Electronic Interconnects in Organic Substrates," *ASME Transactions, Journal of Electronic Packaging*, Vol. 133, September 2011, pp. 1–7.

[49] Lau, J. H., *Solder Joint Reliability: Theory and Applications*, Van Nostrand Reinhold, New York, 1991.

第2章 TSV 技术

2.1 引言

TSV（Through-Silicon Via）技术或硅通孔技术是 3D Si 集成和 3D IC 集成的核心，也是最重要的支撑技术[1,2]。TSV 可以提供芯片到芯片的最短互连、最小焊盘尺寸与节距。与其他互连技术相比，TSV 的优势包括[1~47]：①更好的电性能；②更低的功耗；③更宽的数据位宽，相应地可得到更宽的带宽；④更高的互连密度；⑤更小的外形尺寸；⑥更轻的质量；⑦有望具有更低的成本。

本章将详细介绍制作 TSV 的 6 个关键工艺步：①采用深反应离子刻蚀技术（DRIE）或激光打孔制作 TSV 孔；②采用热氧化工艺（对无源转接板）或等离子增强化学气相沉积（PECVD）制作介电层；③采用物理气相沉积（PVD）制作阻挡层和种子层；④采用电镀 Cu 填充 TSV 通孔，或者对于尺寸非常小的孔，采用溅射方法填充 W；⑤采用化学机械抛光（CMP）去除多余的 Cu；⑥TSV Cu 外露。一般而言，以上各工艺步按照成本从大到小的排序为：PVD＞PECVD＞CMP＞电镀 Cu＞DRIE。

本章首先对 TSV 的发明者进行简单的介绍。

2.2 TSV 的发明

TSV 是由 1956 年诺贝尔物理学奖获得者肖克莱（William Shockley）在 50 多年前[3]发明的，他还发明了晶体管——半导体工业最伟大的发明。1962 年 7 月 17 日，肖克莱于 1958 年 10 月 23 日申请的专利"Semiconductive wafer and method of making the same"在美国获得了专利授权。图 2.1 所示为该专利的关键所有权之一，这也使得现在的半导体工业如此兴奋。从根本上来说，在晶圆上制作"deep pits"（深通孔，今天称作 TSV）能够使信号从顶部直接传递到底部，反过来也一样。"Through-silicon via"（TSV）这个术语最先是由萨瓦斯舒克（Sergey Savastiouk）提出来的，详见发表在 2000 年 1 月的 Solid State Technology 上的文章"Industry Insights：Moore's Law-the Z Dimension"。

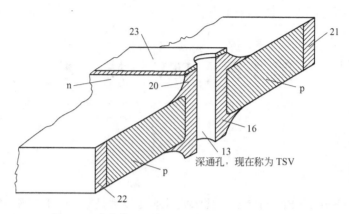

图 2.1　肖克莱（William Shockley）发明了 TSV

（具体注释请见美国专利 3044909，申请日期：1958 年 10 月 23 日）

2.3　采用 TSV 技术的量产产品

为了将 TSV 技术应用于量产商业化产品，惠普公司（Hewlett-Packard，HP）

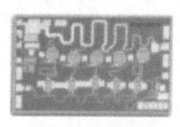
共面 GaAs RF MMIC
（单片微波 IC）

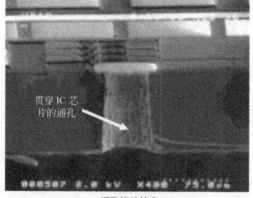
通孔接地技术

图 2.2　HP 的一款采用芯片通孔技术的量产产品（MMIC）(1976)

于 1976 年在其共面 GaAs RF 单片微波集成电路（MMIC）产品中已经使用了通孔接地技术（见图 2.2），但是这并非用于 3D 集成。一般来说，工业界认为东芝（Toshiba）2008 年利用芯片通孔（Through-Chip Vias，TCV）技术生产的互补金属氧化物半导体（CMOS）影像传感器是第一个量产的 3D 集成产品（见图 2.3）。目前还有一些与微机电系统（MEMS）相关的 3D IC 集成产品[1,2]，不过产量都很小。

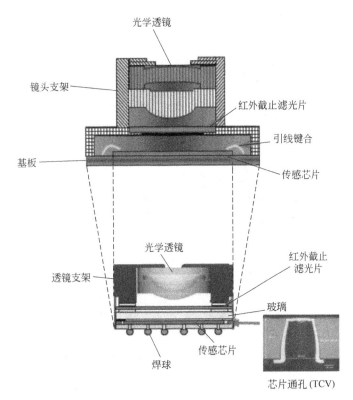

图 2.3　东芝基于 TCV 技术的 CMOS 影像传感器
（从引线键合转换到倒装芯片）（2008）

2.4　TSV 孔的制作

2.4.1　DRIE 与激光打孔

目前至少有两种方法可以制作 TSV 孔，一种是通过激光打孔，另一种是通过深反应离子蚀刻（DRIE）。由于激光打孔是一个单点操作，因此它的性价比只有当芯片上孔的数量比较少时才能体现出来。而 DRIE 工艺是针对一整片晶圆，因此它更适用于制作高密度的 TSV 孔。采用不同类型的激光打孔时，对 TSV 的孔径和节距有所限制：对于 CO_2 激光，孔的直径为顶部 $65\mu m$、底部 $25\mu m$，节距为 $90\mu m$；对于紫外（UV）激光，孔的直径为顶部 $50\mu m$、底部 $25\mu m$，节距为 $125\mu m$；而对于准分子激光，孔的直径为顶部 $18\mu m$、底部 $12\mu m$，节距为 $35\mu m$。

另一方面，采用 DRIE 工艺时，TSV 孔的尺寸和节距可以分别减小至 $1\mu m$ 和 $5\mu m$。TSV 大部分潜在应用的尺寸范围为：$5\mu m \leqslant$ TSV 孔直径 $\leqslant 20\mu m$，孔节距 $\geqslant 20\mu m$。另外需要指出的是，激光打孔形成的孔壁十分粗糙，如图 2.4 所示，需要化学抛光处理获得光滑的壁面，而这又增加了制程的工序及成本。因此本书只讨论 DRIE 工艺。

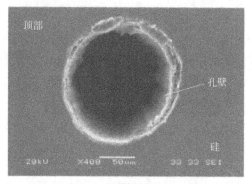

(a) 俯视图

(b) 截面图

图 2.4　利用激光制作的 TSV 孔，可以明显观察到粗糙的孔壁

在大部分 TSV 3D 集成的潜在应用中，无源/有源转接板或芯片的厚度为 $50\sim200\mu m$，堆叠存储芯片的厚度为 $20\sim50\mu m$。因此所有制作的 TSV 都是盲孔[1,2]，而且绝大部分 TSV 的深宽比（深度与直径之比）至少是 2，大的可能达到 50 或更高。

根据已有的文献，目前超过 95% 的 TSV 孔都采用 DRIE 博世（Bosch）工艺制作，它是一种高度各向异性刻蚀工艺。硅通孔可以通过一个基于感应耦合等离子体（ICP）的 DRIE 系统制作，例如应用材料公司（Applied Materials）和 SPTS 公司生产的设备（图 2.5）。虽然 ICP 系统的设计主要是通过刻蚀和钝化交替进行实现对硅基体的深反应离子刻蚀（即所谓的博世工艺），但它也可以用于非博世（non-Bosch）刻蚀加工[14~16]。

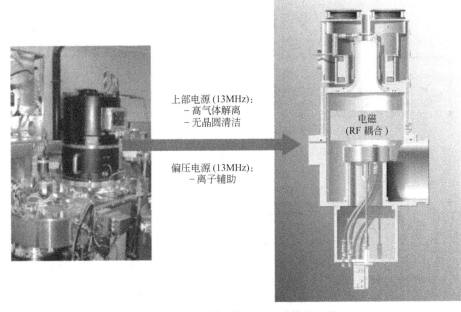

图 2.5　SPTS 公司的 DRIE 系统反应腔

博世工艺是博世公司的科学家们开发的一种改进工艺，用以提高对 DRIE 工艺的控制[14~16]。不同于原来主要依靠离子刻蚀和聚合物钝化之间的平衡工艺，博世工艺采用的是刻蚀和钝化交替进行的方法，如图 2.6 和图 2.7 所示。博世工艺刻蚀过程中，通入的 C_4F_8 气体分解出 CF_x^+ 并沉积为聚合物形成钝化层，然后利用 F^+ 和 SF_x^+ 进行反应离子刻蚀。由于离子轰击的方向性，只有孔底部的聚合物被移除，这样使得刻蚀主要发生在孔的深度方向。每个钝化/刻蚀循环周期只需要几秒的时间，整体的刻蚀速率为 $1\sim 5\mu m/min$，深宽比可以达到 30∶1。DRIE 工艺的主要问题是，刻蚀速率与开孔面积和槽宽有密切关系，因此当制造商采用一种新的光刻掩模布局时，必须先通过晶圆的刻蚀实验对刻蚀速率进行标定[14~16]。

SPTS 公司的 DRIE 系统包括 ICP 等离子源、装载锁和转盘式晶圆加载单元和反应腔，如图 2.5 所示。刻蚀等离子体由一个 13MHz 电源线圈与陶瓷等离子腔内的匹配单元进行电感耦合产生。这种方式能够产生高刻蚀速率所需的高密度等离子体，同时对硅基体的损伤又较小。利用带有自动电源控制和阻抗匹配的底部基片电极上的一个 13MHz RF 偏压电源实现压盘的独立偏置。工艺气体由上部电极组件引入到反应腔中，晶圆由静电卡盘（ESC）固定在与 13MHz 电源相接的底部基片电极上，压盘的温度通过冷却系统的循环去离子（DI）水保持在 10℃。

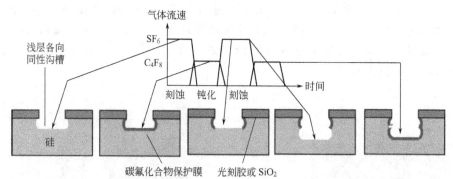

1. 通常，在刻蚀循环中使用 SF_6，而在钝化循环中使用 C_4F_8；
2. 如果刻蚀时间过长而钝化时间过短，所刻的孔会不笔直且壁面也会比较粗糙，出现较大的扇贝形边纹；
3. 另一方面，如果钝化循环时间太长，那么就会降低刻蚀速率和效率；
4. 因此，应当平衡和优化刻蚀质量与刻蚀速率。

图 2.6　TSV DRIE 工艺，SF_6 用于硅刻蚀和钝化，而 C_4F_8 用于钝化

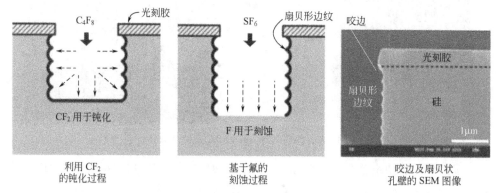

图 2.7　刻蚀步采用 SF_6 气体，钝化（聚合物沉积）步采用 C_4F_8 的工艺循环过程，聚合物保护孔壁从而避免其径向方向受到刻蚀

2.4.2　制作锥形孔的 DRIE 工艺

高深宽比的锥形硅通孔可以通过三个独立控制的工艺步来制作：①利用博世刻蚀工艺制作直孔；②利用可控各向同性刻蚀工艺制作锥形孔；③利用整体各向同性刻蚀工艺进行圆角处理。

上述第一步主要是通过博世刻蚀工艺实现高的刻蚀速率。具有高刻蚀速率的博世工艺一般会得到一个上部稍宽、底部稍窄的竖直孔，如图 2.8 左侧所示。博世工艺中在刻蚀阶段所用气体主要为 SF_6 加上 O_2（氧气），在钝化阶段为 C_4F_8，工艺参数见表 2.1[14~16]。值得注意的是，由于刻蚀/钝化交替循环，因此孔壁呈扇贝形或者比较粗糙。在该工艺步中，孔会被刻蚀到目标深度的 50%~60%。此外，该工艺步也确定了孔底部的直径以及深度。

第二步是采用非博世刻蚀的反应离子刻蚀（RIE）工艺制作符合要求的锥形

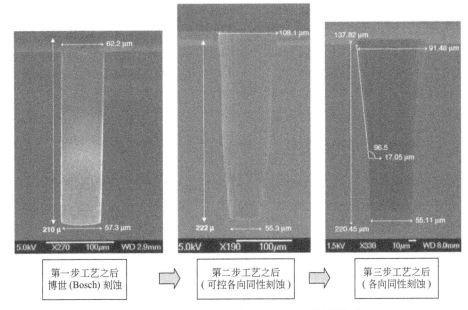

图 2.8 完成第一步工艺之后,孔的直径为 $50\mu m$,刻蚀深度为 $200\sim220\mu m$;第二步工艺为通过可控各向同性刻蚀制作锥形孔;第三步工艺是通过整体各向同性刻蚀对孔顶部进行圆角处理

孔。该工艺基本上是利用 SF_6、O_2 和 Ar(氩气)进行的可控各向同性刻蚀过程(见表 2.1)。O_2 用于辅助壁面钝化并控制横向刻蚀率,调节 SF_6 和 O_2 的比例可以控制锥形孔的锥度。经过该工艺步之后,在孔的顶部会形成尖锐的拐角,如图 2.8 中图所示。完成第一步和第二步工艺后,除了在孔顶部形成尖锐的拐角之外,已经得到了设计的孔深和锥度。

表 2.1 锥形孔三步制作工艺概要

第一步: 直孔(Bosch) 刻蚀工艺	刻蚀循环:APC:77%(26mt);130sccm SF_6;13sccm O_2;线圈功率 600W;压盘功率 20W;时间 6s
	钝化循环:APC:77%(17mt);85sccm C_4F_8;线圈功率 600W;时间 5s
	压盘温度:10℃
	总工艺时长:60min
	刻蚀速率:在 15%~20% 的暴露区域为 $3\sim3.5\mu m/min$
第二步: 锥形孔(RIE) 工艺	APC:78%(30mt);84sccm SF_6;67sccm O_2;59sccm Ar;线圈功率 600W;压盘功率 30W
	压盘温度:10℃
	总工艺时长:60min
	刻蚀速率:在 15%~20% 的暴露区域为 $3.5\sim4.0\mu m/min$
第三步: 孔角圆滑处理 (各向同性刻蚀) 工艺	APC:65%(12~13mt);180sccm SF_6;18sccm O_2;线圈功率 600W;压盘功率 30W
	板温度:10℃
	总工艺时长:10min
	刻蚀速率:在有覆盖层的硅晶圆上为 $1.5\sim2.0\mu m/min$

注:APC=Advanced Pressure Calibrator(压力校准仪);mt=毫托,压强单位;1sccm=$1cm^3/min$,标准流量单位;min=分钟;W=瓦,功率单位。

在完成前面两步工艺之后,去除刻蚀掩膜并清洗,然后开始第三步,即对晶圆进行无掩膜的整体各向同性刻蚀。在该工艺步中,已成形的孔主要受到富含氟自由基的各向同性等离子体刻蚀(见表2.1的工艺参数)。由于这一步主要是自然的化学反应而且扩散有限,所以反应更多地发生在孔的粗糙边缘和顶部区域尖锐的棱角处。这样经过该工艺步后,即可得到圆滑的孔壁,如图2.8右侧所示。目前,大部分文献中的TSV都是直孔,因而只需要进行第一步工艺即可。

2.4.3 制作直孔的DRIE工艺

前面提到过,大部分TSV都是采用DRIE制作的直孔。DRIE工艺的刻蚀速率对于获得高质量的、壁面光滑的TSV孔来说十分重要。高质量的TSV孔意味着最小的扇贝形边纹以及咬边,如图2.6所示[38~43]。通常来讲,在刻蚀和钝化循环中都会用到SF_6,而C_4F_8只用于钝化步,如图2.7所示。如果刻蚀时间太长而钝化时间太短的话,得到的孔将变得不直,同时壁面也会很粗糙,出现较大的扇贝形边纹。另一方面,如果钝化时间太长,则会降低刻蚀速率和效率。因此应当平衡刻蚀速率和刻蚀质量,并对其进行优化。

一般使用实验设计(DoE)确定在200mm和300mm晶圆上制作TSV孔的最优刻蚀速率。设计参数变化范围为:刻蚀速率$1.7\sim18\mu m/min$,TSV直径为$1\mu m$、$10\mu m$、$20\mu m$、$30\mu m$和$50\mu m$,所有实验条件下的背面氦气(BHE)进/出口压强均为2666Pa。为表征刻蚀质量,将刻好的TSV孔沿纵剖面切开,用扫描电子显微镜观察孔壁面的扇贝纹形貌。此外,基于相同的刻蚀参数、掩膜以及9种($5\mu m$、$10\mu m$、$15\mu m$、$20\mu m$、$25\mu m$、$30\mu m$、$40\mu m$、$55\mu m$和$65\mu m$)TSV直径,还比较了200mm和300mm晶圆的刻蚀结果,包括刻蚀速率、TSV深度以及孔壁扇贝形边纹等。最后,给出了一套优化TSV刻蚀工艺的准则和参数[10]。

(1) 300mm晶圆TSV直孔刻蚀的实验设计参数

表2.2列出了300mm晶圆TSV刻蚀的实验设计方案:①共考虑了11组不同的工艺参数;②刻蚀速率范围为$1.7\sim18\mu m/min$;③TSV直径范围为$1\sim50\mu m$;④所有情况下的背面氦气(BHE)进/出口压强均为2666Pa。对于直径为$1\mu m$的TSV,深宽比(AR)为20(深度$20\mu m$);对于直径为$10\mu m$的TSV,深宽比为7.5(深度$75\mu m$);对于直径为$20\mu m$的TSV,深宽比为4.2(深度$85\mu m$);对于直径为$30\mu m$的TSV,深宽比为3(深度$90\mu m$);对于直径为$50\mu m$的TSV,深宽比为2(深度$100\mu m$)。

表2.2 300mm晶圆TSV刻蚀工艺的实验设计

工艺参数	钝化时间/s	刻蚀1时间/s	刻蚀2时间/s	BHE进/出口压强/Pa	CD尺寸/μm	最大扇贝形边纹/nm	刻蚀速率/($\mu m/min$)
1-1st	1.8	1.3	1.0	2666/2666	1	83	2.1
1-2nd	1.6	1.3	1.0	2666/2666	1	68	1.8

续表

工艺参数	钝化时间/s	刻蚀1时间/s	刻蚀2时间/s	BHE进/出口压强/Pa	CD尺寸/μm	最大扇贝形边纹/nm	刻蚀速率/(μm/min)
1-3rd	1.6	1.3	1.0	2666/2666	1	57	1.7
10-1st 20-1st	1.8	1.3	2.0	2666/2666	10 20	278 225	5.8 8.8
10-2nd 20-2nd	1.6	1.3	1.0	2666/2666	10 20	107 93	3.5 4.2
30-1st	1.8	1.3	1.8	2666/2666	30	258	9.55
30-2nd	1.6	1.3	1.3	2666/2666	30	179	9.1
30-3rd	1.8	1.3	1.3	2666/2666	30	146	8.2
30-4th	1.1	1.3	0.6	2666/2666	30	136	5.8
30-5th 50-2nd	1.5	1.3	0.6	2666/2666	30 50	97 99	4.6 5.2
50-1st	1.8	1.3	2.0	2666/2666	50	235	11.0

注：表中第1列第1个数字表示TSV孔直径大小，序列数字表示实验设计次数。例如，"20-2nd"表示直径为20μm TSV孔，第2次实验。

（2）300mm晶圆TSV直孔刻蚀的实验设计结果

图2.9所示为实验设计结果，给出了不同刻蚀速率下以及不同直径TSV孔壁的扇贝形边纹情况。可以看到：①对于所有情况，刻蚀速率越高，孔壁的扇贝形边纹越大，孔壁越粗糙。②对于直径为1μm的TSV孔，刻蚀速率对扇贝形边纹的影

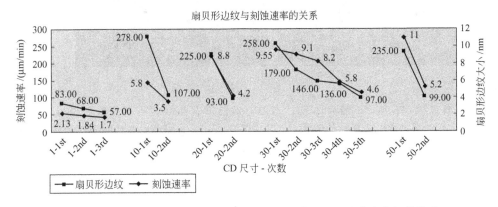

图2.9　刻蚀速率（ER）、TSV直径（CD）和扇贝形边纹大小之间的关系

响很小，当刻蚀速率从1.7μm/min增大至2.13μm/min时，扇贝形边纹的大小从57nm增加到83nm，如图2.10所示。③对于直径为10μm的TSV，刻蚀速率对扇贝形边纹的影响比较明显，当刻蚀速率从3.5μm/min增大至5.8μm/min时，扇贝形边纹的大小从107nm增加到278nm，如图2.11所示。④对于直径为20μm的

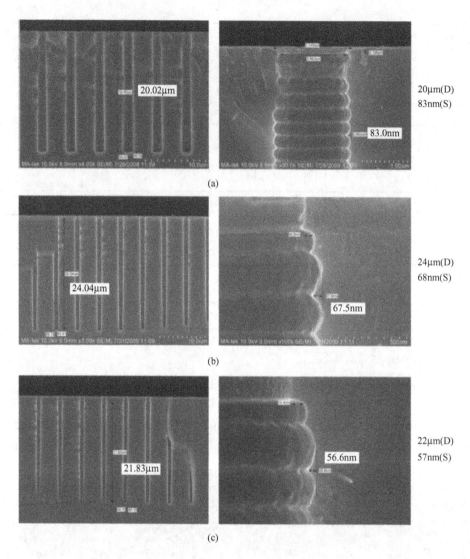

图 2.10　在 300mm 晶圆上刻蚀直径为 1μm 的 TSV 孔：图(a) 孔深度为 20μm，扇贝形边纹为 83nm；图(b) 孔深度为 24μm，扇贝形边纹为 68nm；图(c) 孔深度为 22μm，扇贝形边纹为 57nm

TSV，刻蚀速率对扇贝形边纹的影响更加明显，当刻蚀速率从 4.2μm/min 增大至 8.8μm/min 时，扇贝形边纹的大小从 93nm 增加到 225nm，如图 2.12 所示。⑤对于直径为 30μm 的 TSV，刻蚀速率对扇贝形边纹的影响较大，当刻蚀速率从 4.6μm/min 增大至 9.5μm/min 时，扇贝形边纹的大小从 97nm 增加到 258nm，如图 2.13 所示。⑥对于直径为 50μm 的 TSV，当刻蚀速率从 5.2μm/min 增加到 11μm/min 时，扇贝形边纹的大小从 99nm 增加到 235nm，如图 2.14 所示。

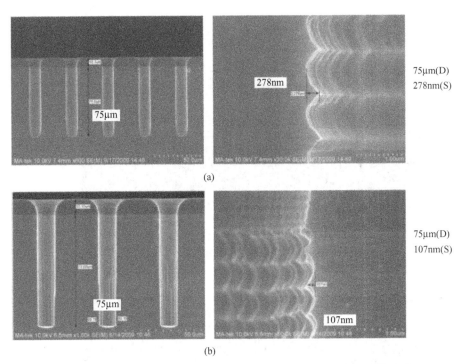

图 2.11 在 300mm 晶圆上刻蚀直径为 $10\mu m$ 的 TSV 孔：图（a）孔深度为 $75\mu m$，扇贝形边纹为 278nm；图（b）孔深度为 $75\mu m$，扇贝形边纹为 107nm

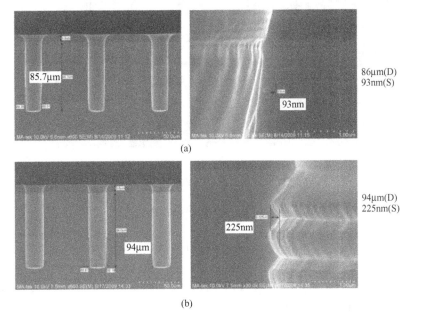

图 2.12 在 300mm 晶圆上刻蚀直径为 $20\mu m$ 的 TSV 孔：图（a）孔深度为 $86\mu m$，扇贝形边纹为 93nm；图（b）孔深度为 $94\mu m$，扇贝形边纹为 225nm

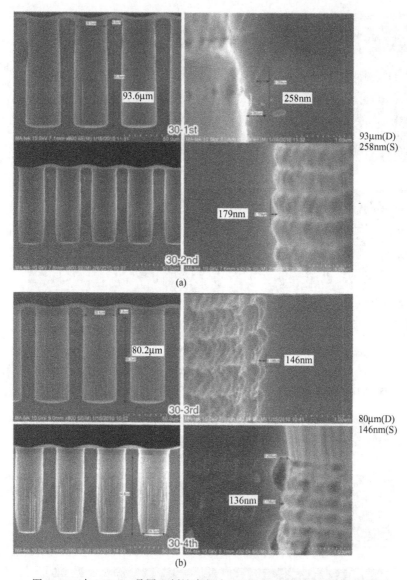

图 2.13 在 300mm 晶圆上刻蚀直径为 30μm 的 TSV 孔：图（a）孔深度为 93μm，扇贝形边纹为 258nm；图（b）孔深度为 80μm，扇贝形边纹为 146nm

（3）直孔刻蚀：200mm 与 300mm 晶圆的对比

200mm 和 300mm 晶圆均在 300mm 反应腔（SPTS-ASE-V300）内进行刻蚀实验，采用相同的光刻掩膜与工艺参数，如表 2.3 所示。可以看出，钝化（C_4F_8）时间为 1.8s，刻蚀 1 的时间（利用 SF_6 去除钝化层）为 1.3s，刻蚀 2 的时间（利用 SF_6 对 Si 进行垂直方向的刻蚀）为 2s。刻蚀时使用了 9 种不同直径的掩膜，分别为 5、10、15、20、25、30、40、55 和 65（μm）。

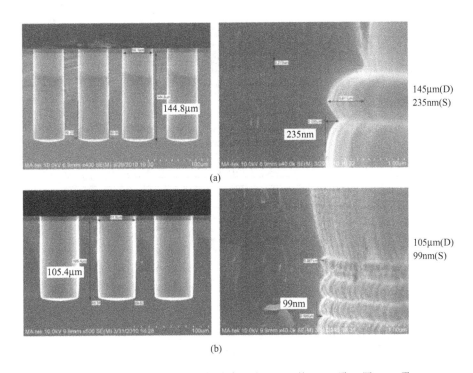

图 2.14 在 300mm 晶圆上刻蚀直径为 50μm 的 TSV 孔：图（a）孔深度为 93μm，扇贝形边纹为 258nm；图（b）孔深度为 80μm，扇贝形边纹为 146nm

表 2.3 200mm 和 300mm 晶圆采用相同的刻蚀工艺参数，只改变刻蚀步时间

钝化时间/s	刻蚀 1 时间/s	刻蚀 2 时间/s	BHE 进/出口压强/Pa
1.8	1.3	2.0	2666/2666

注：1. 200mm 和 300mm 晶圆在同一个 300mm 反应腔内刻蚀，型号为 SPTS-ASE-V300；
2. 采用相同的光刻胶掩膜；
3. 200mm 和 300mm 采用相同的刻蚀工艺参数；
4. 只改变刻蚀步时间。例如，刻蚀直径 5μm、深度 80μm 的 TSV 孔所需时间比直径 5μm、深度 50μm 的孔更长。

200mm 和 300mm 晶圆的刻蚀深度随 TSV 直径的变化情况如图 2.15 所示。可以看出：①对于两种晶圆以及不同的刻蚀循环数而言，一般 TSV 的直径越大，得到的 TSV 深度越深。例如，在刻蚀循环数相同情况下，200mm 和 300mm 晶圆上直径 5μm 的 TSV，其刻蚀深度分别为 51μm 和 49μm；直径 20μm 的 TSV 其刻蚀深度则分别为 83μm 和 81μm。②对于两种晶圆，刻蚀循环数越大或刻蚀时间越长，刻蚀的深度也越深。例如，当刻蚀循环数为 X 时，200mm 和 300mm 晶圆上直径 30μm 的 TSV 其刻蚀深度分别为 97μm 和 94μm；当刻蚀循环数为 $2.25X$ 时，刻蚀

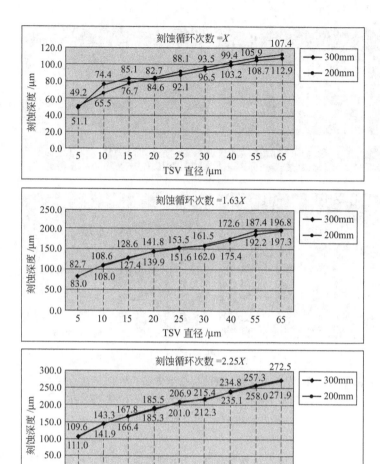

图 2.15 200mm 和 300mm 晶圆在不同刻蚀循环
数下刻蚀深度与 TSV 直径之间的关系示意图

深度则分别为 212μm 和 215μm。③当 TSV 直径和刻蚀循环数相同时，200mm 和 300mm 晶圆的刻蚀深度相差不大。

200mm 和 300mm 晶圆的刻蚀速率随 TSV 直径的变化情况如图 2.16 所示。可以看出：①对于两种晶圆以及不同的刻蚀循环数而言，一般 TSV 的直径越大，刻蚀速率越大。例如，当刻蚀循环数为 X 时，200mm 和 300mm 晶圆上直径 5μm 的 TSV，其刻蚀速率分别为 5μm/min 和 4.8μm/min；当直径为 40μm 时，刻蚀速率则分别为 10μm/min 和 9.7μm/min。②对于两种晶圆，刻蚀循环数越大或刻蚀时间越长，刻蚀速率越小。例如，当刻蚀循环数为 X 时，200mm 和 300mm 晶圆上直径 30μm 的 TSV，其刻蚀速率分别为 9.5μm/min 和 9.2μm/min；当刻蚀循环数为 2.25X 时，刻蚀速率则分别为 6.9μm/min 和 7μm/min。③当 TSV 直径

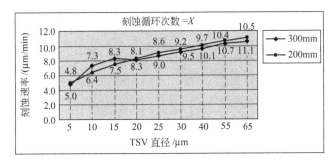

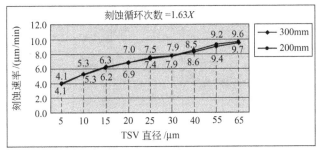

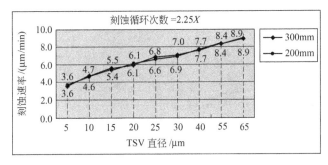

图 2.16 200mm 和 300mm 晶圆在不同刻蚀循环
数下刻蚀速率与 TSV 直径之间的关系

和刻蚀循环数相同时,两种晶圆的刻蚀速率十分接近。

图 2.17 所示为两种晶圆最大扇贝形边纹随 TSV 直径的变化情况。可以看出:①对于两种晶圆以及不同的刻蚀循环数而言,虽然扇贝形边纹大小有所波动,但总体趋势是 TSV 直径越大,最大扇贝纹越大。②对于两种晶圆,刻蚀循环数越大或刻蚀时间越长,产生的扇贝形边纹就越小。③两种晶圆刻蚀的最大扇贝形边纹并不相同,当刻蚀循环数较小时两者差别较小,而当刻蚀循环数较大时两者差别较大。图 2.18 给出了 TSV 直径为 $5\mu m$,刻蚀循环数分别为 X、$1.63X$、$2.25X$ 时所对应的两种晶圆剖面图。

(4) 直孔刻蚀工艺参考

表 2.4 列出了实现高质量(即较小的扇贝形边纹、低成本)TSV 刻蚀需要考虑的关键因素。可以看出:①TSV 刻蚀的关键因素包括刻蚀速率、功率、刻蚀循

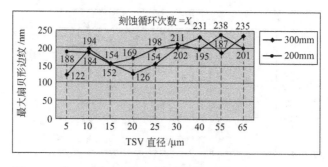

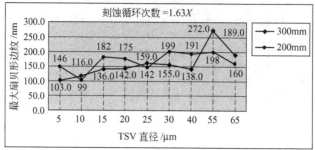

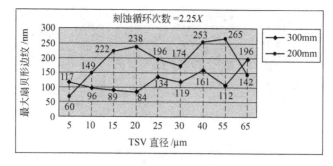

图 2.17 200mm 和 300mm 晶圆在不同刻蚀循环
数下最大扇贝形边纹和 TSV 直径之间的关系

环数、除湿以及扇贝形边纹；②刻蚀速率、功率、刻蚀循环数（时间）之间的平衡是必不可少的；③建议的初始工艺参数为：钝化（形成一个较薄的钝化层）时间为 1.8s，刻蚀 1（去除钝化层）时间为 1.3s，刻蚀 2（对 Si 进行垂直方向的刻蚀）时间为 2s。

（5）总结与建议

本节讨论了刻蚀速率对 TSV 孔壁形貌（扇贝形边纹）的影响。通过实验设计确定了 300mm 晶圆上 TSV 孔的最优刻蚀速率。此外，在相同的刻蚀参数、掩膜以及 TSV 直径条件下，分析比较了 200mm 和 300mm 晶圆的 TSV 刻蚀速率、刻蚀深度以及扇贝形边纹大小等，得到了如下结论和建议[10]。

1) 300mm 晶圆的 TSV 刻蚀：

① 刻蚀速率越大，产生的扇贝形边纹越大。

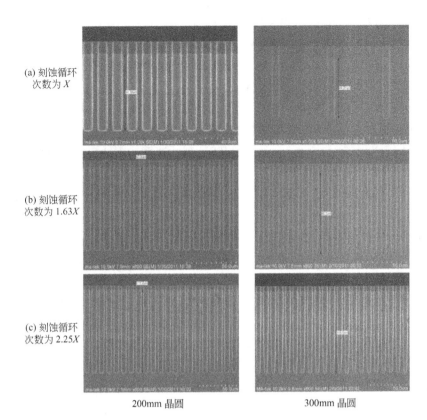

图 2.18 左侧与右侧分别为 200mm 和 300mm
晶圆的剖面图，TSV 直径均为 $5\mu m$

表 2.4 高质量 TSV 刻蚀需要考虑的关键因素

需要考虑的因素	结论
(1) 扇贝形边纹问题	钝化时间(C_4F_8 阶段)增加，扇贝形边纹减小 刻蚀时间(SF_6 阶段)增加，扇贝形边纹增大
(2) 刻蚀速率问题	钝化时间(C_4F_8 阶段)增加，刻蚀速率减小，出现刻蚀条纹和残渣 刻蚀时间(SF_6 阶段)增加，刻蚀速率增大
(3) 功率问题	功率增加，刻蚀速率增大，扇贝形边纹增大，孔壁钝化不充分，导致出现刻蚀条纹 功率减小，刻蚀速率减小，扇贝形边纹减小
(4) 刻蚀循环数问题	刻蚀循环数越大，TSV 直径越大，TSV 深度越深，刻蚀速率越小
(5) 除湿问题	建议在去除光刻胶之后进行除湿 X 截面分析+SEM 检查

② 对于直径为 $1\mu m$ 的 TSV，刻蚀速率对扇贝形边纹的影响很小。当刻蚀速率从 $1.7\mu m/min$ 增大至 $2.13\mu m/min$ 时，扇贝形边纹大小从 57nm 增加到 83nm。

③ 对于直径为 $10\mu m$ 的 TSV，刻蚀速率对扇贝形边纹的影响比较明显。当刻蚀速率从 $3.5\mu m/min$ 增大至 $5.8\mu m/min$ 时，扇贝形边纹大小从 107nm 增加到 278nm。

④ 对于直径为 20μm 的 TSV，当刻蚀速率从 4.2μm/min 增大至 8.8μm/min 时，扇贝形边纹大小从 93nm 增加到 225nm。

⑤ 对于直径为 30μm 的 TSV，当刻蚀速率从 4.6μm/min 增大至 9.5μm/min 时，扇贝形边纹大小从 97nm 增加到 258nm。

⑥ 对于直径为 50μm 的 TSV，当刻蚀速率从 5.295μm/min 增大至 11μm/min 时，扇贝形边纹大小从 99nm 增加到 235nm。

2) 比较 200mm 与 300mm 晶圆的 TSV 刻蚀速率、刻蚀深度以及扇贝形边纹，发现：

① 对于两种晶圆以及不同的刻蚀循环数而言，TSV 直径越大，刻蚀深度越深。

② 对于两种晶圆而言，刻蚀循环数越大，刻蚀深度越深。

③ 当 TSV 直径和刻蚀循环数相同时，两种晶圆上的 TSV 刻蚀深度基本相同。

④ 对于两种晶圆以及不同的刻蚀循环数而言，TSV 直径越大，刻蚀速率越高。

⑤ 对于两种晶圆而言，刻蚀循环数越大，刻蚀速率越低。

⑥ 当 TSV 直径和刻蚀循环数相同时，两种晶圆的刻蚀速率十分接近。

⑦ 对于两种晶圆以及不同的刻蚀循环数而言，TSV 直径越大，扇贝形边纹越大。

⑧ 对于两种晶圆而言，刻蚀循环数越大，扇贝形边纹越小。

⑨ 两种晶圆上的扇贝形边纹并不相同，当刻蚀循环数较小时两者差别较小，而当刻蚀循环数较大时两者差别较大。

2.5 绝缘层制作

由于硅是导电材料，为了防止 TSV 漏电以及 TSV 间的串扰，必须在 TSV 孔壁上制作厚度不小于 $0.1\mu m$ 的 SiO_2 绝缘层。可以采用热氧化法在 TSV 孔壁上形成 SiO_2 层，或者采用等离子增强化学气相沉积（PECVD）工艺沉积出绝缘层。一般来说，热氧化工艺制得的 SiO_2 层比较均匀，但是需要 $>1000℃$ 的高温；而 PECVD 工艺需要经过好几个步骤来制作 SiO_2 层，但其所需温度 $<250℃$。

2.5.1 热氧化法制作锥形孔绝缘层

对于无源转接板，即不包含诸如晶体管之类的任何器件的转接板，可以通过热氧化工艺制作绝缘层。如图 2.19 所示的熔炉（Ellipsiz 公司），可以制作出厚度 $1\mu m$ 的 SiO_2 层。H_2（氢气）和 O_2 在石英腔内混合并在 1050℃ 高温下反应燃烧产

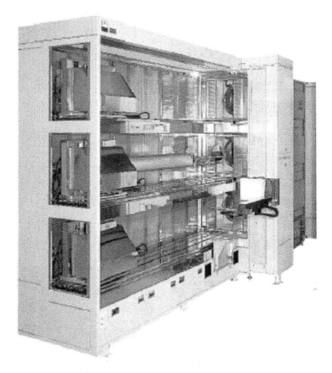

图 2.19 Ellipsiz 公司生产的热氧化系统

生蒸汽,这也就是为什么将该工艺称为湿法工艺的原因。Si 的氧化时间为 3 个半小时,图 2.20 所示为共形 SiO_2,可以看到在 TSV 的顶部、中部和底部都覆盖了

图 2.20 利用热氧化工艺(1080℃)制作的介电绝缘层(SiO_2)

均匀的 SiO_2 层,其均匀度为±5%。由于热氧化过程中 Si 的损耗[14~16],热氧化后 TSV 壁面的粗糙度从 200~250nm 减小到 100nm 以下。对于含有器件的晶圆/转接板,由于其能够承受的最高温度在 400~450℃,因此不能采用热氧化法在 TSV 壁面制作绝缘层。

2.5.2 PECVD 法制作锥形孔绝缘层

也可以通过等离子增强化学气相沉积(PECVD)干法工艺来沉积 SiO_2 层,美国应用材料公司和 SPTS 公司都提供这类的设备,如图 2.21 所示。图 2.22 为采用 PECVD 工艺制作的 SiO_2 层,可以看到,沉积的氧化层厚度在 TSV 的顶部为 1.9~2μm,顶部孔壁上的厚度为 1.3~1.4μm,中部孔壁上的厚度为 0.7~0.8μm,底部孔壁则为 0.35~0.45μm。可见,干法工艺沉积的 SiO_2 层,其均匀度要比湿法工艺差。但这种工艺的温度低于 250℃[16],对于大多数晶圆都适用。

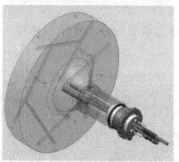

图 2.21 SPTS 公司的等离子增强化学气相沉积(PECVD)设备,
包含一个气冷圆盘用于厚衬膜的低温沉积

2.5.3 PECVD 法制作直孔绝缘层的实验设计

尽管目前已经发展了一些用于制作 SiO_2 的材料,但本节重点关注正硅酸乙酯 [$Si(OC_2H_5)_4$, TEOS] 的研究,因为这种材料是制作半导体层间材料的先驱。TEOS 的重要特性包括具有良好的涂层均匀性、相对惰性以及室温下即具有流动性等特点。另外,TEOS 在 700℃即可热分解并形成 SiO_2,再加上等离子增强,使得沉积温度低于 500℃。TEOS 沉积的机理是表面反应,与硅烷的质量输运机理相比,TEOS 具有较低的黏性系数,使其更容易到达 TSV 的底部区域[5]。

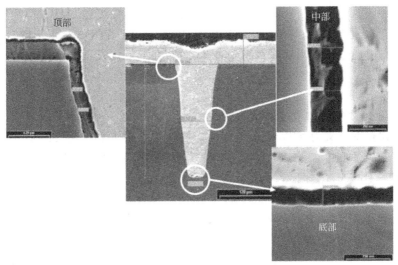

图 2.22 利用等离子增强化学气相沉积（PECVD）工艺制作的介电绝缘层（SiO_2）

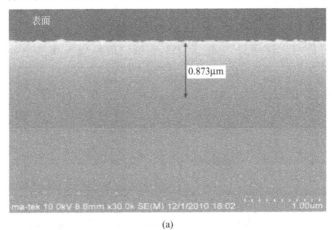

(a)

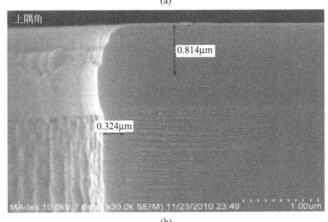

(b)

图 2.23 远离 TSV 孔处的表面氧化层厚度为 8730Å [图(a)]，
TSV 孔附近的表面氧化层厚度为 8140Å [图(b)]

实验中[23]，采用了传统的 300mm AMAT PECVD 设备。以直径为 10μm 和 30μm 的 TSV 孔为对象，对一些关键参数如 RF 功率、TEOS 流量、O_2 流量以及压力等进行实验设计。主要观察区域为 TSV 底部壁面以及下隅角处氧化层最薄地方的阶梯覆盖率（step coverage）[6]。实验中，将温度设定在 180℃，这是由于 3D IC 集成中用于临时键合的大部分粘接材料都不能承受高温。基于这些实验，给出了关于低温 PETEOS 氧化工艺的建议。

阶梯覆盖率定义为 TSV 孔壁上氧化层厚度与晶圆表面氧化层厚度的比率。然而，由于 Si 表面的负载效应，晶圆表面氧化层的厚度在 TSV 孔附近和在远离 TSV 孔的位置是不同的。如图 2.23 所示，图(a)为远离 TSV 孔处的晶圆表面氧化层，其厚度为 8730Å；图(b)中，接近 TSV 孔处的表面氧化层厚度减小至 8140Å。本文中，选取远离 TSV 孔处的表面氧化层厚度作为分母计算阶梯覆盖率[23]。

2.5.4 实验设计结果

图 2.24 中给出了直径 10μm、深度 30μm TSV 孔阶梯覆盖率的实验设计结果。可以看出：①提高阶梯覆盖率的一个直接方法就是增大 RF 功率以及降低 TEOS 流量，这意味着降低沉积速率；②O_2 流量不是影响阶梯覆盖率的主要因素；③压力

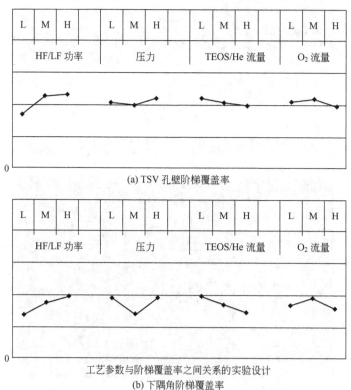

(a) TSV 孔壁阶梯覆盖率

工艺参数与阶梯覆盖率之间关系的实验设计
(b) 下隅角阶梯覆盖率

图 2.24　TSV（直径为 10μm，深度为 30μm）孔壁阶梯覆盖率 [图(a)] 和下隅角阶梯覆盖率 [图(b)] 的实验设计结果

的影响比较复杂，需要进一步的实验研究。

对于直径 $10\mu m$、深度 $60\mu m$ 的高深宽比 TSV，选择更高的 RF 功率、更低的 TEOS 流量以及中等的 O_2 流量等工艺参数，压力参数设为高压（参数 1）和低压（参数 2）两个水平。表 2.5 列出了相应的实验结果，可以看出：①对于高压力参数 1，阶梯覆盖率在底部壁面、下隅角以及底面上分别为 8.67%、7.9% 和 8.3%；②对于低压力参数 2，阶梯覆盖率在底部壁面、下隅角以及底面上分别增加至 9.7%、8.3% 和 11.7%。低压力有助于提高阶梯覆盖率，原因是 TEOS 输运的平均自由路径在低压下更长。

表 2.5 盲孔 TSV（直径为 $10\mu m$，深度为 $60\mu m$）的实验设计结果

项目		参数 1	参数 2	参数 3
HF/LF 功率		H(高)	H(高)	H(高)
压力		H(高)	L(低)	L(低)
TEOS/He 流量		L(低)	L(低)	L(低)
O_2 流量		M(中)	M(中)	M(中)
温度		180℃	180℃	250℃
阶梯覆盖率	上隅角	41.5%	40.0%	37.1%
	顶部孔壁	22.6%	20.2%	34.1%
	中部孔壁	11.7%	13.7%	15.0%
	底部孔壁	8.6%	9.7%	11.7%
	下隅角	7.9%	8.3%	10.3%
	底面	8.3%	11.7%	16.9%

此外，将温度设为参数 3，当温度升高到 250℃ 时（该温度是临时键合胶所能承受的最高温度），可以发现阶梯覆盖率有显著提升，尤其是在底部区域，达到 16.9%。图 2.25 所示为该参数下绝缘层的 SEM 图像，可见温度是影响阶梯覆盖率的关键参数之一。以上结果表明，更高的温度、更高的 RF 功率、更低的压力和更低的 TEOS 流量可以得到更高的 TSV 孔壁氧化层阶梯覆盖率。

2.5.5 总结与建议

① RF 功率、TEOS 流量、压力、温度以及 O_2 流量是影响 TSV 绝缘层制作质量的重要参数。

② 更高的 RF 功率、更低的 TEOS 流量、更高的温度以及更低的压力可以得到更高的 TSV 孔壁氧化层阶梯覆盖率。

③ O_2 流量似乎是一个次要因素。

④ TSV 底部壁面和下隅角处是绝缘层制作的关键区域。

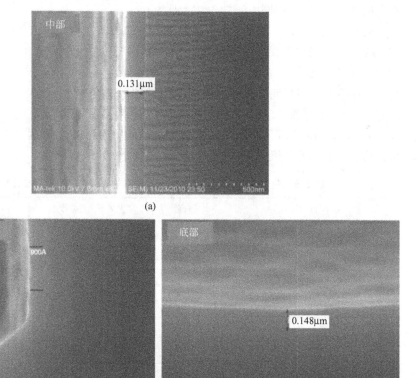

图 2.25 TSV 底部孔壁 [图(a)]、下隅角 [图(b)] 以及底面 [图(c)] 的 SEM 图像（TSV 直径为 10μm，深度为 30μm，参数 3）

2.6 阻挡层与种子层制作

最常用的 TSV 阻挡层材料为钛（Ti）和钽（Ta）。Ti 在元素周期表中的原子序数为 22，它是一种低密度、高强度、有光泽、耐腐蚀的银色过渡金属。Ti 的原子量、电离能、熔点、沸点以及密度分别为 47.88、658kJ/mol、1660℃、3287℃ 和 4.5g/cm³。

另一方面，Ta 在元素周期表中的原子序数为 73，它是一种稀有的、高硬度、蓝灰色、有光泽的金属，具有极高的抗腐蚀性。Ta 的原子量、电离能、熔点、沸点以及密度分别为 180.9497、761kJ/mol、2996℃、5425℃ 和 16.654g/cm³。最常用的 TSV 种子层材料为 Cu，阻挡层和种子层的制作都采用物理气相沉积（PVD）方法，主要设备供应商为探戈（Tango）公司、应用材料公司和 SPTS 公司（见图 2.26）。

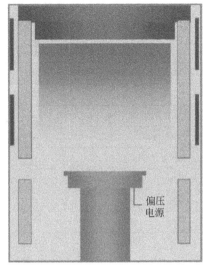

图 2.26 SPTS 的物理气相沉积（PVD）设备

2.6.1 锥形 TSV 孔的 Ti 阻挡层与 Cu 种子层

图 2.22 中的 TSV 孔制作阻挡层和种子层后，其 SEM 图像如图 2.27 所示。Ti 阻挡层厚度为 3000Å，Cu 种子层厚度为 $2\mu m$。制作工艺为 PVD，设备是探戈公司

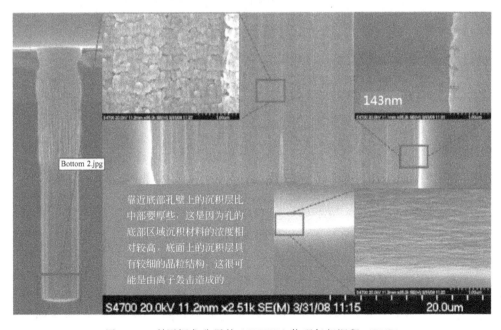

图 2.27 利用探戈公司的 AXCECA 物理气相沉积（PVD）
设备制作的 Ti 阻挡层和 Cu 种子层

的 AXCECA 机台，表 2.6 给出了采用的关键工艺参数。可以看出，孔的底部和壁面上都覆盖良好[15]。

表 2.6 制作 Ti 阻挡层和 Cu 种子层的重要工艺参数

制程	Ti	Cu
厚度	3kÅ	2μm
DC 电源功率	8kW	8kW
线圈功率	600W	600W
偏压电源功率	1000W	1000W
Ar 流量/压力	10sccm/1.5mt	20sccm/2mt
偏置电压	100V	150V
靶电压	550V	630V

Ti 阻挡层和 Cu 种子层同样可以通过湿法工艺制作。高深宽比 TSV 孔的绝缘层和阻挡层沉积可采用 Alchimer 公司的 eG（electrografting）技术，即刚开始采用小电流形成化学键合，接着利用非电镀化学扩散进行沉积。图 2.28 所示即为这种湿法沉积工艺制作的 100nm 厚的 Ti 阻挡层和 1μm 厚的 Cu 种子层，可以看到 TSV 孔壁和底面均覆盖良好。

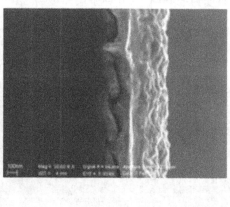

图 2.28 利用 Alchimer 公司的 eG 技术制作的 Ti 阻挡层和 Cu 种子层

2.6.2 直 TSV 孔的 Ta 阻挡层与 Cu 种子层

为了克服深孔沉积的困难，PVD 系统已经从最初的平面二极管溅镀发展到长距离溅射、准直溅射以及射频溅射。最近出现的电离金属等离子（IMP）PVD 系统更进一步增强了溅射工艺的能力。图 2.29 所示为 AMAT PVD 系统的示意图，

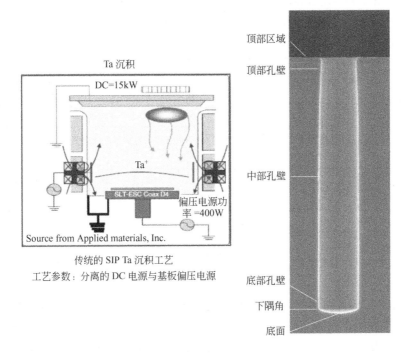

图 2.29 不同 DC 电源功率和基板偏压电源功率的情况下，TSV（直径为 $10\mu m$，深度为 $60\mu m$）孔壁 Ta 阻挡层的阶梯覆盖率

该系统采用了更加先进的自电离等离子（SIP）技术。功率高达数十千瓦的直流（DC）靶电源提高了溅射的效率，并且无需 IMP 系统的射频线圈即可使得靶金属发生自电离。极低的工艺压力和高功率直流靶电源产生的超高等离子体电离能，保证了更好的各向异性沉积。在下面的阻挡层与种子层沉积的阶梯覆盖率实验中，重点研究了对垂直度影响最大的两个参数——DC 靶电源功率和基板偏压电源功率[23]。

2.6.3 直 TSV 孔的 Ta 阻挡层沉积实验与结果

以直径和深度分别为 $10\mu m$ 和 $60\mu m$ 的 TSV 孔为对象，考虑三个水平的 DC 靶电源功率和三个水平的基板功率。带有 SEM 格栅的掩膜布局设计可以确保沿孔中心线进行精确切割，以便得到剖面的 SEM 图像。表 2.7 给出了实验结果，可以看出：①在较低的 DC 靶电源功率水平下，阶梯覆盖率较大；②随着电离功率的增加，等离子具有更高的能量，可以到达深孔的底部；③但是由于 SIP 系统的超高电离率，增大 DC 靶电源功率会减小等离子的平均自由路径，进而降低阶梯覆盖率；④基板偏压电源功率越高，电离金属更容易到达孔的底部；⑤基板偏压电源对于 SIP 系统来说是必需的，否则高度电离的金属会偏离直线并发生随机碰撞，当基板偏压电源功率为零时，阶梯覆盖率几乎为零。

表2.7 基板偏压电源功率和DC电源功率对TSV盲孔Ta阻挡层阶梯覆盖率的影响（TSV孔直径为10μm，深度为60μm）

基板偏压电源功率/W	底部孔壁	底部拐角	参考值
0	0%	0%	13.5%
L(低)	20.2%	12.7%	17.7%
H(高)	25.6%	21.9%	23.5%

不同基板偏压电源功率对Ta阻挡层阶梯覆盖率的影响：更高的基板偏压电源功率对应更高的平均阶梯覆盖率

DC电源功率/W	底部孔壁	底部拐角	参考值
L(低)	21.5%	15.5%	23.2%
M(中)	15.1%	10.5%	15.8%
H(高)	9.2%	8.6%	15.7%

不同DC电源功率对Ta阻挡层阶梯覆盖率的影响：更低的DC电源功率对应更高的平均阶梯覆盖率

通过微调PVD功率参数，当DC靶电源功率较低而基板偏压电源功率较高时，TSV孔下隅角处的阶梯覆盖率高达30%，而该区域往往是采用PVD制作Ta阻挡层时覆盖率最低的地方。图2.30所示为沉积完Ta阻挡层后TSV孔横截面的SEM图像。

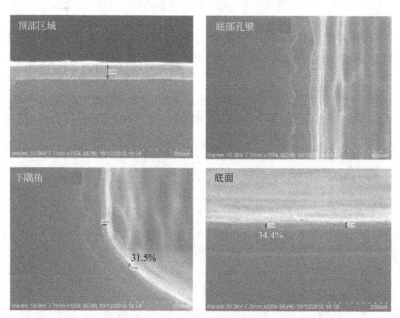

图2.30 优选参数下（较低的DC靶电源功率+较高的基板偏压电源功率），直径10μm、深度60μm TSV孔壁的Ta阻挡层覆盖情况（SEM图像）。Ta层的阶梯覆盖率高达30%

2.6.4 直 TSV 孔的 Cu 种子层沉积实验与结果

Cu 种子层沉积实验以直径为 $10\mu m$、深度为 $30\mu m$ 的 TSV 盲孔为对象，如图 2.31 所示。图 2.32 和图 2.33 所示为实验结果，可以看出：① 与 Ta 阻挡层沉积类似，更高的基板偏压电源功率产生更好的 Cu 种子层阶梯覆盖率，原因是高的基板偏压电源功率对电离金属产生较高的吸引力；② 进一步增大基板偏压电源功率，可能会牺牲沉积效率，但阶梯覆盖率以及覆盖的均匀性（这个更重要）会提高；③ 虽然较高的 DC 电源功率可以使 TSV 孔底部沉积更厚，但较低的 DC 电源功率可以改善深孔关键区域——孔壁和隅角处的阶梯覆盖率。因此，Cu 种子层沉积工艺的最佳参数组合为较低的 DC 电源功率和较高的基板偏压电源功率。

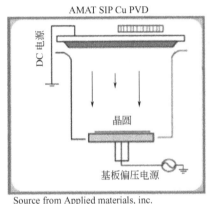

图 2.31 不同 DC 电源功率和基板偏压电源功率组合下，TSV（直径为 $10\mu m$，深度为 $30\mu m$）孔壁 Cu 种子层的阶梯覆盖率

2.6.5 总结与建议

① 制作 TSV 阻挡层（Ta）和种子层（Cu）的重要参数为 DC 靶电源功率和基板偏压电源功率（这个是必需的）。

② 对于 SIP PVD 系统，较低的 DC 靶电源功率可以获得更好的 TSV 阶梯覆盖率，同时，更高的基板偏压电源功率可以使电离金属更容易到达孔的底部。因此，采用较低的 DC 靶电源功率和较高的基板偏压电源功率可以实现较高的 TSV 孔壁阶梯覆盖率。

③ 阻挡层和种子层制作的关键区域与氧化层相同。

④ 2.7.3 节对 TSV 漏电量的测量结果（<50pA）表明，采用这些工艺参数制作的绝缘层和阻挡层性能良好。

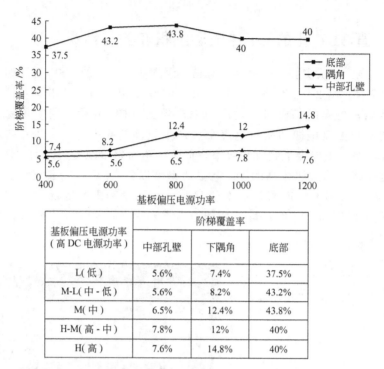

基板偏压电源功率 （高 DC 电源功率）	阶梯覆盖率		
	中部孔壁	下隅角	底部
L（低）	5.6%	7.4%	37.5%
M-L（中-低）	5.6%	8.2%	43.2%
M（中）	6.5%	12.4%	43.8%
H-M（高-中）	7.8%	12%	40%
H（高）	7.6%	14.8%	40%

图 2.32　高 DC 电源功率与不同基板偏压电源功率组合下，TSV（直径 10μm，深度 30μm）孔壁 Cu 种子层的阶梯覆盖率

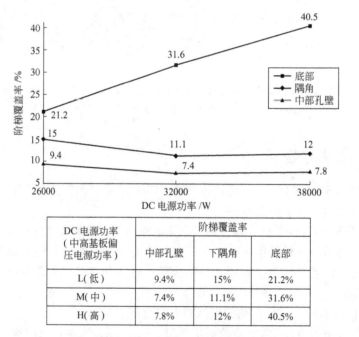

DC 电源功率 （中高基板偏 压电源功率）	阶梯覆盖率		
	中部孔壁	下隅角	底部
L（低）	9.4%	15%	21.2%
M（中）	7.4%	11.1%	31.6%
H（高）	7.8%	12%	40.5%

图 2.33　中高基板偏压电源功率与不同 DC 电源功率组合下，TSV（直径为 10μm，深度为 30μm）孔壁 Cu 种子层的阶梯覆盖率

2.7 TSV 电镀 Cu 填充

TSV 孔可以采用电镀方法填充 Cu、Ti、Al 或焊料，或者采用溅射方法填充 W，或者采用真空印刷方法填充聚合物。根据已有的参考文献，大部分的 TSV 孔都采用电镀 Cu 进行填充，因此本书主要讨论电镀 Cu 填充。大部分的 Cu 电镀设备由 Semitool（Applied Materials 的子公司）提供，如图 2.34 所示。将 TSV 孔用 Cu 填充满对于 3D IC TSV 制程来说至关重要，电镀 Cu 过程中出现的孔隙可能造成潜在的电性能、热性能以及机械可靠性方面的问题。

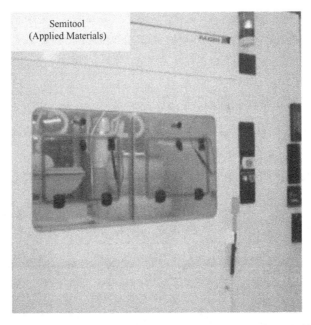

图 2.34 Semitool 公司（Applied Materials 的子公司）的 Cu 电镀设备

深孔和高深宽比 TSV 孔电镀 Cu 时常选用硫酸铜或氰化物电镀液[14~16]，典型的电解液成分包括 $CuSO_4$、H_2SO_4 和 Cl^-，添加剂包括抑制剂、促进剂和整平剂。

2.7.1 电镀 Cu 填充锥形 TSV 孔

即使对于图 2.8 所示的深度达 $300\mu m$ 的 TSV 孔，也可以实现无孔洞填充，如图 2.35 所示[14~16]。对于深孔和高深宽比锥形 TSV 孔，要实现无孔洞填充，电镀液化学成分、低电镀电流以及预处理（如浸润、去离子水冲洗、预吸收促进剂）等都是关键因素。电镀工艺采用反向脉冲电镀，这需要采用包含增亮剂和整平剂的双组分添加剂体系[14~16]。

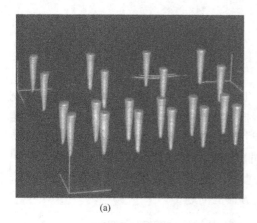
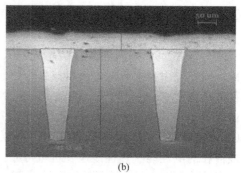

图 2.35 电镀 Cu 填充 TSV 孔，图(a)：填充完成后通过计算机断层扫描（CT）重构得到的 3D X 射线图像；图(b)：填充完成后的无孔洞 TSV（已完成 CMP）

2.7.2 电镀 Cu 填充直 TSV 孔

电镀 Cu 过程中，对填充率进行监控非常重要。一般将填充率定义为填充过程中底部电镀厚度与顶部侧壁电镀厚度的比值。图 2.36～图 2.38 给出了一些 TSV Cu 电镀实验的结果。图 2.36(a) 为部分填充后的图像，按照定义其填充率为 14.3

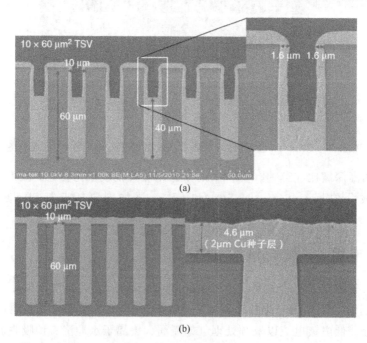

图 2.36 图(a)：自下向上填充，填充率为 14.3；图(b)：填充率为 14.3 时得到的无孔洞 TSV，其深宽比为 6

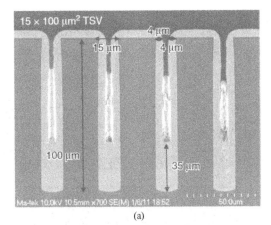

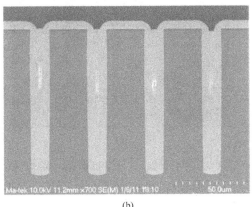

图 2.37　图(a)：自下向上填充，填充率为 4.6；图(b)：填充率为 4.6 时得到的 TSV，其深宽比为 6.7，可以观察到一些细小缝隙

[=40/(1.6−0.2)×2]，该值比所填充 TSV 孔的深宽比 (60/10) 大许多，意味着可以完成填充，而且还有足够的余量对电镀工艺参数进行微调。然而，这也意味着所选工艺参数会增加总的电镀时间。图 2.36(b) 所示为完全填充且无孔隙的 TSV。

图 2.37(a) 为另一组 Cu 电镀实验结果，其填充率为 4.6[=35/(4−0.2)×2]，低于 TSV 孔的深宽比 6.7(100/15)。当电镀时间成比例增加时，会导致孔的中心位置产生缝隙，如图 2.37(b) 所示。

应当指出的是，在电镀液相同的前提下，影响 Cu 电镀效果最重要的工艺参数是电流密度和波形。较大的电流密度可以缩短电镀时间，但是，由于电流在孔顶部集聚以及物质向孔底部输运受到限制，过高的电流密度会导致孔洞的形成[7]。为了避免出现孔洞，同时要缩短总的电镀时间，可以调整电流密度的时间分布，即在自下向上主要填充阶段适当降低电流密度，但保持总电流不变。经过调整之后，直

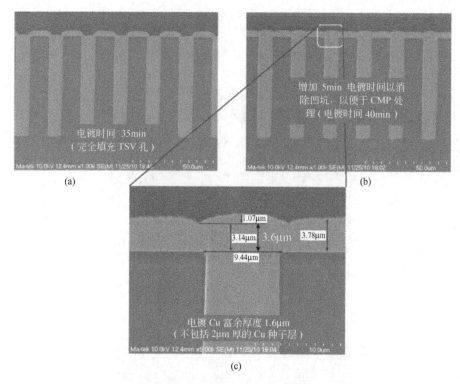

图 2.38　图(a)：电流调整后，35min 即可完成对直径 10μm、深度 60μm TSV 孔的无孔洞填充；图(b)：增加 5min 电镀时间以消除凹坑（总电镀时间为 40min）；图(c) 为放大图

径 10μm、深度 60μm 的 TSV 孔其填充时间由 60min 缩短至 35min，且对填充效果没有影响，如图 2.38(a) 所示。

为后续 CMP 工艺考虑，增加了 5min 电镀时间以使电镀表面更加平坦，如图 2.38(b) 所示。另外，由于电流密度的调整，顶部电镀 Cu 覆盖层富余厚度由 2.6μm 减小到 1.6μm，这也有利于 CMP 工艺。

TSV 电镀 Cu 填充完成后，还需要足够的温度和时间来进行退火处理，否则由于 Cu 晶粒的生长，填充 Cu 会胀出 TSV 表面，可能对后续工艺产生影响。

2.7.3　直 TSV 盲孔的漏电测试

以直径 10μm、深度 60μm 的 TSV 盲孔为对象，采用上述优化工艺参数制作其绝缘层、阻挡层、种子层，并进行电镀 Cu 填充，然后测试其漏电性能。图 2.39 所示为测试结果，可以看出，节距为 150μm 的两个 TSV 之间的漏电电流小于 50pA[23]，表明其绝缘层和阻挡层性能良好。

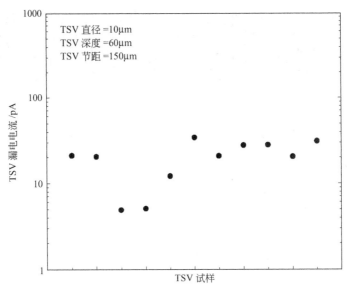

图 2.39　直径 10μm、深度 60μm TSV 盲孔的漏电测试结果

2.7.4　总结与建议

① 当电镀液相同时，影响 TSV 电镀 Cu 填充效果最主要的工艺参数为电流密度和波形。

② 为达到无孔洞填充，在保持总电流不变前提下，需对电流密度的时间分布以及自下向上主要填充阶段的电流密度进行调整。

③ 为后续 CMP 工艺考虑，建议增加几分钟的电镀时间。为了防止 TSV 中填充 Cu 胀出，建议在电镀完成后将 TSV 在足够的温度和时间下进行退火处理。

2.8　残留电镀 Cu 的化学机械抛光（CMP）

2.8.1　锥形 TSV 的化学机械抛光

如图 2.8 和图 2.22 所示深度达 300μm 的锥形 TSV 孔，为了实现无孔洞填充，其电镀时间较长，因而会在晶圆表面形成一层厚度为 30～50μm 的 Cu 覆盖层[14~16]，如图 2.40 所示，这层较厚的 Cu 覆盖层会导致晶圆弯曲。如果采用传统的化学刻蚀去除 Cu 覆盖层，则需要较长时间，并且刻蚀表面不均匀。因此，多采用化学机械抛光（CMP）方法去除晶圆表面的 Cu 覆盖层。对于具有较高应力的 Cu 覆盖层，CMP 设备可选用 Okamoto GNX 200，选用软抛光垫和罗门哈斯（Rohms and Hass）公司的具有强去除率的磨抛液。抛光载荷为 320～350gf/cm²，抛光垫和卡盘的转速为 90r/min，磨抛液的供给速度为 170～200mL/min。

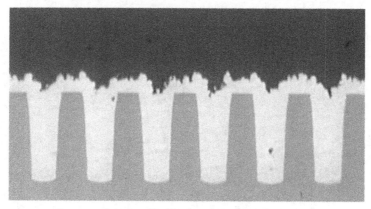

图 2.40 较长的电镀时间在晶圆表面形成 Cu 覆盖层

2.8.2 直 TSV 的化学机械抛光

较深的 TSV 直孔电镀完之后,同样会在晶圆表面形成较厚的 Cu 覆盖层,本节针对 300mm 晶圆表面 Cu 覆盖层的 CMP 去除工艺,讨论工艺优化问题。为了使 TSV Cu 凹坑最小,针对目前的两步 Cu 抛光工艺,提出要选用合适的 Cu 磨抛液。第一步,采用具有较高 Cu 去除率的磨抛液去除大部分的 Cu;第二步,采用具有较强 Cu 钝化能力的磨抛液对 Cu 表面进行整平。经过这两步,300mm 晶圆上直径为 $10\mu m$ 的 TSV 其 Cu 凹坑情况可以改善 97%。通过对 TSV 与 RDL 的电镀 Cu 覆盖层进行优化,可以减小凹坑。

(1) CMP 工艺

近来,基于大马士革技术的低电阻率 Cu 被用作 TSV 导体材料[24]。但是,由于深孔 TSV 的电镀 Cu 时间较长,晶圆表面形成的电镀 Cu 覆盖层必须通过 CMP 去除,如图 2.41 所示。文献 [25] 和 [26] 表明,深孔 TSV 填充完后,晶圆表面的 Cu 覆盖层厚度可达 $30\sim50\mu m$。有两种 CMP 工具可用于去除较厚的 Cu 覆盖层,同时可以保证凹坑最小。一种是采用配有高刻蚀率磨抛液的研磨机/抛光机,另一种是前段 Cu CMP 设备[25]。采用高刻蚀率的磨抛液时,工艺控制至关重要。文献 [27] 的研究表明,通过优化磨抛液和抛光时间,可以实现良好的凹坑控制(凹坑小于 $0.4\mu m$)。

为了得到较高的 Cu 去除率以及最小的凹坑,磨抛液是 CMP 工艺的关键材料。文献 [11] 研究了具有较高 Cu 去除能力和较高 Cu 钝化能力的磨抛液。由于深孔 TSV 电镀填充 Cu 的时间较长,金属覆盖层不仅影响 TSV 的 CMP 效果,而且也影响 RDL 的 CMP 效果。TSV 正面金属的 CMP 工艺(见图 2.41)类似于 CMOS 前段制程(FEOL)中传统的 Cu CMP 工艺。另一方面,对于减薄后的薄晶圆,必须将其与载体(支撑晶圆)临时键合,如图 2.41 所示。因此,还必须考虑背面 CMP 工艺。背面 CMP 工艺是为了露出 TSV Cu 以实现背面 RDL 的连接,与正面 CMP 工艺不同,详见 2.9.1。正面 CMP 是为了去除电镀后晶圆表面带有凹坑的

第 2 章　TSV 技术

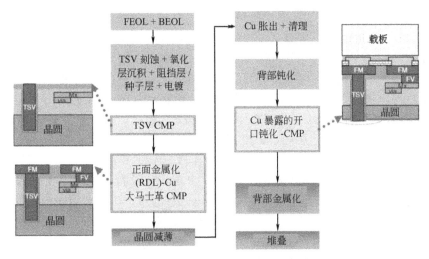

图 2.41　CMP 在 TSV 工艺中的可能应用

Cu 覆盖层，抛光后表面平整且无金属残渣是该工艺的考核指标。而背面 CMP 工艺恰好相反，是为了去除覆盖在 TSV Cu 柱上的绝缘介电层，使得 TSV Cu 的表面露出，同时保证硅表面的介电层损失最小，抛光后 TSV Cu 柱的形貌以及介电层的去除量是该工艺步的考核指标，详见 2.9.1。

本节研究了图 2.41 中与 TSV 相关的前两个 CMP 工艺，包括正面 TSV 和 RDL 的 Cu CMP。提出了一个两步 Cu CMP 工艺，其中包含两种不同的 Cu 磨抛液用以减小 TSV 与 RDL 电镀 Cu 的凹坑。此外，还讨论了 Cu 覆盖层对 CMP 效果的影响。基于研究结果，针对 CMP 磨抛液与 Cu 覆盖层给出了一组工艺指导意见，以优化 TSV 和 RDL 结构上 Cu 的形貌[11]。

(2) 去除 TSV 残留 Cu 的 CMP 工艺

测试模型：对于去除 TSV 和 RDL 上残留 Cu 的 CMP 工艺而言，选择合适的磨抛液是减小凹坑的关键。首先，以含有直径 $10\mu m$、深度 $60\mu m$ TSV 的 300mm 晶圆［见图 2.42(a)］为对象，研究磨抛液对残留 Cu 凹坑的影响。依据电镀电

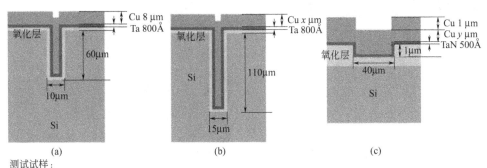

测试试样：
(a) 含有直径为 10μm、深度为 60μmTSV 晶圆，其表面 Cu 覆盖层厚度为 8μm；
(b) 含有直径为 15μm、深度为 110μmTSV 晶圆，其表面 Cu 覆盖层厚度为 x(x=7.2μm、7.8μm、8.6μm、9.0μm)；
(c) 不同厚度 Cu 覆盖层下的 RDL 布局，其凹槽处的线宽为 40μm (y=1.2μm、2μm、4μm、5μm)。

图 2.42　测试模型示意图

流和电镀时间可以计算得到晶圆表面 Cu 覆盖层的厚度大约为 $8\mu m$，阻挡层金属为 Ta，其厚度为 800Å。其次，重点研究 Cu 覆盖层厚度对 TSV 以及 RDL 的影响。图 2.42(b) 所示为 TSV 转接板测试结构，TSV 直径为 $15\mu m$，深度为 $100\mu m$。四种不同的 Cu 覆盖层厚度（$7.2\mu m$、$7.8\mu m$、$8.6\mu m$ 和 $9.0\mu m$）用于确定覆盖层厚度对凹坑的影响，Ta 阻挡层的厚度为 800Å。除此之外，还考察了具有不同布局密度的 RDL，如图 2.42(c) 所示。图 2.42(c) 中，宽度 $40\mu m$、厚度 $1\mu m$ 的 RDL 凹槽上方 Cu 覆盖层的厚度分别为 $1.2\mu m$、$2.0\mu m$、$4.0\mu m$ 和 $5.0\mu m$。

实验设备：采用的 CMP 设备为 AMAT Reflexion，如图 2.43 和图 2.44 所示。机台包括三个抛光压盘，压盘 1 和压盘 2 用于 Cu 磨抛，压盘 3 用于阻挡层（例如 Ta）磨抛和氧化层抛光。研究了用于 Cu 磨抛的磨抛液，用于阻挡层去除的磨抛液等均为商用产品。磨抛液流速设定在 300mL/min，利用高分辨率轮廓仪（HRP）、光学显微镜（OM）以及 SEM 来表征 Cu 表面形貌。

压盘 1(P1) 和压盘 2(P2) 用于 Cu 抛光，压盘 3(P3) 用于阻挡层抛光（例如 Ta）以及氧化层抛光

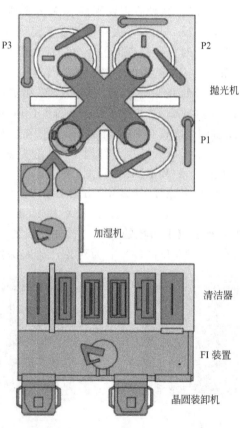

图 2.43　应用材料公司的化学机械抛光机（CMP）

材料和工艺：采用两种 Cu 磨抛液以达到两种不同的抛光目的：一种是 Cu 去除率较高的磨抛液（磨抛液 1），另一种是 Cu 钝化能力较强的磨抛液（磨抛液 2）。

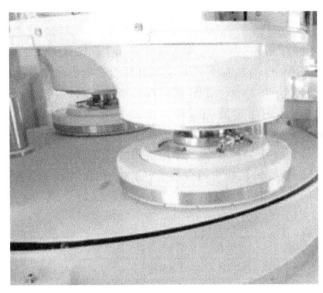

图 2.44 工作中的压盘

图 2.45 给出了两种磨抛液 Cu 去除率随压力的变化情况。当压力为 3psi 时,磨抛液 1 的去除率为 7500Å/min;当压力降至 1psi 时,去除率降到 4700Å/min,与磨抛液 2 在压力为 3psi 时的去除率(4500Å/min)相近。图 2.45 中的表格给出了采用的测试条件。磨抛液 2 与磨抛液 1 相比具有更强的 Cu 钝化能力,因此,在压盘 1 上选用 3psi 压力,并使用具有较高 Cu 去除率的磨抛液 1。在压盘 2 上,则选用磨抛液 1 与 1psi 压力的组合以及磨抛液 2 与 3psi 压力的组合。

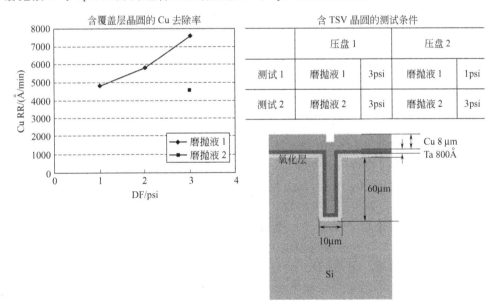

图 2.45 磨抛液 1 与磨抛液 2 的 Cu 去除率

(3) 结果与讨论

Cu 磨抛液的选取：图 2.46 给出了直径 $10\mu m$ 的 TSV 经过 CMP 后的实验结果，其中 TSV 顶部有 $8\mu m$ 厚的 Cu 覆盖层，采用了图 2.42(a) 中的测试模型。当在压盘 1 上使用磨抛液 1 并施加 3psi 压力，在压盘 2 上也使用磨抛液 1 并施加 1psi 压力时，凹坑尺寸均为 $1.2\mu m$，如图 2.46(a) 所示。当将压盘 2 的磨抛液 1 替换为磨抛液 2，同时将压力增加至 3psi 时，则凹坑减小到 250Å，如图 2.46(b) 所示。这个结果表明，对于压盘 2，尽管采用磨抛液 1 与 1psi 压力得到的 Cu 去除率同采用磨抛液 2 与 3psi 压力一样低，但凹坑尺寸却明显不同。另外还表明，在压盘 2 上使用具有较高 Cu 钝化能力的磨抛液（磨抛液 2）对于减缓 Cu 表面凹坑十分有利。这里得到的 250Å 的凹坑结果要比参考文献 [27] 和 [28] 中针对直径为 $10\mu m$ 的 TSV 得到的 1000~4000Å 的测试结果好很多。

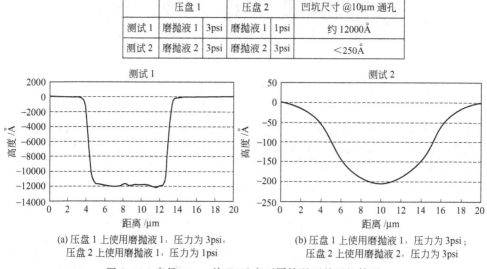

	压盘 1		压盘 2		凹坑尺寸 @10μm 通孔
测试 1	磨抛液 1	3psi	磨抛液 1	1psi	约 12000Å
测试 2	磨抛液 2	3psi	磨抛液 2	3psi	<250Å

(a) 压盘 1 上使用磨抛液 1，压力为 3psi，压盘 2 上使用磨抛液 1，压力为 1psi

(b) 压盘 1 上使用磨抛液 1，压力为 3psi；压盘 2 上使用磨抛液 2，压力为 3psi

图 2.46 直径 $10\mu m$ 的 TSV 在不同情况下的凹坑情况

TSV 覆盖层的影响：基于上一节中关于 Cu 凹坑的实验结果，在压盘 1 上选用磨抛液 1 并施加 3psi 压力，在压盘 2 上选用磨抛液 2 并施加 3psi 压力，以此研究 Cu 覆盖层厚度对 TSV 和 RDL 的影响。TSV 测试结构如图 2.42(b) 所示，其中包含不同的 Cu 覆盖层厚度（如 $7.2\mu m$、$7.8\mu m$、$8.6\mu m$ 和 $9.0\mu m$）。借助激光端点检测仪，当到达 Ta 层时，便自动停止对 Cu 覆盖层的抛光，额外的 60~100s 时间将用于去除 Ta 层。采用高分辨率轮廓仪（HRP）对 CMP 工艺之后的表面形貌进行测量，图 2.47 显示了不同 Cu 覆盖层厚度经过 CMP 处理之后的凹坑结果（500Å）。可以看出，Cu 覆盖层厚度为 $7.2\mu m$ 时的凹坑尺寸超过了 $1.4\mu m$，其原因将在后面给予解释。采用 HRP 对 CMP 后 10 个 TSV（节距为 $30\mu m$）顶部的 Cu 覆盖层进行扫描，其凹坑结果如图 2.48 所示。结果表明，较厚的 Cu 覆盖层不但不会减小凹坑尺寸，反而会增大凹坑尺寸。在压盘 1 上使用的磨抛液具有较低的 Cu 钝化能力，因此更厚的 Cu 覆盖层需要的磨抛时间更长。同时，由于压盘 1 上

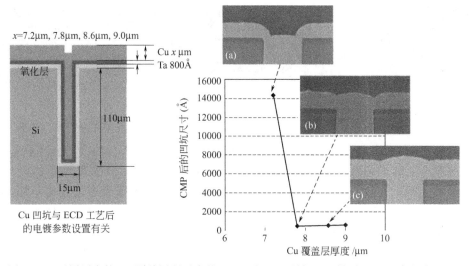

图 2.47　不同厚度的 Cu 覆盖层所对应的 CMP 后 TSV 顶部凹坑情况，TSV 直径为 15μm

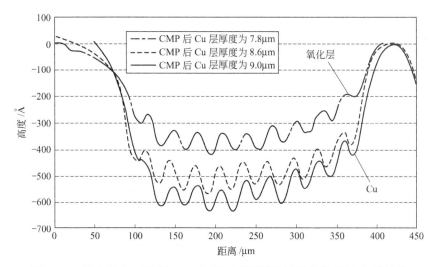

图 2.48　完成 CMP 工艺后，10 个 TSV 表面覆盖层形貌的 HRP 扫描结果

磨抛液在晶圆表面凹坑处的钝化能力较差，抛光时间越长会使得凹坑深度轻微增大。

为什么 7.2μm 厚的 Cu 覆盖层在 CMP 后凹坑更深呢？这是因为电镀时间不足导致最终电镀覆盖层形貌不好，如图 2.49(a) 所示。图 2.47 所示的 CMP 后 Cu 凹坑发生在当 Cu 的电镀形貌和覆盖层与图 2.49(a) 相似的情况下，图 2.49(a) 中 TSV 顶部有一个较深的凹坑。由于在大部分区域磨抛液的 Cu 钝化能力都较低，因此对凹坑处 Cu 的保护能力也较低，从而导致 TSV 顶部受到刻蚀。当该区域的 Cu 覆盖层被去除后，这种刻蚀现象更严重，从而在 TSV 顶部形成凹坑。

另外，对于含有直径 50μm、深度 150μm TSV 的晶圆，其不均匀性使得 Cu 覆

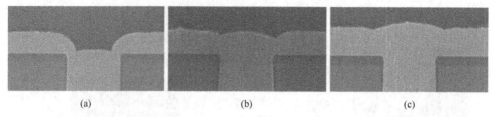

图 2.49　电镀时间 (a)<(b)<(c)，TSV顶部覆盖层由凹坑变为突起

盖层在晶圆的边缘处较厚（12～14μm），而在接近中心的位置较薄（7～10μm）[29]。通过调整 CMP 工艺的压力和速度并不能完全消除覆盖层的不均匀性[29]，因此，完成 CMP 后总会在晶圆表面形成凹坑，依据所处位置的不同，Cu 凹坑的深度为 0.25～1.5μm 不等。根据晶圆电镀 Cu 实验，晶圆中心处的厚度为 8.5μm，边缘处为 10.1μm。幸运的是，本研究中使用的机台可以在不同的区域施加不同的压力。因此，通过在晶圆边缘设置较高的压力达到较高的去除率等参数配置，可以使得整个晶圆的非均匀性对 CMP 后的表面凹坑影响最小。

用于 TSV RDL 的 CMP：由于 RDL 中的金属线路和焊盘密度远大于 TSV，因此有必要研究 Cu 覆盖层对 RDL 凹坑的影响。图 2.42(c) 为 RDL 的测试结构。由于 RDL 在电镀 Cu 时并不像填充 TSV 那样自下而上电镀，而更像是共形电镀过程，因此，RDL 的覆盖层厚度是指图 2.42(c) 中不包括 RDL 阶梯厚度（1μm）在内的厚度 y。图 2.50 所示为线宽 40μm 的 RDL 上厚度分别为 1.2μm 和 2μm 的 Cu 覆盖层经过 CMP 工艺后的 OM 图像。由图 2.50(a) 可以发现，在靠近 RDL 金属线路的边缘处存在一些暗区，此即 RDL 上的金属凹坑位置；而在图 2.50(b) 中并未发现这种情况，说明 RDL 金属线路表面比较平整。

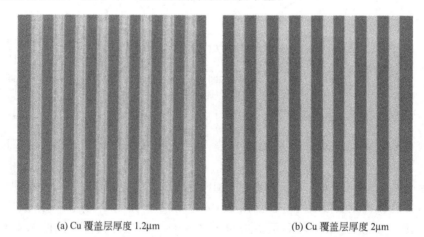

(a) Cu 覆盖层厚度 1.2μm　　　　(b) Cu 覆盖层厚度 2μm

图 2.50　线宽为 40μm 的 RDL 经过 CMP 工艺后的 OM 图像

图 2.51(a) 所示为不同 RDL 金属线宽度/氧化层节距所对应的凹坑情况。以线宽 40μm、间距 40μm 的 RDL 为例，对于 1.2μm 厚的覆盖层，凹坑尺寸为

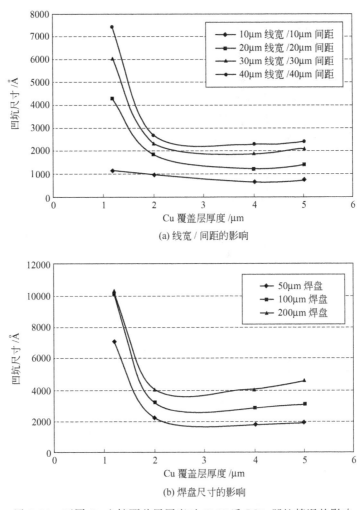

(a) 线宽/间距的影响

(b) 焊盘尺寸的影响

图 2.51　不同 Cu 电镀覆盖层厚度对 CMP 后 RDL 凹坑情况的影响

7400Å；对于 2μm 的覆盖层，凹坑尺寸为 2700Å，所得结果与图 2.50 中的 OM 图像吻合。其他线宽和间距的 RDL，如 10μm、20μm 和 30μm，其凹坑尺寸变化的趋势与 40μm 线宽和间距的 RDL 相同。由此可见，将 Cu 覆盖层的厚度增加至 2μm，即可明显改善 CMP 工艺后金属/氧化层的凹坑情况。图 2.51(b) 和图 2.52 给出了 RDL 金属焊盘的凹坑情况。与 TSV 的情况类似，较厚的 Cu 覆盖层并不能减小金属凹坑的尺寸，尤其对于 200μm 厚的焊盘。所有结果表明，为了达到较好的 Cu CMP 效果，应当考虑对 Cu 电镀覆盖层厚度进行优化。

含有 RDL 的 TSV 晶圆在 CMP 工艺后产生的凹坑对凸点下金属层（UBM）、晶圆微凸点等后续 3D 封装制程有重要影响。通常微凸点允许的高度公差为 5%，即如果微凸点高度为 20μm，那么可接受的高度变化范围不超过 1μm。综合考虑来自 CMP、UBM 制作、掩膜以及晶圆微凸点制作等工艺的平均误差，含有 RDL 的

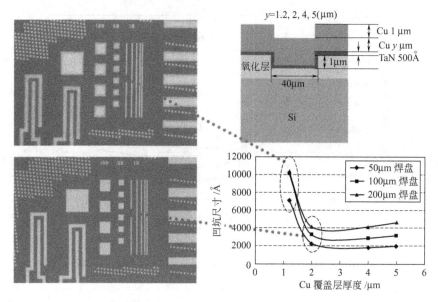

图 2.52　Cu 覆盖层厚度对 CMP 后 RDL 凹坑情况的影响

TSV 其最大允许凹坑尺寸设定为 $0.25\mu m$ 是合理的。另一方面，如果微凸点允许的高度公差为 10%，则相应地，最大允许凹坑尺寸设定为 $0.5\mu m$。该设定值不受 TSV 孔直径、深宽比以及 RDL 尺寸的影响。

（4）直孔 TSV 的 CMP 工艺建议

① 对含有 RDL 和 TSV 的晶圆进行正面 Cu CMP 处理时，建议先采用具有较高 Cu 去除率的磨抛液去除较厚的 Cu 覆盖层，然后改用具有较高 Cu 钝化能力的磨抛液用于清除剩余的 Cu，这样可以减小凹坑尺寸。

② 电镀 Cu 效果对 CMP 工艺后产生的凹坑有重要影响，应尽可能减小电镀 Cu 后形成的凹坑尺寸。电镀后，如果 TSV 顶部的 Cu 覆盖层形貌是凹陷的，则 CMP 后形成的凹坑较大；如果 TSV 顶部的 Cu 覆盖层形貌是凸起的，则 CMP 后形成的凹坑较小。

对 TSV 和 RDL 的 Cu 覆盖层进行 CMP 处理后产生的凹坑，对后续的 3D 封装工艺有重要影响。

2.8.3　总结与建议

① 推荐两步法 CMP 以去除电镀后的 Cu 覆盖层，第一步使用具有较高 Cu 去除率的抛光液，第二步使用具有较高钝化能力的抛光液，可以显著减小 CMP 后的凹坑尺寸。

② 确定了一个适用于 TSV 和 RDL 的最优 Cu 覆盖层：当台阶高度小于 $1\mu m$ 时，凹坑处的电镀覆盖层厚度为 $2\mu m$ 时，可保证 CMP 后凹坑尺寸最小。

2.8.2 节提出了一组有效用于 TSV CMP 工艺的指导意见，涉及磨抛液的选取、减小电镀 Cu 后凹坑、边缘修整以及采用低压力进行背面 CMP 处理。

2.9　TSV Cu 外露

2.9.1　CMP 湿法工艺

（1）测试结构

前面讨论了晶圆正面 CMP 工艺，为了将 TSV Cu 外露，需要对晶圆背面氧化层进行 CMP 处理。文献［11］对晶圆背面 CMP 工艺进行了研究。图 2.53 所示为厚度为 $50\mu m$ 的薄晶圆及其 TSV 结构，其中包含直径 $10\mu m$ 和 $30\mu m$ 两种 TSV，该薄晶圆临时键合在一个较厚的硅晶圆上，然后用于背面氧化层 CMP 测试，同时关注 CMP 工艺前后 Cu 柱的形貌变化。

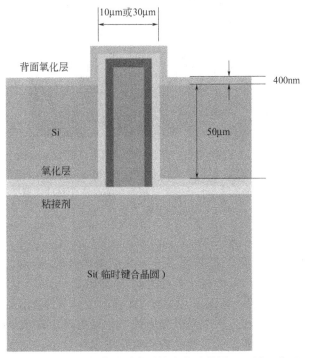

图 2.53　将 $50\mu m$ 厚的薄晶圆临时键合在硅晶圆上，用于背面 CMP
工艺测试，其中包含直径 $10\mu m$ 和 $30\mu m$ 两种 TSV

（2）实验设备

采用的 CMP 设备为 AMAT Reflexion，如图 2.43 所示。只采用压盘 3 去除 TSV 顶部 400nm 厚的氧化层，以露出 TSV Cu 柱。采用的磨抛液、抛光垫以及转盘等均为商用产品。磨抛液流速设置在 300mL/min，抛光压力设置为 1psi，氧化层与 Cu 的去除率分别为 400Å/min 和 200Å/min，利用 HRP 以及 OM 来表征 Cu 的表面形貌。

(3) 临时键合前的边缘修整

在将减薄的 TSV 晶圆临时键合到硅晶圆之前，需在 CMP 机台上对薄晶圆进行传送检测，以确定制程中其崩边的可能性。对厚盲孔 TSV 晶圆边缘完成 0.5mm 的修整后，在传送检测中未发现崩边现象，如图 2.54 所示。另一方面，如果不进行修边，则会出现崩边。基于这些测试结果，建议在临时键合之前，对所有厚盲孔 TSV 晶圆都要进行修边，然后进行背面磨削以露出 TSV Cu，这样可以减小后续制程中薄晶圆崩边可能性。

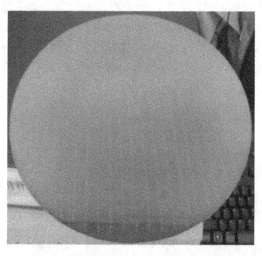

图 2.54 在 CMP 机台上经过传送检测后，将 50μm 厚的薄晶圆背面临时键合到一个硅晶圆上（50μm 厚的薄晶圆背面，在 CMP 机台上经过传送检测后临时键合到硅晶圆上之前，将其边缘修整为 0.5mm）

(4) 结果讨论

将晶圆减薄至 50μm 后，再减薄 2～4μm 以露出 TSV Cu 柱，然后采用背面 CMP 工艺将 TSV 区域的 Cu 外露，最后再将 Si 刻蚀 2～4μm。之后，在晶圆上沉积 4000Å 厚的绝缘氧化层，再次采用 CMP 去除 4000Å 的钝化层（约 60s）。这一步 CMP 主要关注两个指标：一个是硅表面的氧化层损耗应该很小，另一个是去除氧化层后 Cu 柱的表面轮廓。为了使硅表面氧化层损耗最小，将氧化层去除率设置在 400Å/min 左右。

图 2.55 给出了直径为 10μm 的 TSV 在背面 CMP 处理前后的结果。完成氧化层 CMP 之后，在 TSV Cu 柱表面观察到明显的光泽。采用 HRP 对 CMP 前后 Cu 柱的 3D 表面形貌进行观测，Cu 柱的阶梯高度从 2.7μm 变为 1.8μm（见图 2.55），减小了 9000Å。比较 CMP 前后的阶梯高度变化值（9000Å）和氧化层厚度（4000Å），表明 Cu 柱顶部的氧化层已经被去除。由于晶圆表面上 Cu 柱的密度小

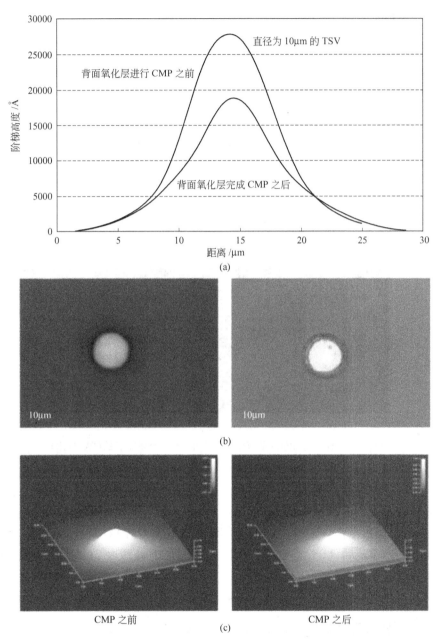

图 2.55 直径为 10μm 的 TSV 在背面 CMP 处理前后的结果对比，
图(a)：2D HRP 测量结果；图(b)：OM 图像；图(c)：3D HRP 云图

于 5%，而且还有些 Cu 柱胀出晶圆表面，所以进行 CMP 时施加的压力主要作用在这些 Cu 柱上，使得阶梯高度有效减小。除此之外，如图 2.56 上图所示，CMP 之前 Cu 柱的形貌类似圆锥，CMP 之后还保持圆锥形。

此外还研究了 TSV 直径对 Cu 柱形貌的影响。图 2.56 所示为直径 30μm 的

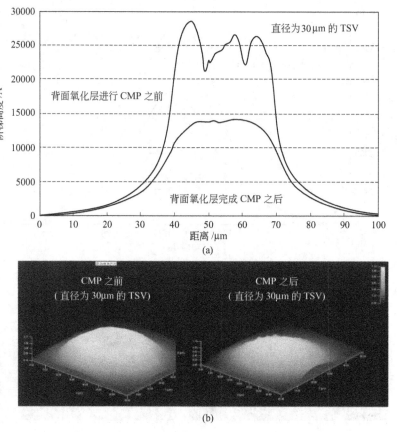

图 2.56 直径为 30μm 的 TSV 在背面 CMP 处理前后的结果对比，
图(a)：2D HRP 测量结果；图(b)：3D HRP 云图

TSV 在 CMP 处理前后 HRP 测量结果。可以看到，直径为 30μm 的 TSV 其 Cu 柱形貌有点像高地。而直径为 10μm 的 TSV 的 Cu 柱形貌是圆锥状的，如图 2.55 所示。图 2.57 给出了两种直径 TSV 在经过 60s 氧化层 CMP 处理后的 Cu 柱 3D 表面形貌，这些结果表明 TSV 直径对于背面氧化层 CMP 处理后 Cu 的胀出轮廓来说是一个重要参数。此外，CMP 工艺不改变 Cu 柱的形貌特征，无论是直径 10μm 或是直径 30μm 的 TSV，CMP 处理前后其 Cu 柱的形貌特征保持不变。

2.9.2 干法刻蚀工艺

(1) 临时键合与背面减薄

参考文献 [44] 中，IMEC 展示了一个非常实用的 TSV Cu 外露工艺，如图 2.58 所示。可以看到，临时键合之后，有源晶圆（设计为 50μm 厚）被减薄到 57μm。减薄后有源晶圆的总厚度变化（TTV）是一个控制的主要参数，因为它影响到 Cu 外露工艺的效果。基于红外（IR）时差测距系统，测得 300mm 键合晶圆和 TSV 晶圆的 TTV 范围均为 1.6μm[44]。

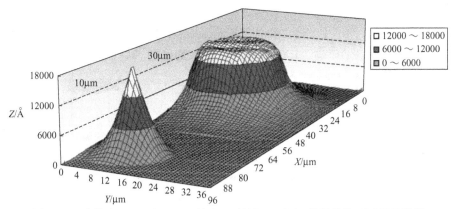

图 2.57　直径为 $10\mu m$ 和 $30\mu m$ 的 TSV 经过 60sCMP 处理后的 3D HRP 结果

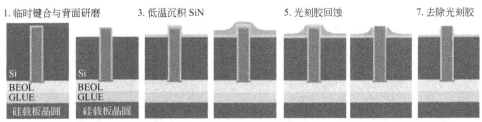

图 2.58　完成 TSV 制程后，IMEC 的晶圆减薄与背面钝化集成工艺

（2）硅的各向同性干法刻蚀

减薄后，采用可选择各向同性干法刻蚀将 TSV 外露出来，同时保留覆盖在 Cu 上的氧化层（如图 2.59 所示），这样可以保证后续工艺步中 Cu 不会被氧化。可以观察到，晶圆上不同的区域，刻蚀量也不同，比如在晶圆中心区域为 $2.9\mu m$，边界区域为 $1.9\mu m$。

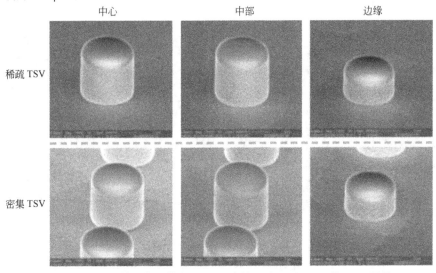

图 2.59　干法刻蚀完凹槽（图 2.58 中第二步）后 TSV 的 SEM 图像

(3) Si_3N_4 钝化层沉积

TSV 外露后，在小于 200℃的温度下将 500nm 厚的 Si_3N_4 钝化层沉积到晶圆背面，防止制作好 RDL 或 UBM 之后、晶圆未堆叠之前，Cu 通过薄晶圆扩散至 FEOL 有源器件层。

(4) 光刻胶涂布与刻蚀、Si_3N_4 和氧化层刻蚀

低温沉积 Si_3N_4 层（厚度均匀度小于 3%）之后，在晶圆背面旋涂一层较厚的光刻胶使表面平坦化，如图 2.60 所示。这种平坦表面允许在没有光刻图形的情况下对其进行回蚀，如图 2.61 所示。通过这种方法，可以将所有的 TSV 外露出来，但仍有一个薄胶层残留在晶圆表面。接下来，通过干法刻蚀去除 TSV Cu 柱顶部上的 Si_3N_4 钝化层和氧化层。最后，通过干法和湿法清除工艺相结合的方法去除残留的光刻胶。这样将 Cu 柱顶部的 Cu 完全外露出来，为后续工艺做准备，如图 2.62 和图 2.63 所示[44]。

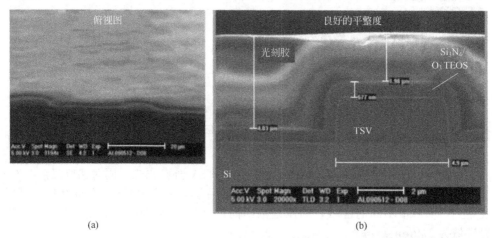

图 2.60　光刻胶涂布后的表面（a）与截面（b）的 SEM 图像

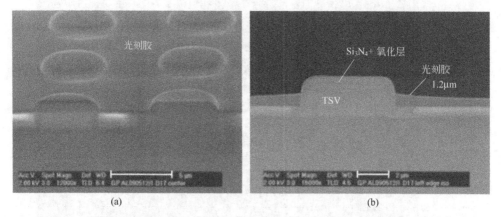

图 2.61　光刻胶回蚀后的表面［图(a)］与截面［图(b)］的 SEM 图像，晶圆表面残留了 1μm 厚的光刻胶，TSV 顶部 500nm 厚的 Si_3N_4 层外露出来

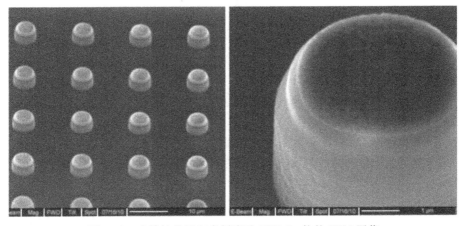

图 2.62　去除钝化层和光刻胶后 TSV Cu 柱的 SEM 图像

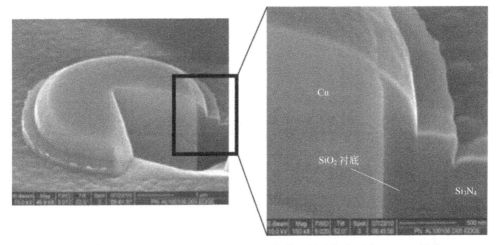

图 2.63　去除钝化层和光刻胶后 TSV Cu 柱截面的聚焦离子束（FIB）图像

2.9.3　总结与建议

① 当采用 CMP 湿法工艺外露 TSV Cu 时，涉及的材料和工艺步都比较少。然而，必须小心防范工艺过程中 Cu 被污染。对于 $50\mu m$ 厚的临时键合晶圆，TSV 直径对外露后 Cu 柱形貌有较大影响：当 TSV 直径较大时，Cu 柱形貌类似于高地；当 TSV 直径较小时，Cu 柱形貌类似于圆锥，而且 CMP 处理前后 Cu 柱形貌不会改变。

② 当采用干法刻蚀工艺外露 TSV Cu 时，涉及的材料和工艺相对复杂一些。但是这种工艺的好处是可以避免外露 Cu 与 Si 直接接触（CMP 湿法工艺中会发生这种情况），结合扩散阻挡层的使用，有效防止了 Cu 污染带来的器件可靠性问题。

2.10　FEOL 与 BEOL

前段制程（FEOL）定义为 IC 制造的第一阶段，在这个阶段诸如晶体管或电阻这样的元器件被制作出来。从裸晶圆到金属层沉积都属于这个阶段。FEOL 通常由晶圆厂完成。

后段制程（BEOL）定义为在晶圆上利用引线将器件进行互连的制造过程。从第一层金属层到钝化的键合焊盘都属于该制程。BEOL 还包括触点与绝缘层的制作以及将晶圆切割为单个的 IC 芯片。

2.11　TSV 工艺

对于 3D Si 集成而言，TSV 孔一般在晶圆到晶圆（W2W）键合工艺之前或之后制作。对于 3D IC 集成来说，TSV 孔可以采用先孔（via-first）、中孔（via-middle）、从晶圆正面的后孔（via-last）以及从晶圆背面的后孔工艺制作，详见表 2.8。

表 2.8　3D 集成工艺、方法及选项

制程	方法/选项		
孔制作	Bosch 深反应离子蚀刻（DRIE）	非 Bosch 深反应离子蚀刻	激光
介电层沉积	SiO_2	Si_3N_4/SiO_2	聚合物
阻挡层/种子层沉积	Ti（或 Ta）/Cu	TiW/Cu W/Cu	W/W
孔填充	Cu	W	导电聚合物、CNT、焊料等
Cu 覆盖层去除	CMP	CMP（两步）	
TSV Cu 外露	湿法刻蚀	干法刻蚀	
TSV 制程	键合前制作 TSV，键合后制作 TSV	先孔工艺，中孔工艺	后孔工艺（正面或背面）
薄晶圆拿持	支撑（载体）晶圆	不需要载体	在堆叠上制作
堆叠	C2C	C2W	W2W
微互连	焊锡凸点	Cu 柱+焊料	Au/Cu 凸点
键合	自然回流，热压键合	直接键合	间接键合（中间层）

2.11.1 键合前制孔工艺

对于采用 Cu 到 Cu 键合的 W2W 键合来说，通常在 W2W 键合之前制作 TSV 孔。图 2.64 所示为 NIMS/AIST/东芝/东京大学（University of Tokyo）等利用 Cu 到 Cu 键合工艺制作的键合晶圆的 SEM 截面图[46]。

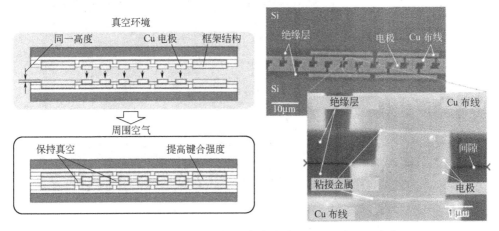

图 2.64　NIMS/AIST/东芝/东京大学基于 Cu 到 Cu（先孔）的 W2W 键合工艺制作的键合晶圆（3D Si 集成）

2.11.2 键合后制孔工艺

对于采用 SiO_2 到 SiO_2 键合的 W2W 键合来说，通常在 W2W 键合后制作 TSV 孔。图 2.65 所示为 MIT 采用 SiO_2 到 SiO_2 键合工艺制作的键合晶圆的 SEM 截面图[47]。

2.11.3 先孔工艺

即在 FEOL 之前制作 TSV 孔，这种工艺只能由晶圆厂完成。但是，即便是在晶圆厂，也很少采用该工艺，因为晶圆上的晶体管要比 TSV 重要得多。

2.11.4 中孔工艺

在 FEOL 结束之后（比如晶体管制作完成后），在小孔和金属线路层制作之前制作 TSV 孔。由于后勤和设备能力，通常这种工艺也由晶圆厂完成。图 2.66 所示为 IEEE/IEDM 采用中孔工艺制作的 TSV，图 2.67 所示为 IBM 采用中孔工艺制作的 TSV[45]。

2.11.5 正面后孔工艺

即在 BEOL 结束之后（制作完钝化层后）制作 TSV 孔。通常是从晶圆的正面

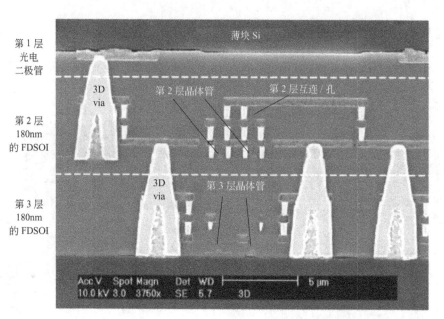

图 2.65 MIT 基于 SiO_2 到 SiO_2（后孔）的 W2W 键合
工艺制作的键合晶圆（3D Si 集成）

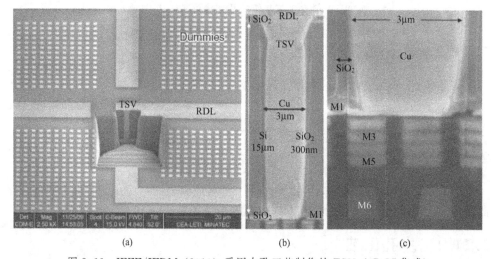

图 2.66 IEEE/IEDM（2010）采用中孔工艺制作的 TSV（3D IC 集成）

制作，一般由晶圆厂制作较小的孔（<$1\mu m$），由外包半导体封装测试厂（OSAT）制作较大的孔（$\geqslant 4\mu m$）。对于 RDL，通常 OSAT 的设备只能制作 $3\mu m$ 线宽和 $3\mu m$ 间距的 RDL，但晶圆厂可以制作亚微米线宽和亚微米间距的 RDL。

2.11.6 背面后孔工艺

除了 TSV 孔是从晶圆的背面制作以外，其余工艺均与 2.11.5 节类似。从图

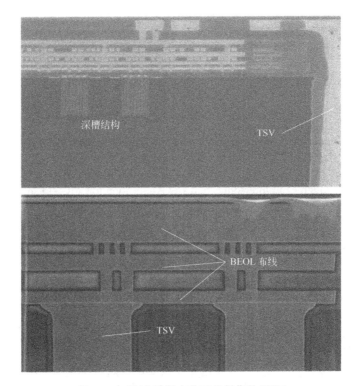

图 2.67 IBM 采用中孔工艺制作的 TSV

2.2（MMIC）和图 2.3（CMOS 影像传感器）可以找到若干例子，不过坦率地讲，它们并不是 3D IC 集成。

2.11.7 无源转接板

当工业界为 3D IC 集成制定 TSV 工艺时，并没有包含无源转接板。由于无源转接板不含器件，因此它不满足之前的任何论述。

2.11.8 总结与建议

① 对于 3D Si 集成，TSV 孔通常都较小（$\leqslant 1\mu m$），均采用 Cu 到 Cu 或者 SiO_2 到 SiO_2 或者混合方式（Cu 到 SiO_2）实现 W2W 键合。

② 对于 3D IC 集成，TSV 孔通常都较大（$\geqslant 1\mu m$），并且采用先孔或者中孔或者后孔工艺来制作。

③ 先孔和中孔工艺通常由晶圆厂完成。

④ 后孔工艺通常可以由晶圆厂完成小孔（$<4\mu m$）和细线宽、细间距（$<3\mu m$）RDL 的制作，由 OSAT 完成大孔（$\geqslant 4\mu m$）和 $3\mu m$ 线宽、$3\mu m$ 间距 RDL 的制作。

2.12 参考文献

[1] Lau, J. H., *Reliability of RoHS-Compliant 2D and 3D IC Interconnects*, McGraw-Hill, New York, 2011.
[2] Lau, J. H., C. K. Lee, C. S. Premachandran, and A. Yu, *Advanced MEMS Packaging*, McGraw-Hill, NewYork, 2010.
[3] Shockley, W., "Semiconductive Wafer and Method of Making the Same," U.S. Patent No. 3,044,909, filed on October 23, 1958, and granted on July 17, 1962.
[4] Sekiguchi, M., H. Numata, N. Sato, T. Shirakawa, M. Matsuo, H. Yoshikawa, M. Yanagida, H. Nakayoshi, and K. Takahashi, "Novel Low Cost Integration of Through Chip Interconnection and Application to CMOS Image Sensor," *IEEE/ECTC Proceedings*, San Diego, CA, May 2006, pp. 1367–1374.
[5] Archard, D., K. Giles, A. Price, S. Burgess, and K. Buchanan, "Low Temperature PECVD Of Dielectric Films For TSV Applications," *IEEE/ECTC Proceedings*, 2010, pp. 764–768.
[6] Kumar Praveen, S., Ho Wai Tsan, and R. Nagarajan, "Conformal Low-Temperature Dielectric Deposition Process Below 200°C For TSV Application," *Proceedings of IEEE Electronics Packaging Technology Conference*, 2010, pp. 27–30.
[7] Beica, R., C. Sharbono, and T. Ritzdorf, "Through Silicon Via Copper Electrodeposition For 3D Integration," *IEEE/ECTC Proceedings*, 2008, pp. 577–583.
[8] Lau, J. H., "Evolution, Outlook, and Challenges of 3D IC/Si Integration," *IEEE/ICEP Proceedings* (keynote), Nara, Japan, April 13, 2011, pp. 1–17.
[9] Lau, J. H., "Evolution, Challenge, and Outlook of TSV, 3D IC Integration and 3D Si Integration," *International Wafer Level Packaging Conference* (plenary), San Jose, CA, October 3–7, 2011, pp. 1–18.
[10] Hsin, Y. C., C. Chen, J. H. Lau, P. Tzeng, S. Shen, Y. Hsu, S. Chen, C. Wn, J. Chen, T. Ku, and M. Kao, "Effects of Etch Rate on Scallop of Through-Silicon Vias (TSVs) in 200-mm and 300-mm Wafers," *IEEE/ECTC Proceedings*, Orlando, FL, June 2011, pp. 1130–1135.
[11] Chen, J. C., P. J. Tzeng, S. C. Chen, C. Y. Wu, J. H. Lau, C. C. Chen, C. H. Lin, Y. C. Hsin, T. K. Ku, and M. J. Kao, "Impact of Slurry in Cu CMP (Chemical Mechanical Polishing) on Cu Topography of Through Silicon Vias (TSVs), Re-distributed Layers, and Cu Exposure," *IEEE/ECTC Proceedings*, Orlando, FL, June 2011, pp. 1389–1394.
[12] Selvanayagam, C., J. H. Lau, X. Zhang, S. Seah, K. Vaidyanathan, and T. Chai, "Nonlinear Thermal Stress/Strain Analysis of Copper Filled TSV (Through Silicon Via) and Their Flip-Chip Microbumps," *IEEE Transactions in Advanced Packaging*, Vol. 32, No. 4, November 2009, pp. 720–728.
[13] Tang, G., O. Navas, D. Pinjala, J. H. Lau, A. Yu, and V. Kripesh, "Integrated Liquid Cooling Systems for 3D Stacked TSV Modules," *IEEE Transactions on Components and Packaging Technologies*, Vol. 33, No. 1, 2010, pp. 184–195.
[14] Khan, N., V. Rao, S. Lim, S. Ho, V. Lee, X. Zhang, R. Yang, E. Liao, Ranganathan, T. Chai, V. Kripesh, and J. H. Lau, "Development of 3D Silicon Module with TSV for System in Packaging," *IEEE Transactions on CPMT*, Vol. 33, No. 1, March 2010, pp. 3–9.
[15] Chai, T., X. Zhang, J. H. Lau, C. Selvanayagam, K. Biswas, S. Liu, D. Pinjala, G. Tang, Y. Ong, S. Vempati, E. Wai, H. Li, B. Liao, N. Ranganathan, V. Kripesh, J. Sun, J. Doricko, and C. Vath, "Development of Large Die Fine-Pitch Cu/Low-k FCBGA Package with Through Silicon Via (TSV) Interposer," *IEEE Transactions on Components, Packaging and Manufacturing Technology*, Vol. 1, No. 5, 2011, pp. 660–672.
[16] Ranganathan, N., L. Ebin, L. Linn, V. Lee, O. Navas, V. Kripesh, and N. Balasubramanian, "Integration of High Aspect Ratio Tapered Silicon Via for Through-Silicon Interconnection," *IEEE/ECTC Proceedings*, Orlando, FL, May 2008, pp. 859–865.
[17] Lau, J. H., "TSV Manufacturing Yield and Hidden Costs for 3D IC Integration," *IEEE/ECTC Proceedings*, Las Vegas, NV, June 2010, pp. 1031–1041.

[18] Chaabouni, H., M. Rousseau, P. Leduc, A. Farcy, R. El Farhane, A. Thuaire, G. Haury, A. Valentian, G. Billiot, M. Assous, F. De Crecy, J. Cluzel, A. Toffoli, D. Bouchu, L. Cadix, T. Lacrevaz, P. Ancey, N. Sillon, and B. Flechet, "Investigation on TSV Impact on 65-nm CMOS Devices and Circuits," *Proceedings of IEEE International Electron Devices Meeting*, 2010, pp. 35.1.1–35.1.4.
[19] Reed, J. D., S. Goodwin, C. Gregory, and D. Temple, "Reliability Testing of High Aspect Ratio Through Silicon Vias Fabricated with Atomic Layer Deposition Barrier, Seed Layer and Direct Plating and Material Properties Characterization of Electrografted Insulator, Barrier and Seed Layer for 3D Integration," *Proceedings of IEEE International 3D Systems Integration Conference*, 2010, pp. 1–8.
[20] Hung Y., C. Hsieh, S. Jeng, H. Tao, C. Min, and Y. Mii, "A New Enhancement Layer To Improve Copper Interconnect Performance," *Proceedings of IEEE International Interconnect Technology Conference*, 2010, pp. 1–3.
[21] Powell, K., S. Burgess, T. Wilby, R. Hyndman, and J. Callahan, "3D IC Process Integration Challenges and Solutions," *Proceedings of IEEE International Interconnect Technology Conference*, 2008, pp. 40–42.
[22] Teh, W. H., R. Caramto, S. Arkalgud, T. Saito, K. Maruyama, and K. Maekawa, "Magnetically Enhanced Capacitively Coupled Plasma Etching For 300-Mm Wafer-Scale Fabrication Of Cu Through-Silicon-Vias For 3D Logic Integration," *Proceedings of IEEE International Interconnect Technology Conference*, 2009, pp. 53–55.
[23] Wu, C., S. Chen, P. Tzeng, J. H. Lau, Y. Hsu, J. Chen, Y. Hsin, C. Chen, S. Shen, C. Lin, T. Ku, and M. Kao, "Oxide Liner, Barrier and Seed Layers, and Cu-Plating of Blind Through Silicon Vias (TSVs) on 300mm Wafers for 3D IC Integration," *Proceedings of IMAPS International Conference*, Long Beach, CA, 2011, pp. 1–7.
[24] Beica, R., C. Sharbono, and T. Ritzdorf, "Through Silicon Via Copper Electrodeposition for 3D Integration," *IEEE/ECTC Proceedings*, 2008, pp. 577–583.
[25] Vempati Srinivasa Rao, Ho Soon Wee, Lee Wen Sheng Vincent, Li Hong Yu, Liao Ebin, Ranganathan Nagarajan, Chai Tai Chong, Xiaowu Zhang, and Pinjala Damaruganath, "TSV Interposer Fabrication for 3D IC Packaging," *IEEE 11th Electronics Packaging Technology Conference*, 2009, pp. 431–437.
[26] Lee Wen Sheng Vincent, Navas Khan, Liao Ebin, S. W. Yoon, and V. Kripesh, "Cu Via Exposure by Backgrinding for TSV Applications," *IEEE 9th Electronics Packaging Technology Conference*, 2007, pp. 233–237.
[27] Takahashi, K., Y. Taguchi, M. Tomisaka, H. Yonemura, M. Hoshino, M. Ueno, Y. Egawa, Y. Nemoto, Y. Yamaji, H. Terao, M. Umemoto, K. Kameyama, A. Suzuki, Y. Okayama, T. Yonezawa, and K. Kondo, "Process Integration of 3D Chip Stack with Vertical Interconnection," *IEEE/ECTC Proceedings*, 2004, pp. 601–609.
[28] Takahashi, K., Y. Taguchi, M. Hoshino, K. Tanida, M. Umemoto, T. Yonezawa, and K. Kondo, "Development of Less Expensive Process Technologies for Three-Dimensional Chip Stacking with Through-Vias," *Electronics and Communications in Japan*, Part 2, Vol. 88, No. 7, pp. 50–60, 2005.
[29] Dean Malta, Christopher Gregory, Dorota Temple, Trevor Knutson, Chen Wang, Thomas Richardson, Yun Zhang, and Robert Rhoades, "Integrated Process for Defect-Free Copper Plating and Chemical-Mechanical Polishing of Through-Silicon Vias for 3D Interconnects," *IEEE/ECTC Proceedings*, 2010, pp. 1769–1775.
[30] Tsai, W., H. H. Chang, C. H. Chien, J. H. Lau, H. C. Fu, C. W. Chiang, T. Y. Kuo, Y. H. Chen, R. Lo, and M. J. Kao, "How to Select Adhesive Materials for Temporary Bonding and De-Bonding of Thin-Wafer Handling in 3D IC Integration?" *IEEE ECTC Proceedings*, Orlando, FL, June 2011, pp. 989–998.
[31] Dang, B., P. Andry, C. Tsang, J. Maria, R. Polastre, R. Trzcinski, A. Prabhakar, and J. Knickerbocker, "CMOS Compatible Thin Wafer Processing Using Temporary Mechanical Wafer, Adhesive and Laser Release of Thin Chips/Wafers for 3D Integration," *IEEE/ECTC Proceedings*, 2010, pp. 1393–1398.
[32] Tamura, K., K. Nakada, N. Taneichi, P. Andry, J. Knickerbocker, and C. Rosenthal, "Novel Adhesive Development for CMOS-Compatible Thin Wafer Handling," *IEEE/ECTC Proceedings*, 2010, pp. 1239–1244.
[33] Zhang, X., A. Kumar, Q. X. Zhang, Y. Y. Ong, S. W. Ho, C. H. Khong, V. Kripesh, J. H. Lau, D.-L. Kwong, V. Sundaram, Rao R. Tummula, and Georg

Meyer-Berg, "Application of Piezoresistive Stress Sensors in Ultra Thin Device Handling and Characterization," *Sensors and Actuators A: Physical*, Vol. 156, 2009, pp. 2–7.

[34] Kwon, W., J. Lee, V. Lee, J. Seetoh, Y. Yeo, Y. Khoo, N. Ranganathan, K. Teo, and S. Gao, "Novel Thinning/Backside Passivation for Substrate Coupling Depression of 3D IC," *IEEE/ECTC Proceedings*, Orlando, FL, 2011, pp. 1395–1399.

[35] Lee, J., V. Lee, J. Seetoh, S. Thew, Y. Yeo, H. Li, K. Teo, and S. Gao, "Advanced Wafer Thinning and Handling for Through Silicon Via Technology," *IEEE/ECTC Proceedings*, Orlando, FL, 2011, pp. 1852–1857.

[36] Jourdain, A., T. Buisson, A. Phommahaxay, A. Redolfi, S. Thangaraju, Y. Travaly, E. Beyne, and B. Swinnen, "Integration of TSVs, Wafer Thinning and Backside Passivation on Full 300-mm CMOS Wafers for 3D Applications," *IEEE/ECTC Proceedings*, Orlando, FL, 2011, pp. 1122–1125.

[37] Halder, S., A. Jourdain, M. Claes, I. Wolf, Y. Travaly, E. Beyne, B. Swinnen, V. Pepper, P. Guittet, G. Savage, and L. Markwort, "Metrology and Inspection for Process Control During Bonding and Thinning of Stacked Wafers for Manufacturing 3D SIC's," *IEEE/ECTC Proceedings*, Orlando, FL, 2011, pp. 999–1002.

[38] Bhardwaj, J., H. Ashraf, and A. McQuarrie, "Dry Silicon Etching for MEMS," *Proceedings of Microstructures and Microfabricated Systems*, Montreal, Quebec, Canada, May 4–9, 1997, pp. 1–13.

[39] C. K. Chung, "Geometrical Pattern Effect on Silicon Deep Etching by an Inductively Coupled Plasma System," *Journal of Micromechanics and Microengineering*, Vol. 14, 2004, pp. 656–662.

[40] Liu, H.-C., Y.-H. Lin, and W. Hsu, "Sidewall Roughness Control in Advanced Silicon Etch Process," *Journal of Microsystem Technologies*, Vol. 10, 2003, pp. 29–34.

[41] Ranganathan, N., K. Prasad, N. Balasubramanian, and K. Pey, "A Study of Thermo-Mechanical Stress and Its Impact on Through-Silicon Vias," *Journal of Micromechanics and Microengineering*, Vol. 18, 2008, pp. 1–13.

[42] Chen, K., A. A. Ayon, X. Zhang, and S. M. Spearing, "Effect of Process Parameters on the Surface Morphology and Mechanical Performance of Silicon Surfaces after Deep Reactive Ion Etching (DRIE)," *Journal of Microelectromechanical Systems*, Vol. 11, 2002, pp. 264–275.

[43] Tian, J., and M. Bartek, "Simultaneous Through-Silicon Via and Large Cavity Formation Using Deep Reactive Ion Etching and Aluminum Etch-Stop Layer," *IEEE/ECTC Proceedings*, Lake Buena Vista, FL, May 27–30, 2008, pp. 1787–1792.

[44] Jourdain, A., T. Buisson, A. Phommahaxay, A. Redolfi, S. Thangaraju, Y. Travaly, E. Beyne, and B. Swinnen, "Integration of TSVs, Wafer Thinning and Backside Passivation on Full 300-mm CMOS Wafers for 3D Applications," *IEEE/ECTC Proceedings*, May 2011, pp. 1122–1125.

[45] Farooq, M. G, T. L. Graves-Abe, W. F. Landers, C. Kothandaraman, B. A. Himmel, P. S. Andry, C. K. Tsang, E. Sprogis, R. P. Volant, K. S. Petrarca, K. R. Winstel, J. M. Safran, T. D. Sullivan, F. Chen, M. J. Shapiro, R. Hannon, R. Liptak, D. Berger, and S. S. Iyer, "3D Copper TSV Integration, Testing and Reliability," *Proceedings of IEEE/IEDM*, Washington, DC, 2011, pp. 7.1.1–7.1.4.

[46] Shigetou, A., T. Itoh, M. Matsuo, N. Hayasaka, K. Okumura, and T. Suga, "Bumpless Interconnect Through Ultrafine Cu Electrodes by Mans of Surface-Activated Bonding (SAB) Method," *IEEE Transaction on Advanced Packaging*, Vol. 29, No. 2, May 2006, pp. 218–226.

[47] Burns, J., B. Aull, C. Keast, C. Chen, C. Chen, J. Knecht, V. Suntharalingam, K. Warner, P. Wyatt, and D. Yost, "A Wafer-Scale 3-D Circuit Integration Technology," *IEEE Transactions on Electron Devices*, Vol. 53, No. 10, October 2006, pp. 2507–2516.

第 3 章 TSV 的力学、热学与电学行为

3.1 引言

本章主要讨论已填充 Cu TSV 的力学、热学与电学行为，给出一些有价值的仿真和测试结果，提出并探讨 TSV 的相关测试方法和设计建议。可以看到，TSV 技术涉及科学和工程的多个学科领域[1~50]。

3.2 SiP 封装中 TSV 的力学行为

已填充 Cu TSV 是 3D IC 集成和 3D Si 集成的一个重要方面，这是因为：①Si 的热膨胀系数（CTE）近似为 $2.5\times10^{-6}/℃$；②Cu 的热膨胀系数大约为 $17.5\times10^{-6}/℃$；③在 Si 和 Cu 之间存在较大的局部热失配，这可能导致 Cu 与 Si 之间或 Cu 与介电层之间的界面发生分层失效；④当 3D 封装结构受热时会发生 TSV Cu 胀出现象。

3.2.1 有源/无源转接板中 TSV 的力学行为

本节采用有限单元法对高性能 3D IC 集成 SiP 器件（见图 3.1）的 TSV 转接板进行可靠性设计（DFR）与分析。尺寸为 25mm×25mm×0.3mm 的 TSV 硅转接板，顶部放置一个包含 9 个 Cu-low-k 层的大尺寸芯片（21mm×21mm）。TSV 转接板双面金属化处理，顶部有 CuNiAu 凸点下金属层（UBM），底面有节距为 150μm 的 SnAgCu 焊锡凸点。基板材料为双马来酰亚胺三嗪（BT），尺寸为 45mm×45mm，采用 1-2-1 层结构（芯板厚度为 0.8mm）[1~3]。基板下有焊盘节距为 1mm 的塑料球栅阵列（PBGA）[4]。详细参数见表 3.1。

图 3.2(a) 所示为用于有限元分析的模型图[2,3]。采用的假设为：①由于 Ta 层的厚度（1000Å）与介电层（1μm）相比小很多，模型中不考虑 Ta 层；②再分布层非常薄，在模型中也不考虑；③Si 和 SiO_2 只发生弹性变形；④TSV Cu 为弹塑性变形；⑤TSV 转接板的无应力状态温度为 125℃。

采用轴对称 1/4 模型，一些关键界面的边界条件和网格划分如图 3.2(b)、(c) 所示。模型中，方向 1 表示 TSV 孔径向，方向 2 表示轴向。温度载荷采用斜坡法

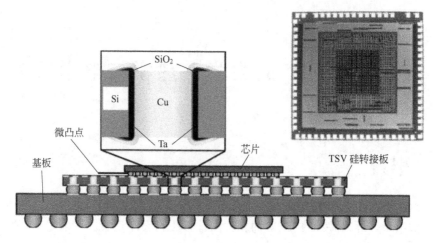

图 3.1 Cu-low-k 芯片（21mm×21mm）与 TSV 转接板（25mm×25mm）（2.5D IC 集成）的示意图

在 125～−40℃ 之间循环，确定该过程中的最大应力。模型的几何尺寸以及材料参数见表 3.2 和表 3.3。

表 3.1 封装结构和尺寸

整体封装	外形尺寸	45mm×45mm
	厚度	1.8mm
顶部芯片	芯片尺寸	21mm×21mm
	IC 技术	Cu-low-k,65nm 技术
	I/O	11000
	节距/焊料	150μm/SnAg
TSV 转接板	外部/内部节距	300μm/600μm
	锥形 TSV	100μm/50μm
Cu	层数	2
	厚度	3μm
	孔直径	40μm
	UBM/焊料	CuNiAu/SnAgCu
有机基板	层数	4
	芯板厚度	800μm
	节距	1mm
	I/O	2000

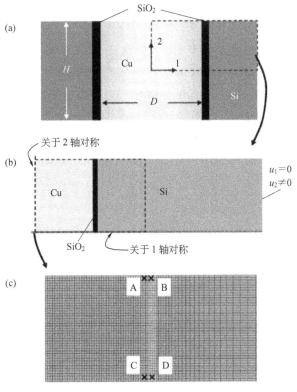

图 3.2 (a) TSV 简化模型；(b) TSV 的 1/4 轴对称模型及其边界条件；(c) 关键区域的网格划分

表 3.2 仿真所用的几何参数

直径(D)	$25\mu m, 50\mu m, 75\mu m$
深宽比(H/D)	$D=25\mu m: 1,2,3,5,7,10,12$ $D=50\mu m: 1,2,3,5,7$ $D=75\mu m: 1,2,3,5$

表 3.3 仿真所用的材料参数

材料	硅(Si)	二氧化硅(SiO_2)	电镀铜(Cu)
弹性模量 /GPa	129.617(25℃) 128.425(150℃)	70	70
泊松比	0.28	0.16	0.34
CTE/(ppm/℃)	2.813(25℃) 3.107(150℃)	0.6	18
应力(MPa)与塑性应变(ε)的关系			240(0ε) 250(0.003ε) 255(0.007ε) 255(0.009ε) 250(0.017ε)

3.2.2　可靠性设计（DFR）结果

在每个温度循环的升温阶段，Cu 的膨胀量为 Si 的 5 倍、SiO_2 的 10 倍。从 $-40℃$ 升温到 $125℃$ 以及从 $125℃$ 冷却到 $-40℃$ 过程中，变形结果如图 3.3(a)、(b) 所示。从图 3.3(a) 中可以看出，SiO_2 层处于高应变状态，Cu 的轴向膨胀使 SiO_2 在轴向受拉，Cu 的径向膨胀使 SiO_2 在径向受压。SiO_2 的变形并没有传递到转接板的 Si 材料区域，因为 Si 与 Cu 以及 SiO_2 相比刚度较大。在降温阶段，界面附近的 A 点和 B 点处的材料处于高应变状态，容易发生失效。

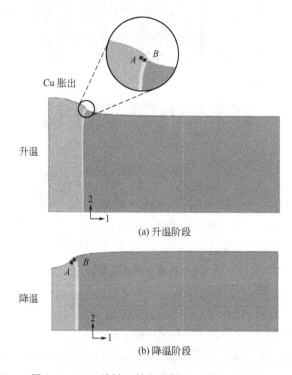

图 3.3　TSV 关键区域变形情况（放大 100 倍）

失效可能发生在两个位置。第一个位置为图 3.2(c) 中的 A 点和 B 点，在降温收缩时，这两个位置处的 Cu 与 SiO_2 界面存在较大的剪切变形；第二个位置为图 3.2(c) 中的 C 点和 D 点，在升温膨胀时，TSV 中间位置的 Cu 和 SiO_2 可能产生裂纹。因此在仿真计算中，重点关注 A、B 点的径向应力和 C、D 点的轴向应变，并将它们作为判断这些区域是否会发生分层和开裂的指标。

图 3.4 给出了不同深宽比情况下关键角点处 Cu 和 SiO_2 的径向应力随 TSV 直径的变化情况。可以看出，径向应力随 TSV 直径增大线性增大，并且当 TSV 的深宽比（H/D）较大时增加较快。SiO_2 中的应力比 Cu 中的大，这是因为 SiO_2 的弹性模量较高。

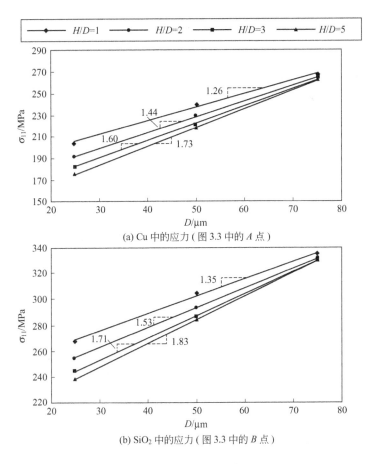

(a) Cu 中的应力（图 3.3 中的 A 点）

(b) SiO_2 中的应力（图 3.3 中的 B 点）

图 3.4 关键角点处的径向应力（σ_{11}）

对于直径较小的 TSV 来说，改变深宽比会引起应力的明显变化。例如，当深宽比从 5 变为 1 时，直径为 $25\mu m$ 的 TSV 其径向应力增大了 15%，而直径为 $75\mu m$ 的 TSV 其径向应力只增大了 2%。图 3.5 所示为 Cu 和 SiO_2 中 C 点和 D 点的轴向应变。和预期的一样，由于两者距离非常近，故 Cu 和 SiO_2 的应变对于所有深宽比和直径都相等。当深宽比为 1 时，轴向应变最大；随着深宽比增加，应变降低；当深宽比超过 5 时，轴向应变不再随深宽比发生变化。由于轴向应变值小于 Cu 的伸长率（2%）以及 SiO_2 的伸长率（30%），因此预计裂纹不会出现在 TSV 的中间位置。

图 3.4 中，不同深宽比时径向应力的变化范围较大；而图 3.5 表明，当深宽比超过 5 时轴向应变值趋于一个稳定值。图 3.6 所示为不同深宽比下 TSV 的轴向应变云图。当深宽比大于 7 时，应变集中在 TSV 顶部较小的区域内，这与深宽比较小时的情况不同。

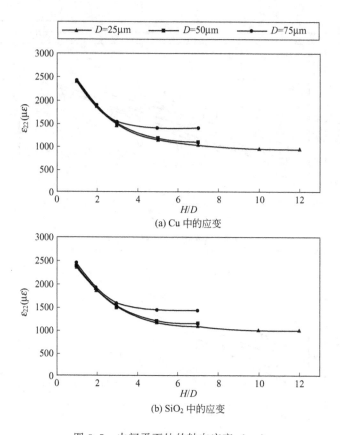

图 3.5 中间平面处的轴向应变（ε_{22}）

3.2.3 含 RDL 层的 TSV

在 TSV 制程中，采用深反应离子刻蚀（DRIE，见 2.4 节）将孔刻好后，需要在 Si 上沉积一层薄的 SiO_2 层，见 2.5 节；接着制作阻挡层和种子层，见 2.6 节；电镀填充，见 2.7 节；晶圆的化学机械抛光（CMP），见 2.8 节；然后沉积另外一层薄的 SiO_2 层。

之后，在室温下溅射再分布层（RDL）Cu。最后将整个结构进行退火处理，以去除工艺中产生的残余应力。因此，在仿真建模时可以忽略工艺残余应力。

建立 RDL 的模型并模拟 RDL 层的应力状态，用于预测 RDL 可能的破坏位置。图 3.7 所示为 TSV 截面图、1/4 有限元模型以及关键区域的网格划分。RDL 层的失效可能发生在与 TSV Cu 表面呈 45°的位置。因此，需要利用方向 1 和方向 2 上的应变得到对角线方向上的应变，即沿方向 1′上的应变 ε_D。

图 3.8 所示为 A 点与 B 点 45°方向的应变。可以看出，对角线应变与孔的直径之间存在线性关系。正如预期的那样，SiO_2 中的应变要比 Cu 中的大，有可能发生

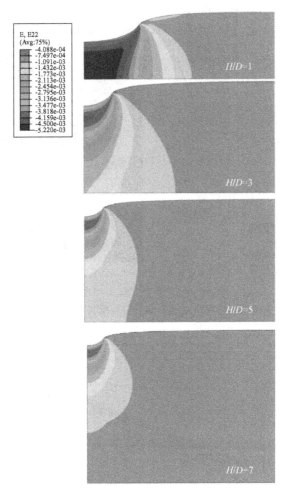

图 3.6　直径为 25μm 的 TSV 在不同深宽比
情况下的轴向应变（ε_{22}）云图

失效，如图 3.7(d) 所示。这是由于该处 Cu（17.5×10^{-6}/℃）和 Si（2.5×10^{-6}/℃）之间的热膨胀系数严重失配造成的，该结果曾经在 2008 年 5 月的 IEEE/ECTC 会议上报道过[3]。较大的 Cu 变形"撕扯"SiO_2，从而导致裂纹产生（见图 3.3）。这种现象称为 Cu 胀出［见图 3.7(d)］，Tezzaron 半导体公司的 Patti 在 2008 年曾经报道过这一现象（源自"Semiconductor International"的特约编辑 Philip Garrou，2009 年 12 月 3 日）。

图 3.9 所示为含 RDL 模型中 C 点的轴向应变，这里没有列出 D 点的轴向应变曲线，因为 C 和 D 的轴向应变相同。注意到这些曲线与图 3.5(a) 相似，图 3.5(a) 为不含 RDL 模型中 Cu 的轴向应变。由此可见，如果只是计算轴向应变，RDL 层是可以忽略的。

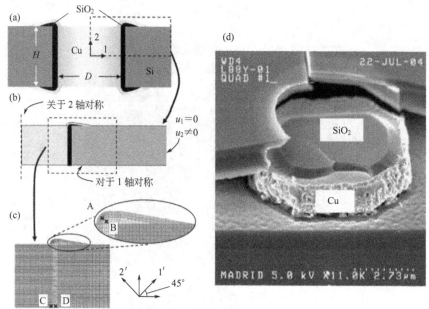

图 3.7 (a) 含 RDL TSV 的示意图;(b) 1/4 轴对称模型与边界条件;
(c) 关键区域的网格划分;(d) 失效模式(来源于 Tezzaron)

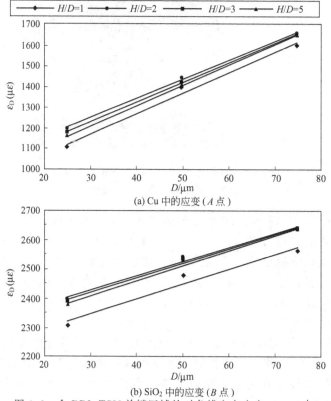

(a) Cu 中的应变(A 点)

(b) SiO_2 中的应变(B 点)

图 3.8 含 RDL TSV 关键区域的对角线方向应变($\varepsilon_D = \varepsilon_1'$)

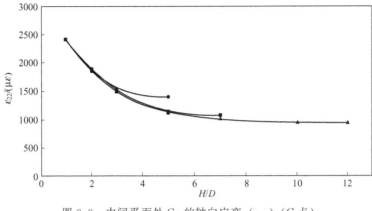

图 3.9　中间平面处 Cu 的轴向应变（ε_{22}）（C 点）

3.2.4　总结与建议

① 一般来讲，当 TSV 的深宽比（TSV 深度与直径之比）超过 5 时，应力和应变几乎不受深宽比的影响。

② 对于不含工艺缺陷的完美 TSV 结构，在 TSV 孔壁界面发生失效的可能性不大，因为模拟发现界面附近单元中的应变较小，不足以导致失效。

③ 对于制程中可能出现的缺陷（例如界面间的键合不良、表面粗糙等，这些会引起应力集中）以及由于局部热膨胀系数不匹配造成的应力等，都可能引起失效。因此，TSV 制程中应尽可能避免缺陷。

④ SiO_2 介电层中的应力比 TSV Cu 中的应力大，因此在制作介电层时要格外小心，以确保 TSV 结构无缺陷。

3.3　存储芯片堆叠中 TSV 的力学行为

与有源/无源转接板不同，同种存储芯片堆叠（例如存储立方）并无 RDL 层，TSV 用于穿过芯片实现 Z 方向互连以增加带宽，如图 3.10 所示。本节采用非线性有限元方法和断裂力学方法分析图 3.10 中所示结构的力学行为，确定最大应力和应变的位置和大小，并采用修正的虚拟裂纹闭合技术（MVCC）[15] 对关键应力/应变区域[23] 的界面能量释放率进行计算。

3.3.1　模型与方法

TSV 的机械失效模式主要是分层，其驱动力来自 TSV 中 Cu/SiO_2 界面、Cu/Si 界面处由热失配引起的最大应力和能量释放率（ERR）。如果最大界面热应力超过许用值或界面处的 ERR 超过临界值，则界面发生分层。

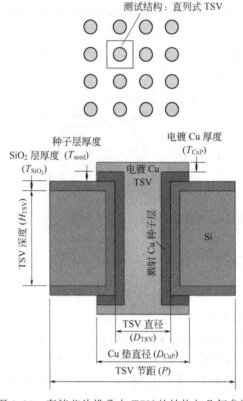

图 3.10　存储芯片堆叠中 TSV 的结构与几何参数

为了计算 TSV 热应力和界面 ERR，建立了单个对称的 TSV 3D 有限元模型（见图 3.10），并在 SiO$_2$ 层与 Cu 种子层之间的界面预埋两种水平裂纹模拟 Cu 垫与 TSV 壁面分层的情况，采用 MVCC 技术计算界面的 ERR[15~17]。除了 TSV Cu 设为非线性材料外，其他材料均设为线弹性材料。为了得到影响 TSV 界面分层的主要几何参数，采用实验设计（DoE）评估裂纹长度、TSV 直径、TSV 节距、TSV 深度、SiO$_2$ 层厚度以及 Cu 种子层厚度等参数的影响。所得结果对于 3D IC 集成中避免 TSV 分层的优化设计很有帮助。

3.3.2　TSV 的非线性热应力分析

热效应不仅影响结构中温度的分布，而且影响应力的分布。热应力不仅是 3D IC 集成器件的可靠性问题，而且是整个半导体工业重要的可靠性问题。为了得到 3D IC 集成中 TSV 结构的热应力，采用 ANSYS 有限元软件对电镀 Cu TSV（EP-Cu TSV）进行仿真分析，评估其热应力分布和界面分层行为。仿真采用的电镀 Cu TSV 结构如图 3.10 所示，包含 SiO$_2$ 层和 Cu 种子层。图 3.11 为对应的有限元模型，采用六面体单元进行网格划分。由于对称性，图中只显示了 1/4 网格模型。在有限元模型中，Cu 被设为温度相关的弹塑性材料[18]，其他所有材料均设为弹性材

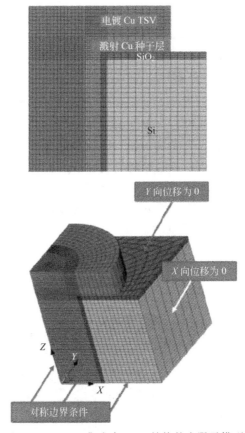

图 3.11　3D IC 集成中 TSV 结构的有限元模型

料,Si 的弹性模量和热膨胀系数设为与温度相关,所用材料参数如表 3.4 和表 3.5 所示。表 3.4 中,E、ν 和 CTE 分别为弹性模量、泊松比及热膨胀系数。

表 3.4　TSV 结构中的材料参数

材　　料	E/GPa	ν	CTE/($\times 10^{-6}$/℃)
芯片	129.617(25℃) 128.425(150℃)	0.28	2.18(25℃) 3.11(150℃)
TSV[18] (电镀 Cu)	70	0.34	18.0
SiO$_2$	70	0.16	0.6
Cu 种子层	110	0.35	17

表 3.5　TSV Cu 的应力应变关系

应　　变	应力(25℃)/MPa	应力(150℃)/MPa
0.001	120	110
0.004	186	179

续表

应 变	应力(25℃)/MPa	应力(150℃)/MPa
0.01	217	214
0.012	234	231
0.04	248	245

仿真计算的温度载荷设置为：整个模型温度均匀分布，并从 125℃ 下降至 $-40℃$。TSV 的深度（H_{TSV}）为 $50\mu m$，SiO_2 层的厚度（T_{SiO_2}）为 $0.5\mu m$，Cu 种子层的厚度（T_{Seed}）为 $1\mu m$，Cu 垫的直径（D_{CuP}）为 $20\mu m$，厚度（T_{CuP}）为 $2\mu m$，TSV 节距（P）为 $40\mu m$，如图 3.10 所示。图 3.12 所示为 3D IC 集成中直径（D_{TSV}）为 $10\mu m$ 的 TSV 热应力分布云图。从图中可以看出，最大应力发生在 Cu 种子层的圆周和 SiO_2 钝化层的表面处，相应的 von Mises 应力为 488MPa。更多情况下的分析结果可以见参考文献 [23]。结果表明，当 TSV 直径增大时，TSV 结构中的最大 von Mises 也随之增大。

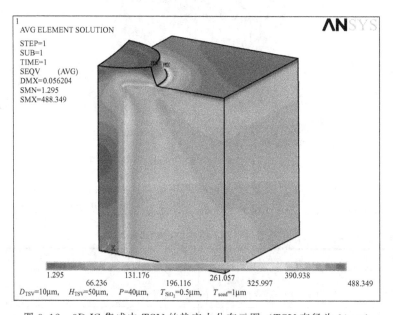

图 3.12　3D IC 集成中 TSV 的热应力分布云图（TSV 直径为 $10\mu m$）

3.3.3　修正的虚拟裂纹闭合技术

断裂力学中，裂纹分为三种基本类型：张开型（Ⅰ型）、滑移型（Ⅱ型）和撕裂型（Ⅲ型）。三种类型的裂纹对应有三个能量释放率 G_I、G_{II} 和 G_{III}，这三个参数需要分别确定。一般的裂纹是上述三种基本类型的混合，因此混合型裂纹的能量释放率（G 值）为三种基本类型所对应的能量释放率 G_I、G_{II} 和 G_{III} 的函数。对于图 3.13 所示的 8 节点实体单元，其三个参数 G_I、G_{II}、G_{III} 可由如下公式计算

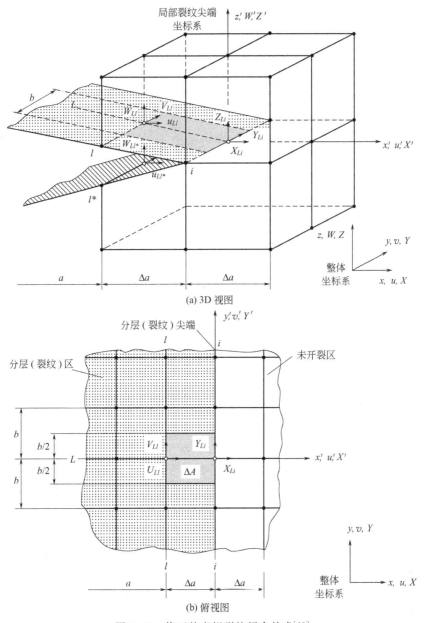

图 3.13 修正的虚拟裂纹闭合技术[15]

得到[15]：

$$G_{\text{I}} = -\frac{1}{2\Delta A} Z_{Li}(w_{Ll} - w_{Ll^*})$$
$$G_{\text{II}} = -\frac{1}{2\Delta A} X_{Li}(u_{Ll} - u_{Ll^*}) \quad (3.1)$$
$$G_{\text{III}} = -\frac{1}{2\Delta A} Y_{Li}(v_{Ll} - v_{Ll^*})$$

式中，$\Delta A = \Delta a \times b$，为虚拟闭合裂纹的面积，$\Delta a$ 为裂纹尖端单元长度，b 为单元宽度，如图 3.13 所示；X_{Li}、Y_{Li} 和 Z_{Li} 为裂纹尖端上第 L 列、第 i 行节点处的力；u_{Ll}、v_{Ll} 和 w_{Ll} 分别为裂纹上表面第 l 行节点的位移；u_{Ll^*}、v_{Ll^*} 和 w_{Ll^*} 分别为裂纹下表面第 l^* 行节点的位移。

在有限元分析中，$G_{\text{I}}/G_{\text{T}}$、$G_{\text{II}}/G_{\text{T}}$ 以及 $G_{\text{III}}/G_{\text{T}}$ 的值通常随着裂纹尖端单元大小的改变而变化，尤其是当单元尺寸较小的时候。G_{T} 为总应变能释放率，$G_{\text{T}} = G_{\text{I}} + G_{\text{II}} + G_{\text{III}}$。但是，在有限元计算中，单元尺寸相对较大，因此 $G_{\text{I}}/G_{\text{T}}$、$G_{\text{II}}/G_{\text{T}}$ 以及 $G_{\text{III}}/G_{\text{T}}$ 的值对单元尺寸大小不敏感。

一般来讲，界面断裂的判据可以表示为

$$\left(\frac{G_{\text{I}}}{G_{\text{IC}}}\right)^l + \left(\frac{G_{\text{II}}}{G_{\text{IIC}}}\right)^m + \left(\frac{G_{\text{III}}}{G_{\text{IIIC}}}\right)^n = 1 \tag{3.2}$$

式中，l、m 和 n 是三个常数；G_{IC}、G_{IIC} 以及 G_{IIIC} 分别代表 I 型裂纹、II 型裂纹与 III 型裂纹的临界能量释放率。在后面的研究中，不考虑 G_{III}。

3.3.4 TSV 界面裂纹的能量释放率

将无应力参考温度设为 25℃，温度载荷设为 −45～125℃。为了研究 TSV 界面分层行为，在 Cu 种子层和 SiO_2 层之间预埋一圆弧状裂纹，上下裂纹面几何上重合，但物理上是分开的。为避免上下裂纹面互相侵入，在上下裂纹面定义接触节点对。模型的几何尺寸为：$T_{\text{CuP}} = 2\mu m$；$T_{\text{Seed}} = 1\mu m$；$T_{SiO_2} = 0.5\mu m$；$D_{\text{CuP}} = 20\mu m$；$D_{\text{TSV}} = 10\mu m$；$H_{\text{TSV}} = 20\mu m$，$P = 25\mu m$。

在 Cu 种子层和 SiO_2 层界面处预埋一个水平裂纹，设计了两种不同开裂方式：一种是沿 Cu 种子层和 SiO_2 层界面从外向内扩展，如图 3.14(a) 所示；一种是沿 Cu 种子层和 SiO_2 层界面由内向外扩展，如图 3.14(b) 所示。首先计算不同长度预埋裂纹的能量释放率，能量释放率最大的裂纹长度即为临界裂纹长度，然后采用临界裂纹长度进行参数分析。

图 3.15 给出了 −45℃ 温度下，预埋裂纹向内扩展时的计算结果。结果表明，该情况下裂纹的能量释放率 G 值与裂纹长度呈反比关系，而且当裂纹尖端接近 Cu 垫边缘时，出现混合的开裂模式。图 3.16 给出了从 −45～125℃ 升温过程中，预埋裂纹向外扩展时的计算结果。当温度为 125℃ 时，裂纹模式为张开型或混合型；当温度为 −45℃ 时，则表现为滑移型。

除此之外，随着裂纹长度的增加，内侧裂纹尖端的 G_{I} 值也随之增加，而 G_{II} 则是先增加后下降，当裂纹长度达到给定裂纹长度的一半时，G_{II} 达到最大值；当裂纹长度为 $3\mu m$ 时，$G_{\text{I}} = G_{\text{II}}$。但是，外侧裂纹尖端能量释放率的变化趋势与内侧裂纹尖端恰好相反。这些裂纹行为主要是 TSV 结构中材料的 CTE 差异导致的。

3.3.5 TSV 界面裂纹能量释放率的参数研究

根据前面的计算结果，确定了向内扩展和向外扩展两种预埋裂纹的临界裂纹长

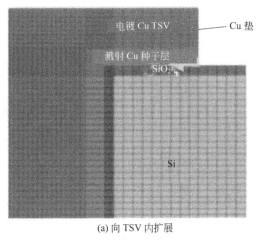

(a) 向 TSV 内扩展

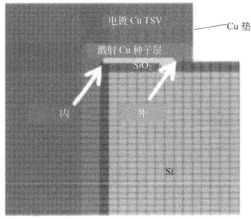

(b) 向 TSV 外扩展

图 3.14　两种裂纹扩展方式

度，如表 3.6 所示。本节采用表中所给的临界裂纹长度，研究 TSV 几何参数对 G 值的影响。首先考虑向内扩展裂纹情况下 SiO_2 层厚度（T_{SiO_2}）的影响，计算结果如图 3.17 所示。可以看出，TSV 直径（D_{TSV}）越大，G 值越大。当 $D_{TSV}=10\mu m$ 和 $20\mu m$ 时，G 值与 TSV 深度（H_{TSV}）、节距以及 T_{SiO_2} 呈反比关系，这表明 Cu 与 Si 的体积百分比对 G 值有重要影响，因为它们的 CTE 差异较大。另外，当 H_{TSV} 超过 $50\mu m$ 时，G 值趋于一个稳定值。如果 T_{SiO_2} 变为 $1.0\mu m$，则当 H_{TSV} 超过 $60\mu m$ 时，G 值趋于稳定；如果 T_{SiO_2} 变为 $0.5\mu m$，则当 H_{TSV} 超过 $70\mu m$ 时，G 值趋于稳定。此外，SiO_2 的弹性模量与其他材料相比较小，使其可以作为 Cu 种子层与 Si 之间的应力缓冲层，这样也就解释了前述 T_{SiO_2} 与 G 值之间的关系。接下来，采用最薄的 SiO_2 层（$T_{SiO_2}=0.5\mu m$）对其他参数进行研究。

图 3.18 给出了其他参数对 G 值的影响。计算时，SiO_2 层的厚度为 $0.5\mu m$，Cu 种子层的厚度为 $1.0\mu m$。可以看出，无论是向内扩展裂纹还是向外扩展裂纹，G 值均与 D_{TSV} 成正比。对于不同的 D_{TSV} 参数，TSV 节距减小都会引起 G 值的快

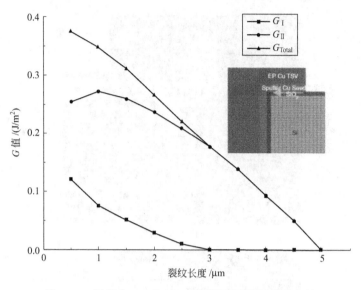

图 3.15 温度为 −45℃时,不同预埋裂纹长度对应的
能量释放率 G (向内扩展情况)

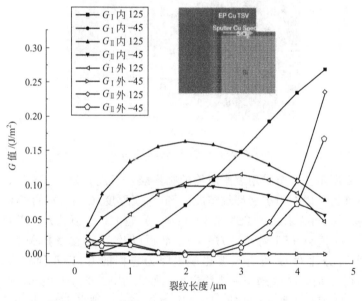

图 3.16 −45℃和 125℃温度下,不同预埋裂纹
长度对应的能量释放率 G (向外扩展情况)

速增大。当 $D_{TSV}=50\mu m$ 时,节距对 G 值的影响特别显著,这表明如果 TSV 直径较大,需要特别关注节距的影响。

除此之外,向内扩展裂纹与向外扩展裂纹相比,其 G 值更大,并且向外扩展裂纹其外侧裂纹尖端的 G 值整体大于内侧裂纹尖端的 G 值。对于所有不同的 D_{TSV}

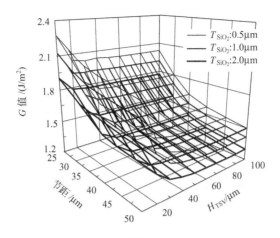

(a) TSV 直径 (D_{TSV})10μm

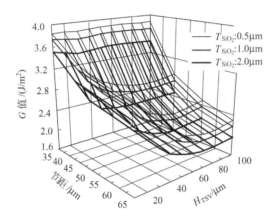

(b) TSV 直径 (D_{TSV})20μm

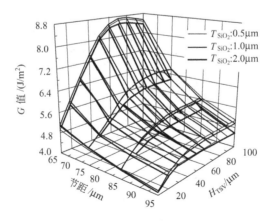

(c) TSV 直径 (D_{TSV})50μm

图 3.17 SiO_2 厚度（T_{SiO_2}）对 G 值的影响（向内扩展情况）

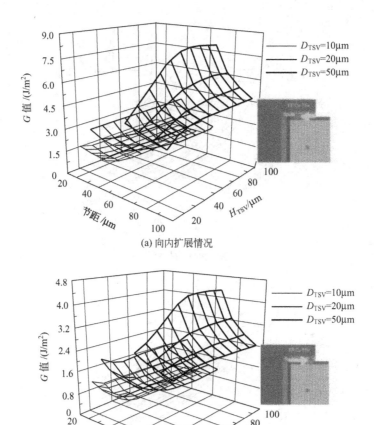

(a) 向内扩展情况

(b) 向外扩展情况（外侧裂纹尖端）

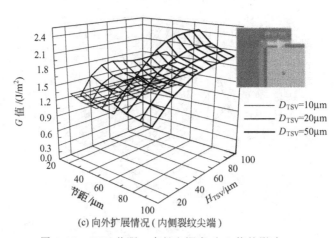

(c) 向外扩展情况（内侧裂纹尖端）

图 3.18　TSV 节距、直径和深度对 G 值的影响

参数，向外扩展裂纹其内侧裂纹尖端处的 G 值最小。这表明裂纹尖端距离外边界越近，其 G 值越高。从图 3.18 可以看出，当 $D_{TSV}=10\mu m$ 时，G 值小于 1.5J/m^2；$D_{TSV}=20\mu m$ 时，G 值小于 2.5J/m^2。当 $D_{TSV}=50\mu m$ 时，向内扩展裂纹的 G 值小于 8.0J/m^2，向外扩展裂纹其外侧裂纹尖端的 G 值小于 4.2J/m^2，其内侧裂纹尖端的 G 值小于 2.4J/m^2。

根据参考文献 [12, 20～22]，Cu 与 SiO_2 的临界应变能释放率（G_c）分别为 10J/m^2 和 8.5J/m^2，因此选用 $G_c=8.5$J/m^2 作为判断是否发生分层的指标。下面将上述分析结果进行总结。

① 对于向内扩展分层问题，如果 TSV 直径为 $10\mu m$ 和 $20\mu m$，不会发生分层。当 TSV 直径为 $50\mu m$ 时，如果增加 TSV 的深度则可能发生分层。

② 无论向内扩展分层或向外扩展分层，TSV 直径为 $10\mu m$ 时，G 值小于 1.5J/m^2；TSV 直径为 $20\mu m$ 时，G 值小于 2.5J/m^2；TSV 直径为 $50\mu m$ 时，G 值小于 8.0J/m^2。因此，不用担心 Cu 垫处发生分层。

表 3.6 参数研究中采用的热加载条件与裂纹长度

分层情况	向外扩展	向内扩展
热加载条件	$-45\sim125°C$	$125\sim-45°C$
临界裂纹长度	$1\mu m$	$4\mu m$

3.3.6 总结与建议

① 当 TSV 直径增大时，最大 von Mises 应力也增大，且最大应力位于 Cu 种子层和 SiO_2 层之间的圆周边界上。

② 对于向内扩展分层情况，G 值与 TSV 直径成正比，与 SiO_2 层厚度和 TSV 节距成反比。对于直径为 $10\mu m$ 的 TSV，当 TSV 的深度大于 $50\mu m$ 时，G 值趋于稳定，不再增大。对于直径为 $20\mu m$ 的 TSV，当 TSV 的深度大于 $80\mu m$ 时，G 值趋于稳定，不再增大。因此，对于直径为 $10\mu m$ 和 $20\mu m$ 的 TSV，增加深度不会引起向内扩展的分层。但是，对于直径为 $50\mu m$ 的 TSV，增加 TSV 的深度可能导致分层的发生。

③ 一般，G 值与 TSV 的直径成正比。Cu 和 Si 的体积百分比对 G 值有影响，当 TSV 直径较大时，节距才对 G 值产生显著影响。此外，向内扩展分层要比向外扩展分层的 G 值高。

④ 无论向内扩展分层或者向外扩展分层，TSV 直径为 $10\mu m$ 时，G 值小于 1.5J/m^2；TSV 直径为 $20\mu m$ 时，G 值小于 2.5J/m^2；TSV 直径为 $50\mu m$ 时，向内扩展裂纹的 G 值小于 8.0J/m^2，向外扩展裂纹其外侧裂纹尖端的 G 值小于 4.2J/m^2，其内侧裂纹尖端的 G 值小于 2.4J/m^2。因此，不需要担心发生分层。以上结论都是在 SiO_2 层厚度为 $0.5\mu m$、Cu 种子层厚度为 $1.0\mu m$ 条件下得到的。

3.4 TSV 的热学行为

3.4.1 TSV 芯片/转接板的等效热导率

3D 集成系统的热管理十分重要，而且 3D 集成的推广应用迫切需要低成本且有效的热管理设计指引和解决方案。应该认识到，即使拥有最先进的软件和高速计算机硬件，仍然不可能对 3D 集成系统中所有的 TSV 进行建模。本节基于热传导理论和计算流体力学（CFD），推导了不同 TSV 直径、节距和深宽比情况下 TSV 转接板的等效热导率。然后，将 TSV 芯片/转接板简化为均匀块体，采用这些等效热导率进行仿真[8]。

图 3.19 所示为一个高性能封装系统的示意图。高性能芯片通过微凸点连接到 TSV 转接板（或芯片）上，TSV 转接板/芯片的底部通过普通焊锡微凸点连接到有机基板上。由于存在较大的热失配，硅转接板/芯片与有机基板之间的焊锡微凸点连接需要下填料填充以提高可靠性。此外，高性能芯片和 TSV 转接板/芯片之间的微凸点互连也需要下填料填充。对于存储芯片堆叠封装，需要同样的处理方法。

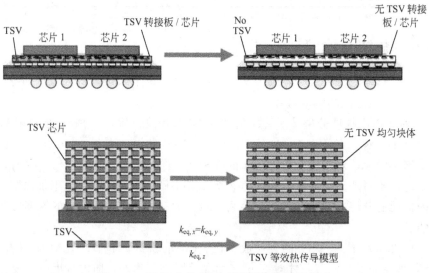

图 3.19　TSV 芯片/转接板的等效热传导模型

TSV 转接板/芯片的等效热导率需要通过详细的 3D CFD 分析来确定。图 3.20 右上角所示为 TSV 芯片/转接板，其热导率与一般的无 TSV 硅芯片不同。无 TSV 硅芯片的热导率是各向同性的，而 TSV 芯片/转接板由于竖直方向的 TSV 孔中填满了 Cu，其热导率是各向异性的，即在 xy 平面方向的等效热导率（$k_{eq,x}=k_{eq,y}$）与 z 方向的等效热导率（$k_{eq,z}$）不相等。

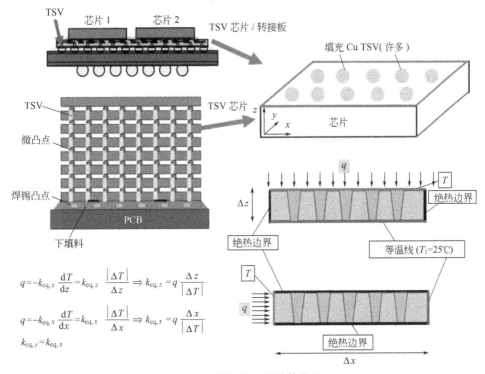

图 3.20 等效热导率计算模型

图 3.20 右侧的中图和下图分别给出了提取 z 方向等效热导率 $k_{eq,z}$ 和 x、y 方向等效热导率 $k_{eq,x}=k_{eq,y}$ 的方法。首先构造包含不同直径、节距和深宽比的填充 Cu TSV 芯片/转接板的几何结构。然后输入各材料的热参数：Si 的热导率为 150W/m/K，Cu 的热导率为 390W/m/K。最后，施加如图所示的热边界条件，计算出温度分布。等效热导率可以通过以下方程得到：

$$q=-k_{eq,z}\frac{dT}{dz}=k_{eq,z}\frac{|\Delta T|}{\Delta z} \Rightarrow k_{eq,z}=q\frac{\Delta z}{|\Delta T|}$$
$$q=-k_{eq,x}\frac{dT}{dx}=k_{eq,x}\frac{|\Delta T|}{\Delta x} \Rightarrow k_{eq,x}=q\frac{\Delta x}{|\Delta T|} \quad (3.3)$$
$$k_{eq,y}=k_{eq,x}$$

例如，为了提取 z 方向的等效热导率，建立了 TSV 芯片/转接板的几何结构，然后在 TSV 芯片顶面施加一个均布热流 q，同时在底面设置一个等温边界（例如 25℃），将周围四个边界设置为绝热边界。利用 Flowtherm 软件可以计算得到芯片顶面的平均温度，同时也得到了 $k_{eq,z}$。

图 3.21 和图 3.22 分别给出了不同直径、节距以及深宽比条件下，填充 Cu TSV 芯片/转接板的等效热导率 $k_{eq,z}$ 与 $k_{eq,x}=k_{eq,y}$。可以看出：①TSV 芯片所有方向上的等效热导率都大于纯硅芯片；②TSV 节距较小的芯片，其所有方向的等效热导率也较大；③TSV 直径较大的芯片，其所有方向的等效热导率也较大。为

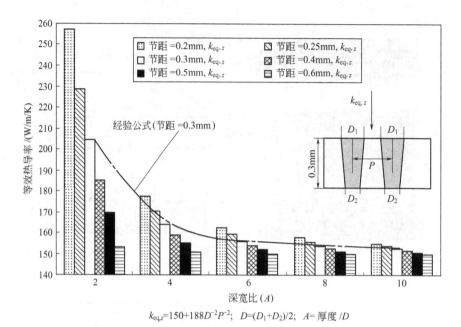

图 3.21　TSV 直径、节距、深宽比与等效热导率（z 方向）的关系

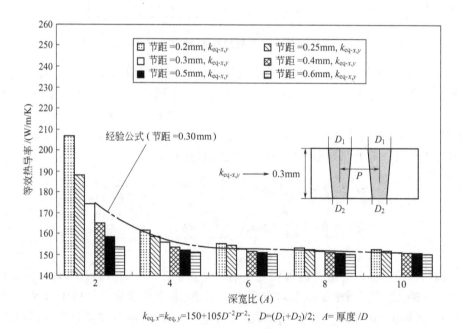

图 3.22　TSV 直径、节距、深宽比与等效热导率（x、y 方向）的关系

方便应用，图 3.21 和图 3.22 中的结果可以拟合为下面的经验公式：

$$k_{eq,z} = 150 + 188 D^2/P^2 \tag{3.4}$$

$$k_{eq,x} = k_{eq,y} = 150 + 105 D^2/P^2 \tag{3.5}$$

式中 P 为 TSV 节距，$D=(D_1+D_2)$ 为 TSV 平均直径，D_1 和 D_2 是锥形 TSV 孔两端的直径。图 3.21 和图 3.32 中的点画线为 $P=0.3\text{mm}$ 时，由上述经验公式得到的结果。通过与 3D CFD 分析结果比较，验证了这些经验公式是精确的。因此，在后面的 3D SiP 热分析中，TSV 芯片/转接板简化为不含 TSV 的均匀块体，其热导率可由经验公式计算得到。表 3.7 列出了用于仿真的材料参数。

表 3.7 仿真采用的封装结构、材料及功率

	芯片	TSV	凸点	下填料	PCB
材料	Si(TSV)	Cu	SnAg	聚合物	FR4
$k/(\text{W}/\text{m}/\text{K})$	经验公式	390	57	0.5	//0.8 ⊥0.3
尺寸/mm	5×5	$\phi 0.05$	$\phi 0.20$ 高度=0.15	5×5×0.15	76×114×1.6
功率/W	0.2W	—	—	—	—

3.4.2　TSV 节距对 TSV 芯片/转接板等效热导率的影响

图 3.23 所示为填充 Cu TSV 节距对 TSV 转接板/芯片等效热导率的影响。TSV 转接板/芯片的厚度为 $300\mu\text{m}$，TSV 平均直径为 $75\mu\text{m}$。可以看出：①当 TSV 节距减小时，即在芯片尺寸不变情况下包含更多的 TSV，这时 TSV 转接板/芯片的等效热导率增大。②TSV 转接板/芯片的等效热导率比纯硅材料大，Si 的热导率为 $150\text{W}/\text{m}/\text{K}$，Cu 为 $390\text{W}/\text{m}/\text{K}$。③$z$ 方向的等效热导率 ($k_{eq,z}$) 比 x、y 方向的等效热导率 ($k_{eq,x}=k_{eq,y}$) 大。④当 TSV 的节距较小时，等效热导率对节距的变化更敏感。

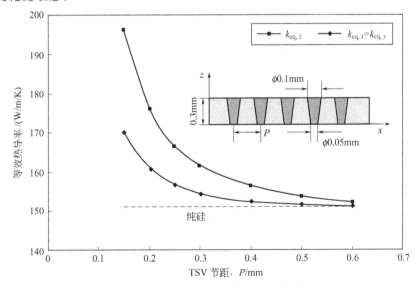

图 3.23　TSV 节距与等效热导率的关系

3.4.3 TSV 填充材料对 TSV 芯片/转接板等效热导率的影响

图 3.24 和图 3.25 分别给出了采用不同材料填充 TSV 孔时，TSV 转接板/芯片 z 方向与 xy 平面方向上的等效热导率。转接板厚度为 $300\mu m$，TSV 孔壁电镀 Cu 厚度为 $10\mu m$，深宽比为 4。可以看出，等效热导率随着填充材料热导率的增大而增大。但是当 TSV 节距较大时（如 $600\mu m$），填充材料对等效热导率的影响可以忽略不计。当填充材料的热导率比纯硅小很多时（如 $4W/m/K$ 和 $25W/m/K$），其对 x 和 y 方向上热导率的影响要远大于 z 方向。这是由于填充的低热导率材料阻碍了 x 和 y 方向上的热传导，而 z 方向上的热传导主要发生在硅材料中，受填充材料的影响较小。

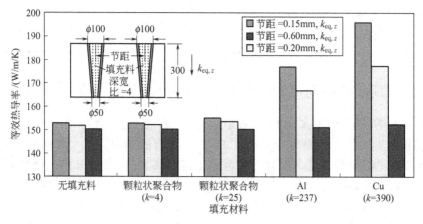

图 3.24 TSV 填充材料与等效热导率（z 方向）的关系

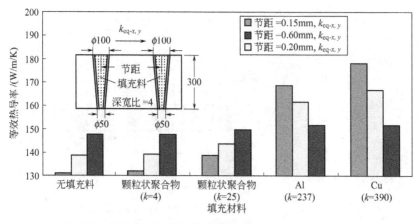

图 3.25 TSV 填充材料与等效热导率（x、y 方向）的关系

3.4.4 TSV Cu 填充率对 TSV 芯片/转接板等效热导率的影响

如图 3.26 所示，将 TSV 孔用 Cu 部分填充，考虑五种不同的 Cu 填充厚度：

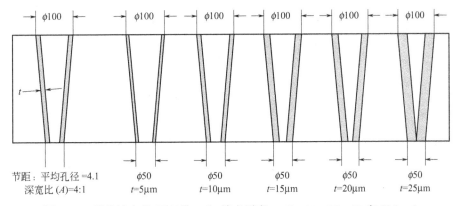

图 3.26 部分填充的 TSV 孔，Cu 填充厚度 $t=5,10,15,20$ 和 $25(\mu m)$

$t=5\mu m$，$t=10\mu m$，$t=15\mu m$，$t=20\mu m$，$t=25\mu m$。转接板厚度为 $300\mu m$，TSV 内无其他填充物。由于 TSV 孔内空间狭小，假设孔内不存在对流。图 3.27 和图 3.28 分别给出了不同节距的 TSV 孔部分填充时，TSV 芯片/转接板 z 方向和 xy 方向的等效热导率计算结果。可以看出，TSV 转接板/芯片的等效热导率随着电镀 Cu 厚度的增大而增大。

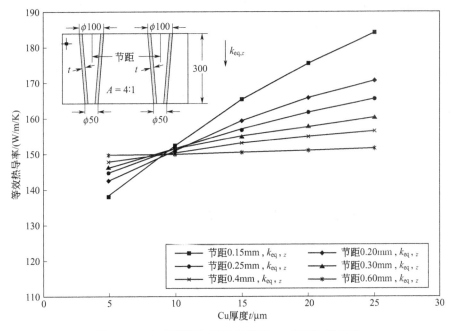

图 3.27 Cu 层厚度与等效热导率（z 方向）的关系

当孔节距较小时，等效热导率对电镀 Cu 的厚度变化较为敏感。例如，当 TSV 的节距为 0.15mm 时，如果电镀 Cu 的厚度从 $5\mu m$ 增加到 $25\mu m$，z 方向的等效热导率从 138W/m/K 增加到 187W/m/K，增大约 40%。当 TSV 的节距为 0.6mm 时，如果电镀 Cu 的厚度从 $5\mu m$ 增加到 $25\mu m$，z 方向的等效热导率从 149W/m/K

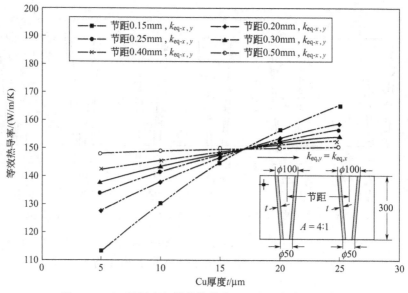

图 3.28　Cu 层厚度与等效热导率（x 和 y 方向）的关系

增加到 153W/m/K，增大不到 3%。

这是为什么呢？从物理上说，部分填充的 TSV 包括空气、电镀 Cu 层和 Si 三种材料，且这三种材料的热导率差别较大，空气的热导率为 0.026W/m/K，Cu 为 390W/m/K，Si 为 150W/m/K。对于一个给定深宽比的 TSV 转接板/芯片，更小的 TSV 节距意味着转接板包含更多的 TSV，也意味着 Cu 和空气的占比更大，而 Si 的占比更小。因此，具有更小节距 TSV 的芯片/转接板其等效热导率对电镀 Cu 厚度更加敏感。

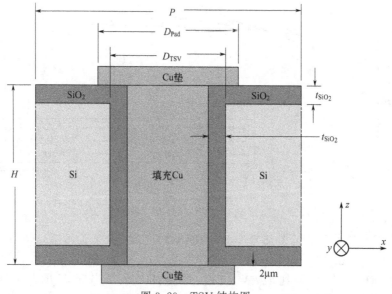

图 3.29　TSV 结构图

3.4.5 更精确的计算模型

前面计算等效热导率时没有考虑 SiO_2 层的影响。图 3.29 为包括 SiO_2 层的 TSV 结构模型[24]，图 3.30～图 3.33 为计算等效热导率所采用的模型。类似地，基于参数研究和曲线拟合[24]，得到如下经验公式。

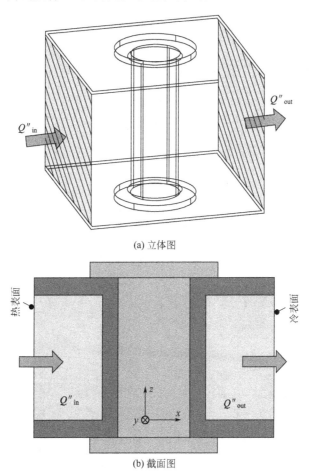

(a) 立体图

(b) 截面图

图 3.30 计算 k_{xy} 的模型

xy 方向的等效热导率 k_{xy}

$$k_{xy} = (90 t_{SiO_2}^{-0.33} - 148) \left(\frac{D_{TSV}}{P} \right) H^{0.1} + 160 t_{SiO_2}^{0.07} \tag{3.6}$$

z 方向的等效热导率 k_z

$$0.002 \leqslant \frac{t_{SiO_2}}{H} \leqslant 0.01 \Rightarrow k_z = 128 \exp\left(\frac{D_{TSV}}{P} \right) \tag{3.7}$$

式中，$0.2 \mu m \leqslant t_{SiO_2} \leqslant 0.5 \mu m$，$10 \mu m \leqslant D_{TSV} \leqslant 50 \mu m$，$H \geqslant 20 \mu m$，$0.1 \leqslant D_{TSV}/P \leqslant 0.77$；$P$ 为 TSV 节距；D_{TSV} 为 TSV 直径；H 为芯片/转接板厚度；

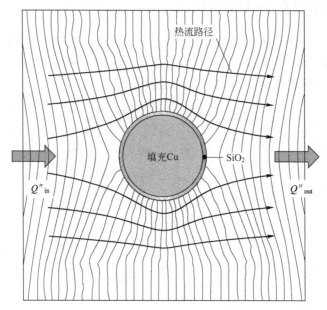

图 3.31　TSV 单元水平方向的热流路径与温度分布图

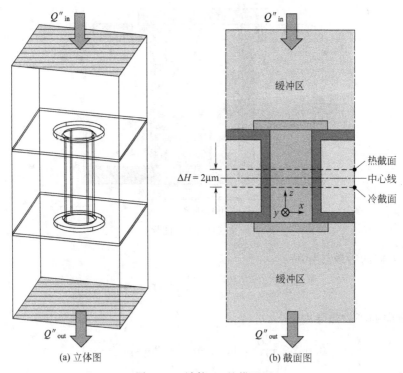

(a) 立体图　　　　　(b) 截面图

图 3.32　计算 k_z 的模型图

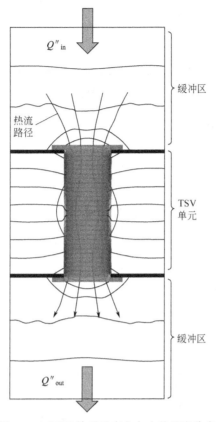

图 3.33 TSV 单元垂直方向上的温度分布

t_{SiO_2} 为 SiO_2 层厚度。

3.4.6 总结与建议

① 对于 3D IC 集成的热性能来说，SiO_2 层相当于 TSV 中的一个热阻，会降低 k_{xy} 和 k_z 的值。另一方面，填充 Cu 又能够提高 k_{xy} 和 k_z 的值。

② 对于所研究的全部情况，k_z 值都比纯硅的要大。也就是说，填充 Cu 增加的热导率大于由于 SiO_2 层存在而减小的热导率。

③ 当 SiO_2 层的厚度为 $0.1\mu m$ 或更薄时，它对热导率的影响不明显。

3.5 TSV 的电学性能

3.5.1 电学结构

TSV 是制作在硅基底上的金属-绝缘体-半导体（MIS）结构[28]；金属为填充

Cu，绝缘体为 TSV 孔壁上的 SiO_2 层，半导体为硅基底（Si）。如图 3.34 所示的 MIS 结构 TSV，为一个信号-接地 TSV（SG-TSV）对。Si 具有有限但非零的电阻率，会影响 TSV 的电学特性。例如，在信号传播和功率传输中会出现慢波和介电准横向电磁波（TEM）模式。通常，商用硅基底的电阻率范围为 5～40Ω·cm，MIS 结构 TSV 的电性能更多地受到上述两种模式的影响。

3.5.2 模型与方程

图 3.35 所示为提出的 MIS 结构 SG-TSV 模型及其参数[28]。模型参数根据 SG-TSV 模型的几何尺寸定义，所用到的几何尺寸主要有：TSV 直径 D（μm）、TSV 长度 L（μm）、两个 TSV 中心之间的节距 P（μm）、绝缘层厚度 T（μm）、Si 的电阻率 ρ_{Si}（Ω·cm），如图 3.34 所示。图 3.35 中模型的参数为

$$R_T = \frac{L\sqrt{1+0.25\times 频率}}{5.8\times 10^7 (D/2)^2 \pi} \tag{3.8}$$

$$L_S = \left(\frac{0.8P}{50}+1\right)\frac{\mu_0 L}{2\pi}\left[\ln\left(\frac{2L}{D/2}\right)-1\right] \tag{3.9}$$

$$M = \left(\frac{0.8P}{50}+1\right)\frac{\mu_0 L}{2\pi}\left\{\ln\left[\frac{L}{P}+\sqrt{1+\left(\frac{L}{P}\right)^2}\right]-\sqrt{1+\left(\frac{P}{L}\right)^2}+\frac{P}{L}\right\} \tag{3.10}$$

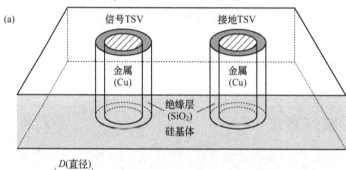

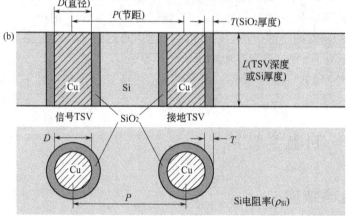

图 3.34 （a）MIS 结构 SG-TSV；（b）MIS 结构 SG-TSV 模型的几何参数定义

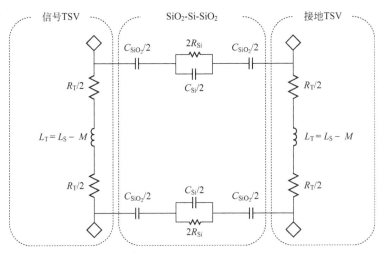

图 3.35 MIS 结构 SG-TSV 模型和式(3.8)~式(3.13)中的参数定义

$$C_{SiO_2} = \frac{3.9\varepsilon_0 L}{T}(2\pi D/2) \tag{3.11}$$

$$C_{Si} = \frac{11.9\varepsilon_0 \pi L}{\ln\left[\frac{P}{D} + \sqrt{\left(\frac{P}{D}\right)^2 - 1}\right]} \tag{3.12}$$

$$R_{Si} = \frac{11.9\varepsilon_0 \rho_{Si}}{100 C_{Si}} = \frac{1}{G_{Si}} \tag{3.13}$$

式中，R 为电阻；L 为电感；C 为电容；M 为两个 TSV 之间的交互区域电感；μ_0 和 ε_0 分别为真空磁导率及介电常数。式(3.8)中，频率的单位为 GHz。式(3.8)~式(3.13)中左端量的单位分别为 Ω/TSV、H/TSV、H/SG-TSV、F/TSV、F/SG-TSV 和 Ω/SG-TSV。

式(3.8)中，$\sqrt{1+0.25\times}$ 频率项反映了 TSV Cu 柱的集肤深度效应，式(3.9)与式(3.10)中的 $(1+0.8P/50)$ 项分别为自我修正项和依赖于 TSV 节距的 TSV Cu 柱交互电感项。式(3.11)可以通过考虑慢波模式下 SiO_2 层中的集中电场分布以及 TSV Cu 柱与硅基底之间的金属-绝缘体-金属（MIM）电容获得。式(3.12)和式(3.13)分别来自介电准 TEM 模式的一个双传输线电容和关系式 $G_{Si} = 2\pi f_e C_{Si}$[50]。

3.5.3 总结与建议

这里给出以下一些重要结果和建议[28]。

① 在低频范围，当信号在 SG-TSV 中传输时，决定信号传输速率的有效介电常数和磁导率受到两个 TSV 周围 SiO_2 层的厚度及其节距的影响。当 SiO_2 层的厚度较薄、节距较大时，有效介电常数增大，慢波模式效应增强，使得电容增大，导致严重的信号损耗。

② 在高频范围，介电常数和磁导率与 Si 相同。因此，MIS 结构的 SG-TSV 的电特性与埋入有损介电材料的双传输线相同。因此，用于信号传输应用的 MIS 结构其 SG-TSV 应当采用较小的直径、较小的节距、较短的长度、较薄的 SiO_2 层厚度以及较大的硅基底电阻率以使电容最小。但是，如果高频时信号损耗很重要，则应当对 SG-TSV 的节距进行优化，以平衡电容和插入损耗。

3.6 盲孔 TSV 的电测试

3.6.1 测试目的

第 2 章提到过，在制作 TSV 的工艺中，TSV 均为盲孔，而且需要许多工艺步才能完成。如果某一工艺步制作的 TSV 质量不能满足要求，那么后续工艺纯属浪费。因此，制程中需要对 TSV 质量进行评估，以决定后续制程是否必要。本节的主要目标是提出 TSV 电特性的测试方法，并将 TSV 测试装置与高频耦合测试集成在一起，用于测试 TSV 耦合参数并对 TSV 制程进行监测。

3.6.2 测试原理与仪器

图 3.36 所示为测试装置示意图，包含两个有一定间距的 TSV，分别与信号输入和输出电路相连。虽然这种设计方法没有将晶圆基底与最低电势连接，但是在寄生参数 R_{sub} 和 C_{sub} 之间仍然会发生交互耦合。

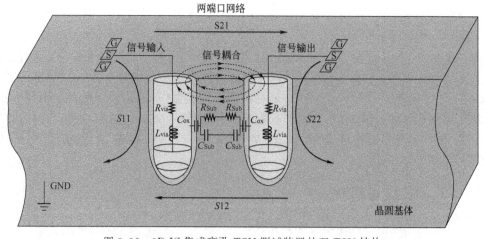

图 3.36 3D IC 集成盲孔 TSV 测试装置的双 TSV 结构

当将低频信号输入到任何一个 TSV，由于两个 TSV 之间的寄生电容导致绝缘，信号不会传输到另一个 TSV，这两个 TSV 就好像是一个断路的两个末端。但是，当将高频信号输入到其中一个 TSV 时，通过电容耦合信号可以到达另一个

TSV。因此，耦合的信号能量可以在另外一个 TSV 中进行测量。由于信号能量与两个 TSV 之间的耦合参数有关，而且由于耦合参数包含寄生参数，所以 TSV 的寄生参数可以被提取出来。通过与正常 TSV 的寄生参数进行比较，便可以确定 TSV 的状态。为了提高精度，可以使用高频 S 参数测试。TSV 的四个 S 参数，S11、S21、S22 和 S12 可以通过使用接地-信号-接地（GSG）探测的方法精确确定。

图 3.37 所示为盲孔 TSV 的 GSG 平面布局图，利用这种布局可以对 S11 和 S12 参数的频率进行分析和仿真。

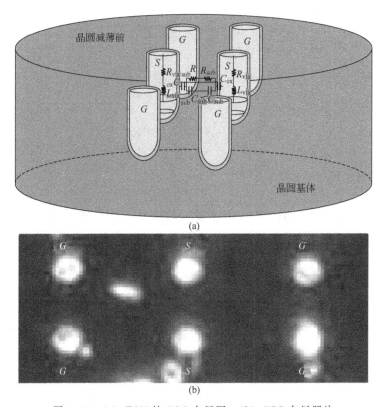

图 3.37 （a）TSV 的 GSG 布局图；（b）GSG 布局照片

图 3.38 给出了在 TSV 孔壁制作过程中 S 参数的仿真结果。如果制作的孔壁厚度不稳定，C_{ox} 会随孔壁厚度的变化增大或减小。当 TSV 孔壁较薄时，C_{ox} 增加，由于耦合效应，从输入端输入的信号在输出端的强度较强。当孔壁较厚时，C_{ox} 下降，耦合效应后的信号较弱。如果孔壁制作失败，导致 TSV 和基底间短路，$C_{ox}=0f$，可能导致一个无穷大的耦合频率。因此，采用这种 3D IC TSV 测试装置结合高频耦合效应完全可以检测 TSV 的制作质量。

为了使客户快速判断 TSV 的制作质量，将电特性参数 C_{ox} 与制程中孔壁的厚度相联系。首先，假设 $H \gg D$，只考虑圆柱电容公式并将其用于 TSV，得到式 (3.14)[49]：

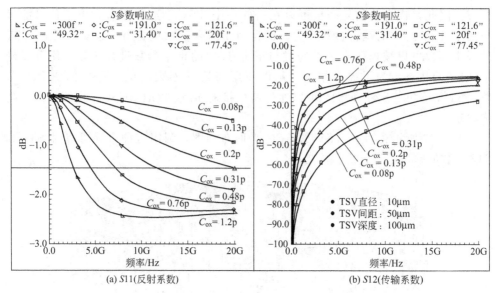

(a) $S11$(反射系数)　　(b) $S12$(传输系数)

图 3.38　不同 S 参数的等效电容分布

$$C_{ox} = \frac{2\pi\varepsilon H}{\ln\left(\dfrac{D}{D-2d}\right)} \tag{3.14}$$

式中，$C_{ox}(f)$ 为孔壁的最大电容量，是频率 f 的函数；H 为 TSV 的高度；D 为 TSV 的直径；ε 为介电常数；d 为孔壁厚度，如图 3.36 所示。式（3.14）可改写为：

$$d > \frac{D}{2}(1 - e^{\frac{2\pi\varepsilon H}{C_{ox}(f)}}) \tag{3.15}$$

例如，依据电性能来判断孔壁厚度是否符合规范，可分两部分定义限制条件。首先是孔壁制作的最小限制取决于直流电流（DC）的条件，这是采用漏电电流的观点，漏电量必须要小于一个特定值。一般而言，SiO_2 在孔壁上的等效阻抗要大于 $1G\Omega$[47]。为了确保这些条件处在一个安全的范围内，这里推荐采用的安全因子为 2。换句话说，等效阻抗应当大于 $2G\Omega$，因此最小孔壁厚度应当大于 $0.2\mu m$。

除此以外，必须获得不同工作频率所需的孔壁厚度。根据不同的系统应用，可以定义穿过 TSV 的信号传输能量。在本研究中，定义能量低于穿过 TSV 传输能量一半的信号可能会丢失，具体数值为 3dB[48]。回波损耗（反射能量）必须小于 3dB，考虑到安全因子为 2，那么回波损耗必须小于 1.5dB。插入损耗（传导能量）必须大于 20dB，同时一个更大的值代表更低的损耗[48]。这些规范都用来确定 SiO_2 的厚度，根据这些规范以及一对 GSG 布局测试，用户可以很容易地诊断 TSV 是否含有一个明显的电学缺陷。因此，可以确定后续 TSV 制程是继续（合格或通过）还是停止（失效或没通过）。

图 3.39 给出了盲孔 TSV 晶圆的电测试流程图，包括三个主要部分。第一部分

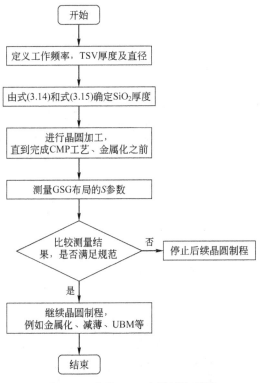

图 3.39 盲孔 TSV 电测试流程图

是基于应用频率和式(3.14)、式(3.15)设计 GSG 测试布局图、确定 TSV 直径和 SiO_2 层厚度。测试位置建议设在晶圆边角区域（即不制作芯片的区域）。第二部分为晶圆加工，直到 CMP 工艺后、金属化之前，然后测试 GSG 布局的 S 参数。最后一部分，将测试结果与规范值比较，决定是否进行后续制程。

3.6.3 测试方法与结果

首先选择 TSV 的直径 D 及其深度 H 用于确定 GSG 测试布局。采用电阻-电容 (RC) 模型仿真，可以得到不同频率下的 C_{ox} 值，如图 3.40 上图所示。对于特定的 TSV 工作频率，得到 C_{ox} 值后利用式(3.14)和式(3.15)可以确定 TSV SiO_2 层的厚度，如图 3.40 下图所示。

图 3.37(b) 给出了 GSG 测试布局的俯视图。这些 TSV 制作在一个 300mm 后段制程（BEOL）工艺（正面金属化之前）晶圆上，图 3.41 所示为该晶圆的测试装置。图 3.42 给出了该晶圆盲孔 TSV 电学测试的方法，图中 8 个测试模块位于晶圆的边角区域，每个测试模块拥有 9 个测试通道（GSG 测试布局）。对每个测试通道在 0~20GHz 范围内测量 10 个数据，然后取这些数据的平均值作为测试结果。最后，将测得的回波损耗与由方程计算得到的标准值进行对比，这样用户就可以很容易地确定 TSV 是否含有一个明显的电性能缺陷。

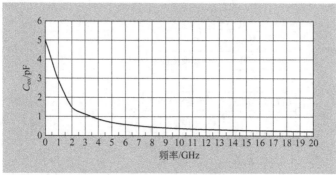

SiO₂层厚度	$d > \dfrac{D}{2}[1-e^{-\dfrac{2\pi \varepsilon H}{C_{ox}(f)}}]\ \&\ d > 0.2\mu m$			
工作频率	0～5GHz	0～10GHz	0～15GHz	0～20GHz
$D=10\mu m, H=50\mu m$	d＞0.20mm	d＞0.20mm	d＞0.21mm	d＞0.26mm
$D=10\mu m, H=100\mu m$	d＞0.20mm	d＞0.29mm	d＞0.42mm	d＞0.51mm
$D=10\mu m, H=200\mu m$	d＞0.30mm	d＞0.57mm	d＞0.80mm	d＞0.98mm

图 3.40　TSV SiO₂ 层厚度的标准值

图 3.41　12 寸（300mm）晶圆的测试装置

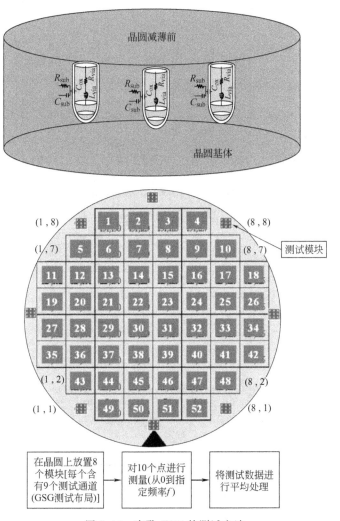

图 3.42 盲孔 TSV 的测试方法

图 3.43 和图 3.44 分别给出了两个 TSV 结构的 S11 和 S12 参数的测试结果。可以看出：①当工作频率小于 3.5GHz 时，回波损耗小于 -1.5dB；②当频率增加到 20GHz 时，插入损耗大于 -20dB。如果想让它在高于 3.5GHz 的频率下工作，就需要改进 TSV 工艺。通过本方法，可以在盲孔 TSV 制程中很容易地实现对 3D IC 工艺质量的监控。

3.6.4 盲孔 TSV 电测试指引

图 3.45 给出了盲孔 TSV 电测试的基本步骤。第一步定义 TSV 模块和 TSV SiO_2 厚度标准。第二步测量晶圆上 TSV 模块的 S 参数，并将这些测量数据进行平均处理。第三步，将测得的回波损耗与由方程计算得到的标准值进行比较，用户可以轻易地确定 TSV 是否含有一个明显的电性能缺陷。

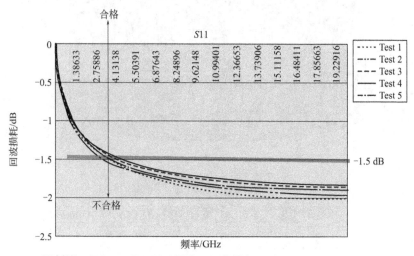

图 3.43 基于 $S11$ 的盲孔 TSV 通过测试的条件

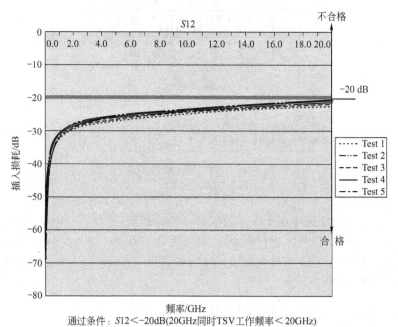

图 3.44 基于 $S12$ 的盲孔 TSV 通过测试的条件

第3章 TSV的力学、热学与电学行为

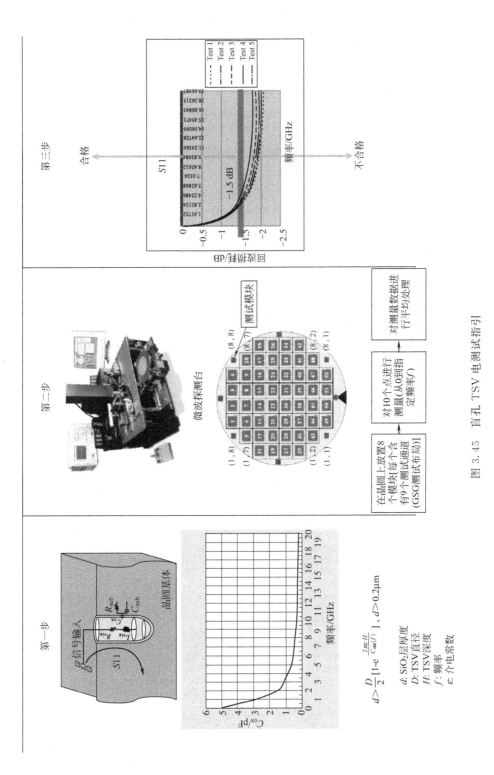

图3.45 盲孔TSV电测试指引

3.6.5 总结与建议

本节提出并且讨论了用于确定盲孔 TSV 电学完整性的一种新颖的测试方法和规范。由于电学测试是在盲孔 TSV 的 CMP 之后和金属化/UBM 之前进行的，可以预先将不合格的晶圆筛选出来，从而避免制造成本和时间上的浪费。主要结果和建议总结如下[46]。

① 由式(3.14) 和式(3.15) 并根据应用频率和 TSV 的深宽比，可以确定 TSV SiO_2 层的厚度。

② 推荐采用具有 GSG 结构的一对 TSV 作为 TSV SiO_2 层厚度的检测设计。

③ 采用 S 参数分析技术和矢量网络分析仪（包含一对 GSG 结构图案），提出了一种回波和传输损耗的测试方法。

④ 根据回波损耗（必须小于 1.5dB）和插入损耗（必须大于 20dB），可以确定 TSV SiO_2 层的厚度。

⑤ 将测量的回波损耗与由式(3.14) 和式(3.15) 计算得到的厚度标准进行比较，可以很容易地判断出 TSV 是否含有一个明显的电性能缺陷。

⑥ 提供了一套用于 3D IC 集成中盲孔 TSV 电学测试的指引。

3.7 参考文献

[1] Lau, J. H., *Reliability of RoHS-Compliant 2D and 3D IC Interconnects*, McGraw-Hill, New York, 2011.

[2] Chai, T., X. Zhang, J. H. Lau, C. Selvanayagam, K. Biswas, S. Liu, D. Pinjala, G. Tang, Y. Ong, S. Vempati, E. Wai, H. Li, B. Liao, N. Ranganathan, V. Kripesh, J. Sun, J. Doricko, and C. Vath, "Development of Large Die Fine-Pitch Cu/Low-*k* FCBGA Package with Through-Silicon Via (TSV) Interposer," *IEEE Transactions on Components, Packaging and Manufacturing Technology*, Vol. 1, No. 5, 2011, pp. 660–672.

[3] Selvanayagam, C., J. H. Lau, X. Zhang, S. Seah, K. Vaidyanathan, and T. Chai, "Nonlinear Thermal Stress/Strain Analysis of Copper Filled TSV (Through-Silicon Via) and Their Flip-Chip Microbumps," *IEEE Transactions in Advanced Packaging*, Vol. 32, No. 4, November 2009, pp. 720–728.

[4] Lau, J. H., *Ball Grid Array Technology*, McGraw-Hill, New York, 1996.

[5] Shen, Y. L., "Thermomechanical Stresses in Copper Interconnects: A Modeling Analysis," *Microelectronic Engineering*, Vol. 83, 2006, pp. 446–459.

[6] Vandevelde, B., C. Okoro, M. Gonzalez, B. Swinnen, and E. Beyne, "Thermomechanics of 3D-Wafer Level and 3D Stacked IC Packaging Technologies," *International Conference on Thermal, Mechanical and Multi-Physics Simulation and Experiments in Microelectronics and Micro-Systems*, 2008, pp. 1–7.

[7] Kuo, T. Y., S. Chang, Y. Shih, C. Chiang, C. Hsu, C. Lee, C. Lin, Y. Chen, and W. Lo, "Reliability Tests for a Three-Dimensional Chip Stacking Structure with Through Silicon Via Connections and Low Cost," *IEEE/ECTC Proceedings*, 2008, pp. 853–858.

[8] Lau, J. H., and G. Y. Tang, "Thermal Management of 3D-IC Integration with TSV (Through-Silicon Via)," *IEEE/ECTC Proceedings*, 2009, pp. 635–640.

[9] Hsieh, M. C., C. K. Yu, and S. T. Wu, "Thermomechanical Simulative Study for 3D Vertical Stacked IC Packages with Spacer Structures," *Semiconductor Thermal Measurement, Modeling and Management Symposium*, 2010.

[10] Hsieh, M. C., C. K. Yu, and W. Lee, "Effects of Geometry and Material Properties for Stacked IC Package with Spacer Structure," *International Conference on Thermal, Mechanical and Multi-Physics Simulation and Experiments in Microelectronics and Micro-Systems*, 2009, pp. 239–244.

[11] Ranganathan, N., K. Prasad, N. Balasubramanian, and K. L. Pey, "A Study of Thermomechanical Stress and Its Impact on Through-Silicon Vias," *Journal of Micromechanics and Microengineering*, Vol. 18, 2008.

[12] Liu, X., Q. Chen, P. Dixit, R. Chatterjee, R. R. Tummala, and S. K. Sitaraman, "Failure Mechanisms and Optimum Design for Electroplated Copper Through-Silicon Vias (TSVs)," *IEEE/ECTC Proceedings*, 2009, pp. 624–629.

[13] Hsieh, M. C., and C. K. Yu, "Thermomechanical Simulations for 4-Layer Stacked IC Packages," *International Conference on Thermal, Mechanical and Multi-Physics Simulation and Experiments in Microelectronics and Micro-Systems*, 2008, pp. 254–260.

[14] Ladani, L. J., "Stress Analysis of Three-Dimensional IC Package as Function of Structural Design Parameters," *Microelectronic Engineering*, Vol. 87, 2010, pp. 208–215.

[15] Krueger, R., "The Virtual Crack Closure Technique: History, Approach and Application," ICASE Report No. 2002–10, NASA, April 2002.

[16] Wu, C. J., M. C. Hsieh, C. C. Chiu, M. C. Yew, and K. N. Chiang, "Interfacial Delamination Investigation Between Copper Bumps in 3D Chip Stacking Package by Using the Modified Virtual Crack Closure Technique," *Microelectronic Engineering*, 2011 (in press).

[17] Hsieh, M. C., S. T. Wu, R. M. Tain, C. H. Chen, and C. K. Lin, "Thermomechanical Analysis and Interfacial Energy Release Rate Estimation for Metal-Insulator-Metal Capacitor Device," *Microelectronic Engineering*, 2011 (in press).

[18] Read, D. T., Y. W. Cheng, and R. Geiss, "Morphology, Microstructure, and Mechanical Properties of a Copper Electrodeposit," *Microelectronic Engineering*, Vol. 75, 2004, pp. 63–70.

[19] Iannuzzelli, R., "Predicting Plated-Through-Hole Reliability in High-Temperature Manufacturing Process," *IEEE/ECTC Proceedings*, 1991, pp. 410–421.

[20] Wang, H. W., and H. W. Ji, "Size Effect and Microscopic Experimental Analysis of Copper Foils in Fracture," *Key Engineering Materials*, Vol. 324–325, 2006, pp. 253–266.

[21] *AZoM: The A to Z of Materials and AZojomo*.

[22] Volinsky, A. A., N. R. Moody, and W. W. Gerberich, "Interfacial Toughness Measurements for Thin Films on Substrates," *Acta Materialia*, Vol. 50, 2002, pp. 441–466.

[23] Hsieh, M. C., S. Wu, C. Wu, J. H. Lau, R. Tain, and R. Lo, "Investigation of Energy Release Rate for Through-Silicon Vias (TSVs) in 3D IC Integration," *EuroSimE Proceedings*, April 2011, pp. 1/7–7/7.

[24] Chien, J., Y. Chao, J. H. Lau, M. Dai, R. Tain, M. Dai, P. Tzeng, C. Lin, Y. Hsin, S. Chen, J. Chen, C. Chen, C. Ho, R. Lo, T. Ku, and M. Kao, "A Thermal Performance Measurement Method for Blind Through-Silicon Vias (TSVs) in a 300-mm Wafer," *IEEE ECTC Proceedings*, Orlando, FL, June 2011, pp. 1204–1210.

[25] Akasaka, Y., "Three-Dimensional IC Trends," *Proceedings of the IEEE*, Vol. 74, No. 12, December 1986, pp. 1703–1714.

[26] Akasaka, Y., and T. Nishimura, "Concept and Basic Technologies for 3D IC Structure," *IEEE Proceedings of International Electron Devices Meetings*, Vol. 32, 1986, pp. 488–491.

[27] Gat, A., L. Gerzberg, J. Gibbons, T. Mages, J. Peng, and J. Hong, "CW Laser of Polycrystalline Silicon: Crystalline Structure and Electrical Properties," *Applied Physics Letters*, Vol. 33, No. 8, October 1978, pp. 775–778.

[28] Pak, J., J. Cho, J. Kim, J. Lee, H. Lee, K. Park, and J. Kim, "Slow Wave and Dielectric Quasi-TEM Modes of Metal-Insulator-Semiconductor (MIS) Structure Through-Silicon Via (TSV) in Signal Propagation and Power Delivery in 3D Chip Package," *IEEE/ECTC Proceedings*, May 2011, pp. 667–672.

[29] Lau, J. H., C. K. Lee, C. S. Premachandran, and A. Yu, *Advanced MEMS Packaging*, McGraw-Hill, New York, 2010.

[30] Lau, J. H., "Evolution, Outlook, and Challenges of 3D IC/Si Integration," *IEEE/ICEP Proceedings* (keynote), Nara, Japan, April 13, 2011, pp. 1–17.

[31] Lau, J. H., "Evolution, Challenge, and Outlook of TSV, 3D IC Integration and

[31] 3D Si Integration," *International Wafer Level Packaging Conference* (plenary), San Jose, CA, October 3–7, 2011, pp. 1–18.
[32] Roullard, J., S. Capraro, A. Farcy, T. Lacrevaz, C. Bermond, P. Leduc, J. Charbonnier, C. Ferrandon, C. Fuchs, and B. Flechet, "Electrical Characterization And Impact On Signal Integrity Of New Basic Interconnection Elements Inside 3D Integrated Circuits," *IEEE/ECTC Proceedings*, 2011, pp. 1176–1182.
[33] Cadix, L., M. Rousseau, C. Fuchs, P. Leduc, A. Thuaire, R. El Farhane, H. Chaabouni, R. Anciant, J.-L. Huguenin, P. Coudrain, A. Farcy, C. Bermond, N. Sillon, B. Flechet, and P. Ancey, "Integration and Frequency Dependent Electrical Modeling of Through-Silicon Vias (TSV) for High-Density 3DICs," *IEEE/ECTC Proceedings*, 2010, pp. 1–3.
[34] Sheng, F., S. Chakravarty, and D. Jiao, "An Efficient 3D-to-2D Reduction Technique for Frequency-Domain Layered Finite Element Analysis of Large-Scale High-Frequency Integrated Circuits," *IEEE Proceedings of Electrical Performance of Electronic Packaging*, 2007, pp. 295–298.
[35] Thorolfsson, T., and P. D. Franzon, "System Design for 3D Multi-FPGA Packaging," *IEEE Proceedings of Electrical Performance of Electronic Packaging*, 2007, pp. 171–174.
[36] Sun, J., J. Lu, D. Giuliano, P. Chow, and J. Gutmann, "3D Power Delivery for Microprocessors and High-Performance ASICs," *IEEE Proceedings of Applied Power Electronics Conference*, 2007, pp. 127–133.
[37] Khan, M. A., and A. Q. Ansari, "Quadrant-Based XYZ Dimension Order Routing Algorithm for 3D Asymmetric Torus Routing Chip (ATRC)," *Proceedings of International Conference on Emerging Trends in Networks and Computer Communications*, 2011, pp. 121–124.
[38] Geng, F., D. Xiaoyun, and L. Luo, "Trends in Networks and Computer Communications," *International Conference on Electronic Packaging Technology & High Density Packaging*, 2009, pp. 85–90.
[39] Cadix, L., M. Rousseau, C. Fuchs, P. Leduc, A. Thuaire, R. El Farhane, H. Chaabouni, R. Anciant, J.-L. Huguenin, P. Coudrain, A. Farcy, C. Bermond, N. Sillon, B. Flechet, and P. Ancey, "Integration and Frequency Dependent Electrical Modeling of Through-Silicon Vias (TSV) for High-Density 3DICs," *International Interconnect Technology Conference*, 2010, pp. 1–3.
[40] Gu, X., B. Wu, M. Ritter, and L. Tsang, "Efficient Full-Wave Modeling of High Density TSVs for 3D Integration," *IEEE/ECTC Proceedings*, 2010, pp. 663–666.
[41] Soon, W., W. Seung, Z. Qiaoer, K. Pasad, V. Kripesh, and J. H. Lau, "High RF Performance TSV Silicon Carrier for High-Frequency Application," *IEEE/ECTC Proceedings*, 2008, pp. 1946–1952.
[42] Lamy, Y. P. R., K. B. Jinesh, F. Roozeboom, D. J. Gravesteijn, and W. F. A. Besling, "RF Characterization and Analytical Modeling of Through-Silicon Vias and Coplanar Waveguides for 3D Integration," *Transactions on Advanced Packaging*, Vol. 33, No. 4, 2010, pp. 1072–1079.
[43] Chaabouni, H., M. Rousseau, P. Leduc, A. Farcy, R. El Farhane, A. Thuaire, G. Haury, A. Valentian, G. Billiot, M. Assous, F. De Crecy, J. Cluzel, A. Toffoli, D. Bouchu, L. Cadix, T. Lacrevaz, P. Ancey, N. Sillon, and B. Flechet, "Investigation on TSV Impact on 65-nm CMOS Devices and Circuits," *IEEE Proceedings of International Electron Devices Meeting*, 2010, pp. 35.1.1–35.1.4.
[44] Wu, J. H., J. Scholvin, and J. A. del Alamo, "Through-Wafer Interconnect in Silicon for RFICs," *IEEE Transactions on Electronic Devices*, Vol. 51, No. 11, November 2004, pp. 1765–1771.
[45] Leung, L. L. W., and K. J. Chen, "Microwave Characterization and Modeling of High Aspect Ratio Through-Wafer Interconnect Vias in Silicon Substrates," *IEEE Transactions of Microwave Theory and Techniques*, Vol. 53, No. 8, August 2005, pp. 2472–2480.
[46] Sheu, S., Z. Lin, J. Hung, J. H. Lau, P. Chen, S. Wu, K. Su, C. Lin, S. Lai, T. Ku, W. Lo, and M. Kao, "An Electrical Testing Method for Blind Through Silicon Vias (TSVs) for 3D IC Integration", *Proceedings of IMAPS International Conference*, Long Beach, CA, October 2011, pp. 208–214.
[47] Sunohara, M., T. Tokunaga, T. Kurihara, and M. Higashi, "Silicon Interposer with TSVs (Through-Silicon Vias) and Fine Multilayer Wiring," *IEEE/ECTC Proceedings*, 2008, pp 847–852.
[48] Ziolkowski, R. W., "An Efficient, Electrically Small Antenna Designed for VHF

and UHF Applications," *IEEE Antennas and Wireless Propagation Letters*, Vol. 7, August 2008, pp. 217–220.
[49] Pozar, D., *Microwave Engineering*, 3rd ed., Wiley, New York, 2004.
[50] Shibata, T., and E. Sano, "Characterization of MIS Structure Coplanar Transmission Lines for Investigation of Signal Propagation in Integrated Circuits," *IEEE Transactions on Microwave Theory and Techniques*, Vol. 38, No. 7, July 1990, pp. 881–890.

第4章 薄晶圆的强度测量

4.1 引言

晶圆减薄以及薄晶圆的拿持是3D IC集成和3D Si集成需要解决的第二个关键技术。3D IC集成和3D Si集成备受关注的一个原因就是堆叠起来的薄晶圆/芯片的总厚度比一个普通的晶圆/芯片还要薄,而更小的体积是诸如智能手机和平板电脑等移动电子产品最受青睐的特性。薄晶圆的缺点在于其容易发生翘曲且强度极低。晶圆越薄,翘曲越大,强度越低。本章讨论采用压阻应力传感器测量薄晶圆的强度和翘曲的方法,还讨论晶圆背面磨削减薄工艺对多层堆叠Cu-low-k芯片纳观力学行为的影响。

4.2 用于薄晶圆强度测量的压阻应力传感器

4.2.1 压阻应力传感器及其应用

压阻应力传感器是原位应力测量的有力工具[1~10],在半导体器件封装应力测量中极具潜力[1~4]。Edwards等[4]使用n型压阻应力传感器定性评估了塑料封装中的应力状态。定量测量时,需要首先确定压阻系数。目前已经提出了几种压阻系数的校准方法[3,5~9]。校准后的压阻应力传感器被广泛用于评估芯片粘贴、底部填充以及塑封等封装工艺后硅芯片表面的应力[1,3,10]。硅芯片与其他封装材料(如环氧树脂、基板或PCB板)之间的热膨胀系数(CTE)存在较大不匹配,在经过几个封装工艺步后,硅芯片表面会产生100~200MPa的残余应力[1,3,10]。

在3D集成中,压阻应力传感器可以用来测量晶圆减薄、SiO_2沉积、金属化层制作以及电镀等工艺过程中器件和转接板的强度。本节讨论用于减薄后晶圆强度和翘曲测量的压阻应力传感器的设计、制作和校准方法。

4.2.2 压阻应力传感器的设计与制作

n型压阻应力传感器的设计布局如图4.1和图4.2所示。传感器R_1、R_2、R_3、R_4分别沿着p型[110]硅晶圆的[110]、[$\bar{1}$00]、[$\bar{1}$10]、[010]方向制作。

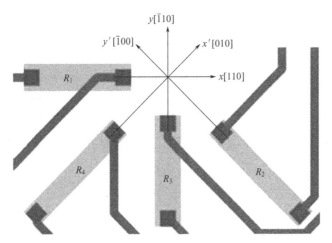

图 4.1 压阻应力传感花的布局

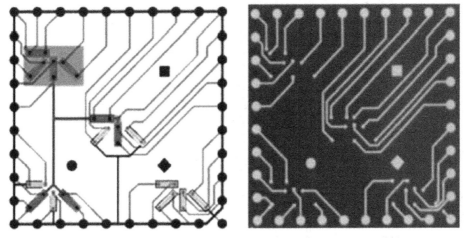

图 4.2 压阻应力传感器芯片内部金属线及探测焊盘的光学图像

R_1 和 R_3 的阻值大约为 $0.625\text{k}\Omega$，R_2 和 R_4 约为 $0.932\text{k}\Omega$。连接线的电阻大约为 178Ω。传感器阻值的差别来自于它们具有不同的尺寸：R_1 和 R_3 的尺寸为 $350\mu\text{m}\times100\mu\text{m}$，$R_2$ 和 R_4 为 $420\mu\text{m}\times80\mu\text{m}$。

n 型应力传感器采用常规工艺制作在晶圆上。首先，在 p 型 [100] 晶圆上通过热氧化生长出厚度为 100Å 的 SiO_2 层；然后，在晶圆上涂覆一层光刻胶（PR），在要制作传感器的地方去除光刻胶。移除光刻胶后，将砷（As）植入 p 型硅晶圆以形成 n 型电阻。最后，将光刻胶从晶圆上整体剥离。光刻胶被剥离后，将 4000Å 的 SiO_2 层通过等离子增强化学气相沉积（PECVD）方法沉积到晶圆上，在氧化层上开电阻接触窗，沉积 TaN 和制作 Al 金属化层以形成金属触点、金属镀膜线和探测焊盘。完成金属化层后，再次用 PECVD 沉积 5000Å 厚的 SiO_2 层，并制作焊盘图案。图 4.2 为传感器芯片的光学图像，图中显示了金属镀膜线和探测焊盘。可以看到，在一个 5mm×5mm 芯片的不同位置制作了 4 个传感花。

4.2.3 压阻应力传感器的校准

为达到测量目的，压阻应力传感器要满足两个基本要求：①在载荷下具有足够的、可测量到的阻值变化；②通过校准确定传感器的压阻系数。传感器校准包括根据已知加载应力测量阻值的改变量，并通过压阻理论确定压阻系数[11]。根据压阻理论，如果给图 4.1 所示的传感花的 [110] 方向施加单轴应力 σ_x，那么压阻系数可由下式给出：

$$\Pi_{11} + \Pi_{12} = \frac{1}{\sigma_x}\left(\frac{\Delta R_1}{R_{10}} + \frac{\Delta R_3}{R_{30}}\right) \tag{4.1}$$

$$\Pi_{44} = \frac{1}{\sigma_x}\left(\frac{\Delta R_1}{R_{10}} - \frac{\Delta R_3}{R_{30}}\right) \tag{4.2}$$

式中，ΔR_i 和 R_{i0} 分别为应力引起的第 i 个传感器的阻值改变量和原阻值。知道了压阻系数后，通过测量 ΔR_i 即可确定平面内应力分量 σ_x 和 σ_y 的大小（假设为平面应力状态，即 $\sigma_z=0$）：

$$\sigma_x = \frac{\Pi_{44}\left(\frac{\Delta R_1}{R_{10}} + \frac{\Delta R_3}{R_{30}}\right) + (\Pi_{11} + \Pi_{12})\left(\frac{\Delta R_1}{R_{10}} - \frac{\Delta R_3}{R_{30}}\right)}{2\Pi_{44}(\Pi_{11} + \Pi_{12})} \tag{4.3}$$

$$\sigma_y = \frac{\Pi_{44}\left(\frac{\Delta R_1}{R_{10}} + \frac{\Delta R_3}{R_{30}}\right) - (\Pi_{11} + \Pi_{12})\left(\frac{\Delta R_1}{R_{10}} - \frac{\Delta R_3}{R_{30}}\right)}{2\Pi_{44}(\Pi_{11} + \Pi_{12})} \tag{4.4}$$

校准压阻传感器的方法多种多样。例如，可以从传感器芯片上取出一条进行四点弯曲（4PB）试验[3,5~10]进行校准，也可以对整个传感器芯片进行四点弯曲试验[8]，还可以采用晶圆真空吸盘法[9]以及静水压力载荷法[3]等。本节采用含多个传感器芯片的长条试样进行四点弯曲校准，这种方法设置简单且有较多的文献数据可供参考[3,5~10]。图 4.3 所示为校准压阻应力传感器的条状试样，尺寸为 70mm×10mm，包含两排共 28 个传感器芯片，但在校准时只用到位于试样中心的传感花。

如图 4.3 所示，校准时采用的主要设备为：四点弯曲夹具、加载设备、显微镜和万用表。显微镜用于定位传感器焊盘，以便采用万用表的微型探针从焊盘测量传感器的阻值变化。采用 Instron 公司的微力测试仪，用其 100N 的加载单元给试样施加一个给定的力（2F），该力通过图 4.3 右下图中的装置分为 2 个力施加在试样上，试样的受力情况如图 4.4 所示。这种加载方式使得试样的传感器一侧承受均匀的单向拉应力 σ，其计算公式为：

$$\sigma = \frac{3F(L-d)}{t^2 h} \tag{4.5}$$

式中，L 是加载跨度；d 是支撑跨度；h 是试样宽度；t 是试样厚度。当试样的弯曲变形较小且 t 和 h 与 d 和 L 比较均较小时，式（4.5）精确度较高[8]。本章采用的尺寸分别为 $L=50$mm、$d=20$mm、$h=10$mm、$t=0.73$mm。

校准时，需要在加载前先测量传感器的电阻值，然后采用增量加载法（单步增

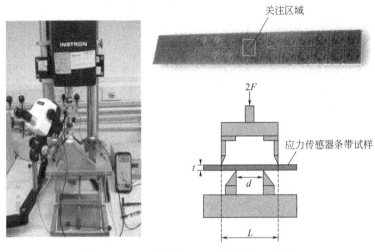

图 4.3 四点弯曲（4PB）试验装置与试样

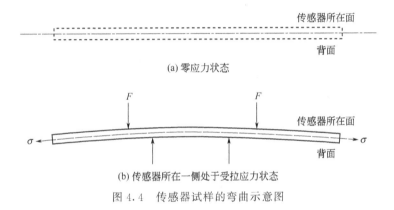

图 4.4 传感器试样的弯曲示意图

量为 4N）依次增加直到 24N。每个载荷增量步，均由微探针和万用表测量阻值。考虑到数据的分散性，需要对至少 8 个试样进行测试，然后取平均。这 8 个试样还要从两个不同的晶圆上截取。

图 4.5 给出了传感器阻值随所施加应力的变化情况。可以看出，n 型传感器的电阻值随着应力的增大而减小。为了由式（4.1）和式（4.2）计算传感器的压阻系数，将图 4.5 中的阻值正则化处理，得到图 4.6，图中两条直线的斜率即为压阻系数，其值列于表 4.1 中。

表 4.1 n 型应力传感器的压阻系数

$\Pi_{11}+\Pi_{12}$	$-1.98\times10^{-4}\,\mathrm{MPa}^{-1}$
Π_{44}	$-1.06\times10^{-4}\,\mathrm{MPa}^{-1}$

最后得到的 n 型传感器的压阻系数 $\Pi_{11}+\Pi_{12}$ 和 Π_{44} 分别为 -1.98×10^{-4} MPa^{-1} 和 $-1.06\times10^{-4}\,\mathrm{MPa}^{-1}$，其值与之前的报道[2]比较一致。

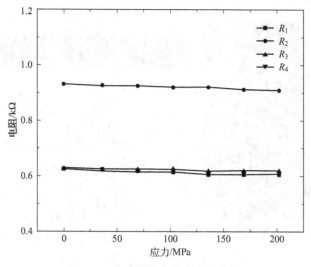

图 4.5 传感器阻值随应力的变化

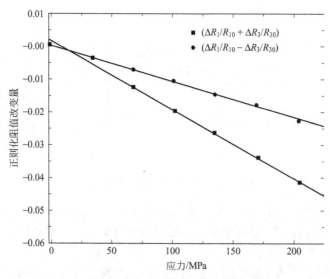

图 4.6 传感器正则化阻值改变量与应力的关系

一旦确定了 n 型传感器的压阻系数 $\Pi_{11}+\Pi_{12}$ 和 Π_{44}，σ_x 和 σ_y 就可通过式（4.3）和式（4.4）求得。下面就采用这种方法评估经过背面磨削减薄后晶圆上的应力，以及与胶带键合后晶圆上的应力。

4.2.4 背面磨削后晶圆的应力

为了研究减薄磨削对晶圆应力的影响，初始厚度相同（均为 $730\mu m$）的传感器晶圆被减薄到 $400\mu m$、$200\mu m$ 和 $100\mu m$ 三个不同的厚度[12]。采用商用减薄机（Okamono GNX200）分别用粗磨、细磨和抛光对晶圆背面进行磨削减薄。粗磨阶段的工艺参数为：磨削砂轮的粒度为 600、主轴转速为 3000r/min、卡盘转速为 230r/min、进给速度 180~190μm/min。细磨阶段的工艺参数为：磨削砂轮的粒度为 2000、主轴转速为 3400r/min、卡盘转速为 230r/min、进给速度 12~16μm/min。抛光工艺参数为：磨抛液流量为 200mL/min，压力为 150gf/cm², 磨抛垫转速为 230r/min，真空吸盘转速为 210r/min。

晶圆减薄后，9 个测量位置（如图 4.7 所示）与减薄前相同。测得电阻后，应力分量 σ_x 和 σ_y 通过式(4.3)和式(4.4)得到。减薄到 100μm 厚晶圆的应力结果列于表 4.2 中。可以发现，减薄后晶圆上主要为压应力。图 4.8 说明平均应力与晶圆厚度有关，而且面内压应力随着晶圆厚度的减小呈指数增大。

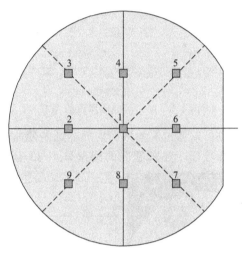

图 4.7 应力测量的位置

表 4.2 减薄后晶圆的面内应力分量 σ_x 和 σ_y MPa

位置编号 \ 晶圆厚度	100μm		200μm		400μm	
	应力 σ_x	应力 σ_y	应力 σ_x	应力 σ_y	应力 σ_x	应力 σ_y
1	−16.0	−23.6	0.1	−7.4	12.8	20.3
2	−47.3	−24.6	−31.1	−8.5	−11.5	−3.9
3	−112.0	−89.2	−60.0	−52.3	−35.7	−28.1
4	−71.6	−48.8	−27.6	−20.0	−35.7	−28.1
5	−51.9	−44.3	−19.5	−12.0	−11.5	−3.9
6	−47.3	−24.6	−11.5	−3.9	−15.0	7.1
7	−51.9	−44.3	0.1	−7.4	12.8	20.3
8	−79.7	−56.9	−50.8	−13.0	1.2	23.8
9	−27.6	−20.0	−3.4	4.2	9.3	31.9
平均值	−56.1	−41.8	−22.6	−13.4	−8.1	4.4
标准差	28.6	22.2	21.9	16.0	18.9	22.1

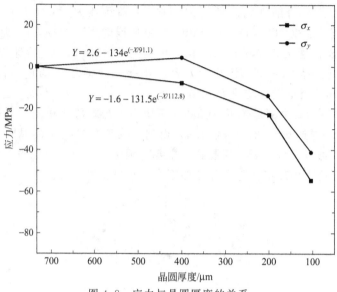

图 4.8 应力与晶圆厚度的关系

晶圆压应力的大小也可以通过测量减薄前后晶圆的翘曲得到，这样可以与晶圆应力传感器测量结果互相验证。晶圆翘曲的测量采用 33 点测量法，结果如图 4.9

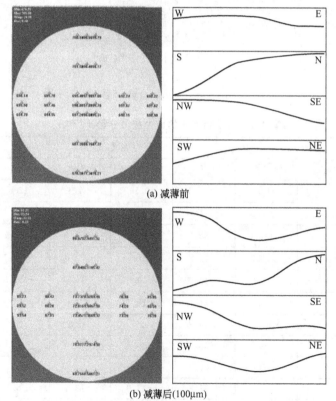

图 4.9 晶圆的翘曲测点布置与翘曲轮廓线

所示，并列于表4.3中。随着减薄晶圆厚度的下降，翘曲由正值变为负值，表明晶圆平面内的应力状态由拉应力变为压应力。这种应力状态的转变与晶圆材料和减薄工艺有关。

晶圆包含了两层材料，第一层（底层）为单晶硅层，第二层（顶层）为SiO_2，在400℃左右温度下通过PECVD工艺沉积到硅晶圆上，厚度为9000Å。当晶圆冷却到室温时，由于SiO_2的热膨胀系数（CTE=0.54×10^{-6}/K[13]）与硅的热膨胀系数（CTE=2.33×10^{-6}/K[13]）不匹配，使得晶圆发生向上凸的弯曲变形，在晶圆顶部SiO_2层内形成拉应力。

图4.10为晶圆在SiO_2层沉积工艺后形成的应力示意图，在室温下晶圆正面受拉应力，图4.9(a)中减薄前的初始晶圆的弯曲轮廓验证了晶圆正面确实存在这样的拉应力。但是，在采用应力传感器测量时，将室温下减薄前晶圆正面标定为"无应力状态"，并设其为测量参考点。所以，晶圆减薄后用应力传感器检测到的晶圆正面的压应力实际上是原有拉应力的减小量。晶圆被减薄后，导致其正面压应力增大的因素来自：①晶圆从背面被减薄后，硅层厚度减小，晶圆弯曲刚度减小，释放了部分晶圆正面的拉应力，致使测量到的压应力增大；②晶圆背面磨削导致晶圆背面的亚表面损伤；③晶圆背面磨削时产生的热应力。

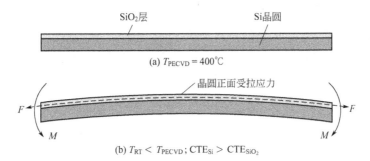

图4.10 （a）在400℃温度下PECVD沉积SiO_2；
（b）当温度降至室温时晶圆弯曲，产生弯曲应力

表4.3 晶圆厚度与翘曲值的关系

晶圆厚度/μm	翘曲/μm
730（初始厚度）	8.1
400	9.9
200	−3.9
100	−9.2

背面磨削后的硅晶圆亚表面损伤可采用透射电子显微镜（TEM）观察到，如图4.11所示。TEM观察分析发现，背面磨削造成晶圆在距表面约200nm薄层内的损伤，损伤层包含无定形硅和多晶硅，且含有较多位错和堆垛层错。这些亚表面损伤导致磨削后硅晶圆的本征应力[13]。

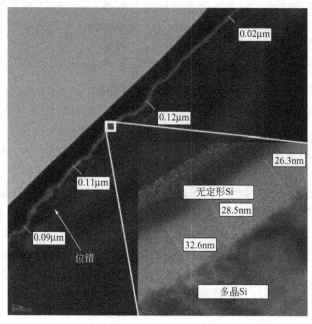

图 4.11 背面磨削硅晶圆亚表面损伤层的 TEM 图像

为了进一步理解背面磨削后晶圆产生残余应力的机理,采用著名的斯通(Stoney)公式[14]进行简单的定量分析。根据斯通公式,当圆形膜-基体复合板的挠度为 δ 时,其残余应力可表示为[13]

$$\sigma = \frac{\delta E d^2}{3(1-\mu)tl^2} \tag{4.6}$$

式中,E 为基体材料的弹性模量;μ 为基体材料的泊松比;d 为基体厚度;t 为膜厚度;l 为板半径。将晶圆半径、厚度、翘曲数据和物理常量($E=107\text{GPa}$ 和 $\mu=0.42$)[15]代入式(4.6),即可得到背面磨削后的晶圆应力,其值列于表 4.4 中。计算结果表明,减薄前晶圆正面表面为拉应力,经过背面磨削拉应力减小并变为压应力,与前面的结论一致。

表 4.4 根据斯通公式和翘曲值计算得到的背面磨削晶圆正面上的应力

晶圆厚度/μm	计算应力/MPa	正则化的计算应力/MPa
730(减薄前)	29.4	0
400	10.8	−18.6
200	−1.1	−30.5
100	−0.6	−30.0

为了将上述计算得到的应力与压阻应力传感器测量得到的平均应力进行比较,需将计算应力值正则化,这是因为测量时将减薄前(初始)晶圆上的应力设为 0。图 4.12 为比较的结果,明显看到,当晶圆厚度大于 200μm 时,随着晶圆厚度的增

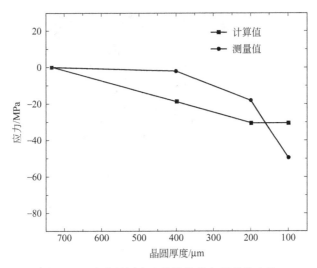

图 4.12 减薄晶圆应力的计算值与测量值比较

加,两种方法得到的压应力呈现几乎同样的增大趋势。但是,当晶圆减薄到 $200\mu m$ 以下时,计算值与测量值开始偏离。

除此之外,还可以看到,在相同的晶圆厚度下,测量应力与计算应力的大小也有差别。其原因可能来自两个方面。第一,斯通公式计算的是基体上薄膜中的应力,而应力传感器测量的是晶圆表层传感器所在层的应力。尽管这两个层很接近,但它们的位置以及材料却不尽相同。第二,晶圆中的应力实际上是晶圆亚表面在磨削时发生内部损伤的结果(见图 4.11),其性质是磨削引起的硅晶圆内部本征应力,严格讲已不能采用斯通公式计算了。可见,尽管斯通公式可以提供一些关于背面磨削后晶圆应力的信息,却不能完全揭示背面磨削工艺对薄晶圆应力的精确影响。特别是厚度小于 $100\mu m$ 的薄晶圆更是如此,因为在如此薄的晶圆里,亚表面损伤层已经十分接近晶圆的正面。这时,采用本章中的压阻式应力传感器测量薄晶圆的应力就十分必要。

4.2.5 切割胶带上晶圆的应力

晶圆与胶带黏合压力为 $0.3MPa$,速度 $10mm/s$,胶带厚度 $70\mu m$,且在室温下完成黏合。与胶带黏合后,在 $70℃$ 下烘烤 $15min$ 以增强切割胶带和晶圆间的黏合强度。烘烤后,需将多余胶带(没有包裹住晶圆的那部分胶带)从晶圆上除去。图 4.13 所示为一个黏合到切割胶带上的 $400\mu m$ 厚的晶圆[12]。

在晶圆表面上 9 个位置进行测量,如图 4.14 所示。面内应力分量 σ_x 和 σ_y 由式(4.3)和式(4.4)计算得到,它们的平均值如图 4.13 所示。可以看出,黏合在切割胶带上晶圆的拉应力增大了,$400\mu m$ 厚晶圆上均为拉应力,而 $100\mu m$ 厚晶圆上则为压应力,但应力值却小了很多。图 4.14 和表 4.5 中由 33 点测量法得到的晶圆弯曲轮廓线和翘曲值也进一步验证了上述结果。

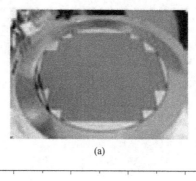

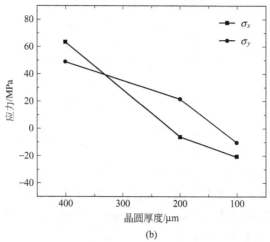

图 4.13 （a）安装在切割胶带上的 $400\mu m$ 厚薄晶圆的光学图；
（b）安装到切割胶带上的薄晶圆表面应力

表 4.5　安装在切割胶带上的不同厚度晶圆的翘曲值

晶圆厚度/μm	翘曲值/μm
400	21.5
200	−20.8
100	−29.5

晶圆应力状态从压应力到拉应力转换的原因为：切割胶带与晶圆的黏合过程实际上是一个层压贴合过程，即切割胶带（高分子材料）被拉伸然后与晶圆背面进行贴合；之后，由于切割胶带本身的黏弹性性质，具有恢复其原来形状的趋势。所以，贴合后切割胶带的收缩变形使晶圆上表面受拉应力。这种现象与热失配情况类似，只是这种现象中的应力是由伸长的高分子胶带材料的回缩引起的。

4.2.6　总结与建议

① 现阶段已经成功地设计、制造并且校准了压阻式应力传感器，测量得到的压阻系数与文献中的结果一致。

② 应力传感器的测量结果表明，背面磨削会在薄晶圆的表面产生压应力，且这个

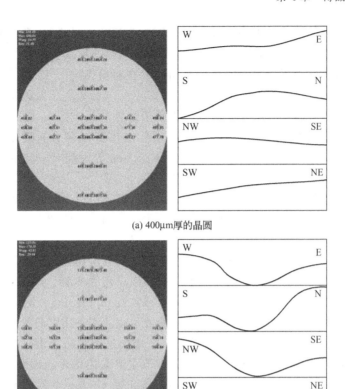

图 4.14 安装到切割胶带上的两种薄晶圆的翘曲数据和弯曲轮廓

压应力会随着晶圆厚度的减小而呈指数增大趋势。对于厚度为 $100\mu m$ 的薄晶圆,当平均压应力为 56MPa 时,局部的压应力可能达到 112MPa。因此,应力传感器可以用来监测晶圆背面磨削时应力的变化情况,并可以依据监测结果选择恰当的工艺参数和材料,如选应力松弛的等离子蚀刻工艺和低温 PECVD 氧化物作为介电层材料等。

③ 将薄晶圆安装到切割胶带后,发现晶圆的应力状态由黏合前的受压转变为受拉。厚度为 $400\mu m$ 的薄晶圆,由于切割胶带的黏弹性行为,晶圆上的应力由 8MPa(受压)转变为 64MPa(受拉),表明晶圆在黏合到切割胶带后承受了剧烈的应力变化。因此,需要采用恰当的处理方法克服这些问题,如使用支撑晶圆等。

4.3 晶圆背面磨削对 Cu-low-k 芯片力学行为的影响

4.3.1 实验方法

背面磨削后,晶圆表面出现划痕、晶体缺陷和应力等[16],使得减薄后晶圆的强度较差。迄今为止，通过评估磨削面质量，已经对晶圆减薄/背面磨削工艺中晶圆的强度问题进行了较多研究[16~21]。除了研究晶圆厚度减小引起的晶圆强度下降外，一些学者还发现了背面磨削造成晶圆的亚表面损伤[22]。Blech等[23]研究了晶圆背面机械磨削对晶圆变形的影响。Hoh Huey Jiun等研究了减薄晶圆的表面粗糙度以及通过应力释放法去除表面层对晶圆断裂强度的影响[21]。除了减薄晶圆的表面粗糙度外，磨削方向对晶圆强度也有较大影响[24]。减薄后晶圆的切割也是个难题，因为切割时会产生崩边，形成粗糙边缘。Chen等发现对于超薄晶圆，可以在磨削前进行切割，这样能将晶圆强度提高10%~15%[20]。在背面磨削过程中，想让晶圆保持初始强度是不现实的，但是，在磨削后采用应力释放法可以在一定程度改善晶圆的强度。应力释放的方法包括湿法化学蚀刻、干法蚀刻或等离子蚀刻、干法抛光和化学机械抛光（CMP）[25,26]。

近些年来，许多研究人员通过评价晶圆强度的方法评估晶圆背面磨削工艺的质量。晶圆强度的评价方法有：三点弯曲实验、四点弯曲实验、球-环（ball-on-ring）测试法、球碎（ball-breaker）测试法以及环-环（ring-on-ring）测试法等。但这些测试方法受表面粗糙度、减薄程度、应力释放过程、切割边缘质量等工艺参数和材料参数的影响较大[20~23,25,26]。注意到，关于磨削过程对晶圆正面电路层影响的文献十分有限，有必要对该问题进行研究。本节主要采用纳米压痕及纳米划痕技术，并结合TEM技术，分析背面磨削工艺对晶圆正面Cu-low-k层的影响。通常，晶圆正面的电路层厚度只有几个微米，因此，常规的晶圆强度评价法已无法使用。将采用纳米压痕和纳米划痕技术进行研究。

采用纳米压痕和纳米划痕法，分别对没有经过背面磨削和经过背面磨削晶圆的Cu-low-k层进行测试，研究其断裂强度、弹性模量、硬度及黏合强度。一般而言，经过背面磨削减薄后晶圆的强度会明显下降。然而，在本研究中注意到，背面磨削后晶圆Cu-low-k层的力学行为却有所改善。因此，可以认为：一方面，背面磨削在一定程度上降低了晶圆的强度；另一方面，背面磨削却增强了Cu-low-k层的结构完整性。

4.3.2 实验过程

(1) 试样制备

采用的试样[17]是由15层不同薄膜组成的多层low-k堆叠。各层组分为Si_3N_4、USG、Blok（SiC）以及黑钻石（BD low-k），如图4.15所示。所有试样基于8英寸[100]硅晶圆制备，制备环境为1000级半导体工艺无尘车间。所采用的测试结构专门用来研究BD（low-k）结构的完整性，这种BD结构与具有三个金属化层的Cu-low-k结构相似。因此，根据后段制程（BEOL）互连设计要求，所测试结构具有三个BD low-k层，且没有铜金属线。试样总厚度大约为3400nm。制备不含金属线测试试样的主要原因是因为不带金属线的low-k层的强度比含金属线的更弱。这项研究对于了解背面磨削对晶圆low-k层的影响很有帮助[17]。

(2) 背面磨削工艺

使用日本迪斯科（Disco）公司的商业化背面磨削系统对测试的 low-k 晶圆进行减薄处理。减薄时首先使用 300 号粒度砂轮进行粗磨，然后用 2000 号粒度的砂轮进行精磨，最后通过干抛光来去除亚表面损伤。干抛光是 Disco 公司发明的一种特殊的应力释放工艺，采用 8000 号粒度的砂轮，且不使用任何化学物质。这种磨抛方式能够保证磨削后晶圆的稳定性[27]。

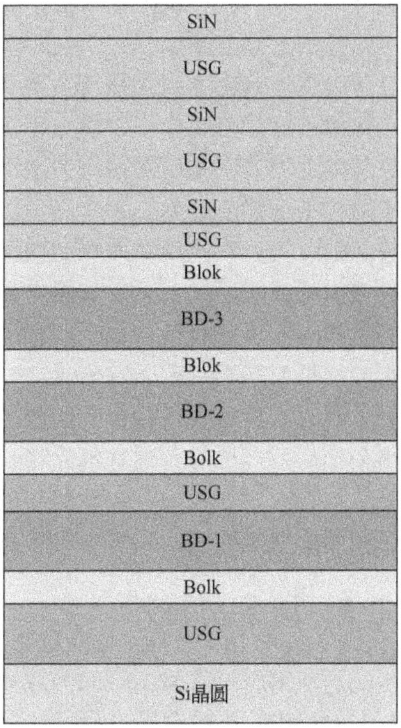

(3) 纳米压痕实验

与传统的硬度测试方法类似，纳米压痕实验也是一种很有用的材料力学性能表征技术，但纳米压痕技术采用纳米压头可以在更小的尺度上表征材料的力学性能。纳米压痕的基本原理是用一个已知力学性能的压头去压感兴趣的另一种力学性能未知的材料，借助纳米压痕曲线可以同时获得材料的硬度和弹性模量[28]。此外，纳米压痕技术还能用于测试残余应力、材料的弹塑性力学行为、蠕变与松弛性能、疲劳和断裂韧性等。

图 4.15 具有 15 层结构的 Cu-low-k 层测试试样

本研究中，采用美国 MTS 公司生产的 XP 纳米压痕仪，带有连续刚度测量（CSM）功能和 Berkovich 压头。连续刚度测量法可以得到力学性能与压入深度之间的关系曲线，并能避免材料蠕变、黏弹性和热漂移等因素对测量结果的影响。

在纳米压痕测试中，给试样施加一个载荷，记录相应的压入深度的实时变化情况，并作出载荷位移曲线，然后采用 Oliver 和 Pharr 方法[28]直接从载荷位移曲线得到被测试样的硬度和弹性模量。载荷位移曲线的三个主要参数在评定材料力学性能时起关键作用，这三个参数是：①最大载荷 P_{max}；②最大位移 h_{max}；③由初始载荷曲线计算得到的接触刚度 $S=dP/dh$。测试结果的精度与可重复性主要取决于这三个参数的测试是否准确。硬度 H 和弹性模量 E 可由下面的 Oliver 和 Pharr 公式进行计算[28]：

$$\frac{E}{1-\upsilon^2}=\frac{\sqrt{\pi}}{2}\times\frac{1}{\sqrt{A_{max}}}\times\frac{dP}{dh} \tag{4.7}$$

以及

$$H=\frac{P_{max}}{A_{max}} \tag{4.8}$$

式中，υ 为泊松比；dP/dh 为压痕卸载曲线初始斜率；P_{max} 为最大载荷；A_{max} 为最大载荷下的接触区域投影面积。对于理想 Berkovich 压头，投影区面积 A

$=24.56h_c^2$，h_c 是实际接触深度。失效时的载荷定义为临界载荷（L_C）。

(4) 纳米划痕测试

薄膜基体结构的黏合强度主要是指界面结合强度，与界面性能有关。界面间的相互作用可能是化学作用、静电作用或范德华力作用。根据经验，黏合强度是指薄膜从基体上剥离下来所需的应力或载荷的大小。目前，已经发展了大量的测量薄膜基体结构黏合强度的方法[29]。但是，使用较多的是纳米划痕方法[30]。本节中所有的纳米划痕实验均使用前面提到的 MTS 公司的纳米压痕仪进行，金刚石划痕压头以斜坡加载方式划过 low-k 晶圆，直到发生失效。本研究中，黏合强度采用纳米划痕实验的临界法向载荷（L_C）表征。对于每个试样，至少要做 5 次实验，且所有的实验结果要基本一致才算有效。对于纳米划痕实验，要使用直径 $5\mu m$ 的圆锥形压头，因为这种压头在各个方向的压入面基本相等。圆锥压头的主要优点在于它对于哪个轴都是对称的，这就消除了倒金字塔形压头的方向性影响[31]。金刚石压头以斜坡加载方式划过 low-k 晶圆，划痕长 $500\mu m$，载荷为 $0\sim250mN$，所有的实验均在相同的划痕速度（$1\mu m/s$）下进行。

4.3.3 结果与讨论

(1) 纳米压痕实验测得的失效载荷、硬度 H、弹性模量 E

本研究中，失效载荷、硬度 H 和弹性模量 E 通过分析纳米压痕的载荷位移曲线得到。图 4.16 给出了未背面磨削和经过背面磨削晶圆（背面磨削至 500、300、150 和 $75\mu m$ 厚）的纳米压痕载荷位移曲线。每个试样至少进行了 10 次测试，一致性良好。low-k 晶圆的泊松比为 0.2 左右。从图中可以看出，与磨削后的晶圆比较，未磨削晶圆的载荷位移曲线在较小的载荷和压入深度下即发生拐点，这是因为

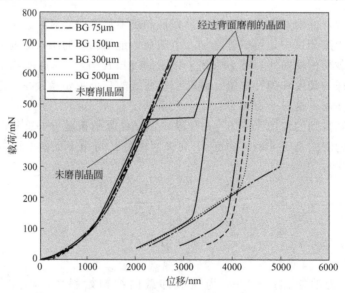

图 4.16 背面磨削至不同厚度晶圆试样的典型位移载荷曲线

薄膜突然开裂和分层引起的[32]。未磨削晶圆在载荷为（456.25±21.22）mN、压入深度为（2422.41±58.53）nm 时失效，而经过背面磨削的晶圆失效时的载荷范围为 482~661mN、压入深度为 2405~2979nm。表 4.6 和图 4.17 给出了所有试样的失效载荷和压入深度值。图 4.18 给出了未磨削晶圆和磨削后晶圆在压痕实验后的光学显微图像。从纳米压痕曲线和光学图像分析可知，未磨削晶圆和磨削后晶圆的失效载荷和压入深度均不同。未磨削晶圆和磨削至 500μm 厚的晶圆在压痕后出现了大量的分层和崩边，而磨削至其他厚度（300μm、150μm 和 75μm）的晶圆则发生分层，与载荷位移曲线上的拐点吻合。磨削至 500μm 的晶圆呈现出混合响应，即在实验中出现了崩边，同时失效载荷为 482.17mN，处于中等水平。原因可能是背面磨削的程度也处于中等水平。其他磨削厚度（300μm、150μm 和 75μm）晶圆，即使提高压入载荷，也不会引起损伤/崩边，而只发生界面分层。

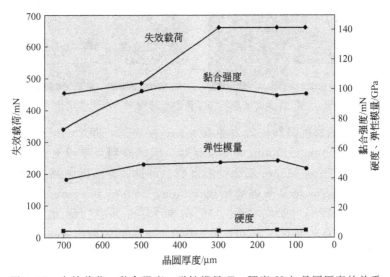

图 4.17　失效载荷、黏合强度、弹性模量 E、硬度 H 与晶圆厚度的关系

表 4.6　未磨削和背面磨削至不同厚度晶圆的力学性能

试样	失效载荷和位移		low-k 区域中部对应的性质/GPa		临界载荷 L_c/mN
	载荷/mN	压入深度/nm	硬度（2000nm）	弹性模量（500nm）	
未磨削	456.25±21.22	2422.42±58.53	4.27±0.10	41.81±2.10	72.80±3.98
BG500μm	482.17±25.25	2405.86±70.47	4.94±0.36	51.55±3.25	98.65±7.50
BG300μm	661.20±7.57	2979.79±21.58	5.10±0.27	53.01±2.47	100.73±3.56
BG150μm	658.45±4.74	2809.01±30.00	5.10±0.64	53.64±1.95	96.00±3.76
BG75μm	658.60±12.21	2942.52±71.20	5.06±0.71	49.75±2.62	96.30±2.22

未磨削和背面磨削晶圆共同的特征是失效都发生在 low-k 层。未磨削晶圆和磨削至 500μm 晶圆的失效发生在 low-k 层中部，即 2400nm 处的 BD2 层（见图

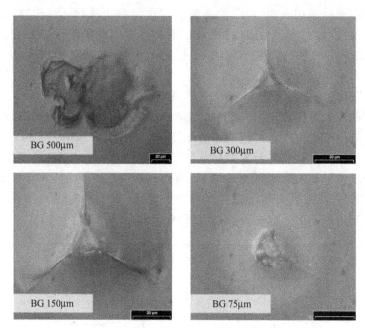

图 4.18 背面磨削至不同厚度晶圆的纳米压痕光学图像

4.15）；其余厚度的磨削晶圆失效发生在 low-k 层底部，即 2800～2900nm 的 BD1 层。背面磨削至 300μm、150μm 和 75μm 厚晶圆的断裂强度没有大的差别，但都比未磨削晶圆和磨削至 500μm 晶圆的强度高。可见，背面磨削晶圆越薄，失效载荷有增大趋势。但是，当晶圆被磨削到 300μm、150μm 和 75μm 时，失效载荷并没有大的变化，原因是：①300μm、150μm 和 75μm 晶圆的背面磨削工艺基本相同；②纳米压痕设备的精度不足以分辨这些背面磨削间的细小差别；③也可能是在对三种厚度晶圆的减薄过程中采用了一样的压力。本研究表明，经过背面磨削后，晶圆正面 low-k 层强度有所增大，这可能是磨削过程中施加的压力和热应力作用的结果。背面磨削时施加的压力或载荷有可能提高了 low-k 层的界面黏合强度，尤其是增大了多层界面间的范德华力，导致层内薄膜的密度变大。许多研究者正在研究背面磨削过程是否会降低芯片的强度，但本研究的结论正好相反。

图 4.19 和图 4.20 分别给出了弹性模量和硬度的纳米压痕测试结果。可以看到，所有试样的弹性模量和硬度均不是常量，而是和接触深度密切相关，这主要是由于被测试样由具有不同物理性质的薄膜和 15 个界面组成。从图中还可以看出，开始时由于在 low-k 层顶端存在 Si_3N_4 层，测得的所有试样的硬度和弹性模量都较大。当压入深度达到 1500nm 之前，背面磨削晶圆的 low-k 层的硬度均明显高于未磨削晶圆，如图 4.20 所示。即便是失效载荷偏低的 500μm 磨削晶圆，其硬度也仍然比未磨削晶圆高。磨削后的晶圆之间的硬度差异不大。

如图 4.19 所示，弹性模量随压入深度的变化情况比较复杂。开始时，经过背面磨削晶圆的弹性模量较大，当压入深度达到 1000nm 时，150μm 厚晶圆弹性模量的变化趋势与未磨削晶圆类似，75μm 和 300μm 厚晶圆的弹性模量则相对较低。所

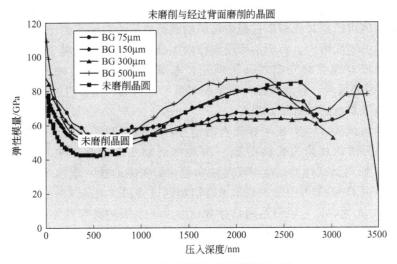

图 4.19 弹性模量随压入深度的变化

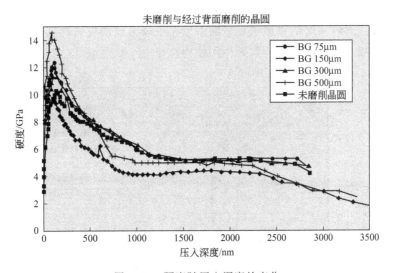

图 4.20 硬度随压入深度的变化

有试样的弹性模量均出现这种复杂趋势，原因主要是：弹性模量是一种高度敏感的材料属性，受到被测膜下面薄膜层和基体的影响较大。

未磨削晶圆与磨削晶圆相比，其硬度差距要比弹性模量明显得多。测得的弹性模量和硬度的最小值表征了 low-k 层 BD 区的性能，因为 BD 层的力学性能比 Si_3N_4、USG 和 Blok 等层要弱。因此，硬度的最小值发生在压入深度 2000nm 处，位于 low-k 层中间；而弹性模量的最小值发生在压入深度 500nm 附近。最小值分别为：未磨削晶圆 $H=4.27GPa$、$E=41.81GPa$；经过背面磨削的晶圆 $H=4.94\sim5.10GPa$，$E=49.75\sim51.55GPa$。这些值反映了 low-k 层的性能。

而对于弹性模量的测量结果，不能像硬度那样与压入深度建立联系。总体而言，经过背面磨削的晶圆具有更高的硬度和弹性模量，这种现象在 low-k 区域尤为明显。其原因可能是背面磨削过程中施加的载荷或压力对层界面产生了影响，并导致各层薄膜的密度增大。由于纳米压痕技术本身比较复杂，且对于多层结构的纳米压痕技术尚没有建立起完整的分析理论，本章对未磨削和背面磨削晶圆的正面 low-k 层的断裂强度、硬度和弹性模量的测量与分析也只是初步的结果。

（2）纳米划痕法测试黏合强度

临界法向载荷（L_C）用来表征未磨削和经过背面磨削晶圆正面 low-k 层的黏合强度。临界法向载荷为实验时第一次观察到划痕深度突然增加时的载荷。图 4.21 给出了所测试试样的划痕深度与法向载荷的关系曲线，图 4.22 给出了对应的光学图像。由图 4.21 可明确识别出未磨削和经过背面磨削晶圆的临界载荷 L_C，该值与图 4.22 互相验证，说明达到临界载荷时 low-k 层的确与基体发生了分离（分层）。图 4.22 中，横排的三个图分别为划痕轨迹的开始部分、中间部分和结束部分。在达到临界载荷之前，划痕周围观察到少量裂纹等损伤，在即将达到临界载荷时，划痕呈现小的转折。图中箭头所指的位置为临界载荷的位置，从这点开始，发生了显著分层或崩裂。从临界载荷点开始，由于压头尖端突然碰到硅基底，划痕深度突然增大。划痕深度的突然增大意味着整个 low-k 层的灾难性崩裂，也意味着纳米压头尖端犁入硅基底。这种犁入作用使得整个 low-k 层破坏。

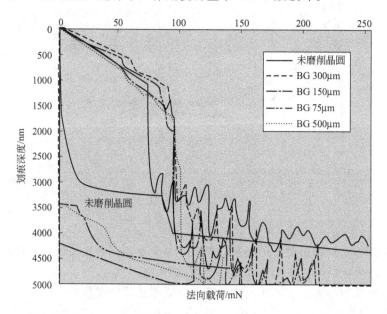

图 4.21 纳米划痕深度与法向载荷的关系曲线（未磨削晶圆与经过背面磨削的晶圆）

与纳米压痕测试类似，经过背面磨削晶圆的 low-k 层也呈现出较高的临界法向载荷值（100.73～96.30mN），而未磨削晶圆的 low-k 层只有 72.80mN。但是，对于经过不同背面磨削的晶圆，它们的临界载荷差别不大。例如，背面磨削至 500μm 厚晶圆的 L_C=98.65mN，背面磨削至 300μm 晶圆的 L_C=100.73mN，背

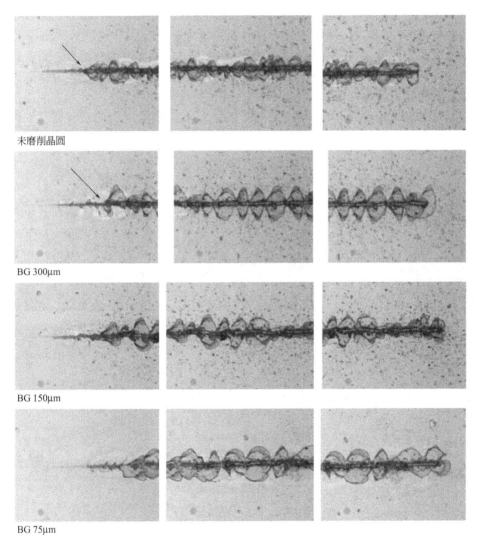

图 4.22 Cu-low-k 层纳米划痕轨迹的光学显微图像
（箭头指示出临界载荷点，从这点开始发生明显分层和崩裂）

面磨削至 150μm 晶圆的 L_C=96.00mN，背面磨削至 75μm 晶圆的 L_C=96.30mN。这些结果列在表 4.6 中，并在图 4.17 中进行了对比。在纳米划痕测试中，未磨削和经过背面磨削晶圆的 low-k 层都表现出了明显的黏合失效，从临界载荷点及其以后的划痕形貌可以观察到。尽管背面磨削程度对 low-k 层的影响有可能是不同的，但由于纳米划痕实验方法所限，观察不到这种差异，因此，测得的所有背面磨削晶圆的临界载荷几乎相同。

（3）TEM 分析

如前面讨论的那样，纳米压痕和纳米划痕实验结果都说明 low-k 层的力学性能在经过背面磨削后得到提高。为了确认这种影响，对未磨削和背面磨削至 75μm 和

150μm 晶圆的 low-k 层横截面进行 TEM 分析，结果如图 4.23 所示。从 TEM 横截面图分析可知，磨削前后晶圆 low-k 层界面的结构变化较大，而其他区域的界面几乎没有变化。如图 4.23(a) 所示，未磨削晶圆 low-k 层界面为波浪形；而经过背面磨削晶圆（75μm 和 150μm）low-k 层的界面是光滑的，如图 4.23(b)、(c) 所示。这说明背面磨削时施加的力对 low-k 层界面有影响。此外，与未磨削晶圆比较，经过背面磨削晶圆的 low-k 层中的 BD1、BD2 和 BD3 层的密度增大 2%～13%，原因可能是由于 low-k 层具有较低的强度。

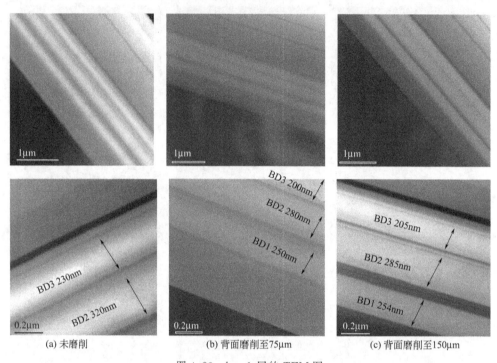

图 4.23　low-k 层的 TEM 图

4.3.4　总结与建议

① 未磨削晶圆正面 low-k 层在载荷为 456.25mN 时发生失效，压入深度为 2422.41nm；而经过背面磨削晶圆的 low-k 层失效载荷为 482.17～661.20mN，对应的压入深度为 2405.86～2979.79nm。

② 未磨削和经过背面磨削晶圆正面 low-k 层的共同特征是失效都发生在 low-k 层内。

③ 硬度最小值出现在压入深度 2000nm 处，弹性模量最小值发生在压入深度 500nm 处。未磨削晶圆 low-k 层的最小硬度 $H=4.27\mathrm{GPa}$，最小弹性模量 $E=41.81\mathrm{GPa}$；经过背面磨削晶圆 low-k 层的最小硬度 $H=4.94\sim5.10\mathrm{GPa}$，最小弹性模量 $E=49.75\sim53.64\mathrm{GPa}$。

④ 与纳米压痕实验类似，纳米划痕实验发现，经过背面磨削晶圆的 low-k 层

具有更高的临界载荷值（100.73~96.30mN），而未磨削晶圆只有72.80mN。

⑤ 在纳米划痕实验中，背面磨削程度不同的晶圆，其 low-k 层的临界载荷没有明显差别。

⑥ 经过背面磨削，晶圆正面 low-k 层的力学性能有所增强。

4.4 参考文献

[1] Miura, H., A. Nishimura, S. Kawai, and G. Murakami, "Structural Effect of IC Plastic Package on Residual Stress in Silicon Chips," *Proc 40th IEEE Electron. Comp. Technol. Conf.*, Vol. 1, Las Vegas, May 1990, pp. 316–321.

[2] Sweet, J. N. "Die Stress Measurement Using Piezoresistive Stress Sensor," in J. H. Lau, ed., *Thermal Stress and Strain in Microelectronics Packaging*, New York: Van Nostrand & Reinhold, 1993, pp. 221–268.

[3] Rahim, M. K., J. C. Suhling, D. S. Copeland, M. S. Islam, R. C. Jaeger, P. Lall, and R. W. Johnson, "Die Stress Characterization in Flip Chip on Laminate Assemblies," *IEEE Trans. Comp. Packag. Technol.*, Vol. 28, No. 3, September 2005, pp. 415–429.

[4] Edwards, D., K. Heinen, S. Groothuis, and J. Martinez, "Shear Stress Evaluation of Plastic Packages," *IEEE Trans. Comp. Hyb. Manuf. Technol.*, Vol. 10, No. 4, December 1987, pp. 618–627.

[5] Lo, T. C. P., and P. C. H. Chan, "Design and Calibration of a 3D Microstrain Gauge for in situ on Chip Stress Measurements," in *Proc. ICSE*, Malaysia, November 1996, pp. 252–255.

[6] Gee, S. A., V. R. Akylas, and W. F. Bogert, "The Design and Calibration of a Semiconductor Strain Gauge Array," in *Proc. IEEE Int. Conf. Microelectron. Test Struct.*, Vol. 1, February 1988, pp. 185–191.

[7] Zhong, Z. W., X. Zhang, B. H. Sim, E. H. Wong, P. S. Teo, and M. K. Iyer, "Calibration of a Piezoresistive Stress Sensor in [100] Silicon Test Chips," in *Proc. 4th Electron. Packag. Technol. Conf.*, December 2002, pp. 323–326.

[8] Beaty, R. E., R. C. Jaeger, J. C. Suhling, R. W. Johnson, and R. D. Butler, "Evaluation of Piezoresistive Coefficient Variation in Silicon Stress Sensors Using a Four-Point Bending Test Fixture," *IEEE Trans. Comp. Hyb. Manuf. Technol.*, Vol. 15, No. 5, October 1992, pp. 904–914.

[9] Suhling, J. C., R. A. Cordes, Y. L. Kang, and R. C. Jaeger, "Wafer-Level Calibration of Stress-Sensing Test Chips," *IEE/ECTC Proceedings*, Washington, DC, May 1994, pp. 1058–1070.

[10] Suhling, J. C., and R. C. Jaeger, "Silicon Piezoresistive Stress Sensors and Their Application in Electronic Packaging," *IEEE Sens. J.*, Vol. 1, No. 1, June 2001, pp. 14–30.

[11] Bittle, D. A., J. C. Suhling, R. E. Beaty, R. C. Jaeger, and R. W. Johnson, "Piezoresistive Stress Sensors for Structural Analysis of Electronic Packages," *J. Electron. Packag.*, Vol. 113, No. 3, September 1991, pp. 203–215.

[12] Kumar, A., X. Zhang, Q. Zhang, M. Jong, G. Huang. V. Lee, V. Kripesh, C. Lee, J. H. Lau, D. Kwong, V. Sundaram, R. Tummula, and G. Meyer-Berg, "Residual Stress Analysis in Thin Device Wafer Using Piezoresistive Stress Sensor", *IEEE Trans. Comp. Packag. Technol.*, Vol. 1, No. 6, June 2011, pp. 841-850.

[13] Ohring, M., *The Materials Science of Thin Films*, Academic Press, San Diego, 1991, pp. 325–552.

[14] Stoney, G. G., "The Tension of Metallic Films Deposited by Electrolysis," *Proc. R. Soc. London A*, Vol. 82, No. 553, May 1909, pp. 172–175.

[15] Sze, S. M., *VLSI Technology*, McGraw-Hill, New Delhi, India, 2003, p. 657.

[16] Takahashi, K., H. Terao, Y. Tomita, Y. Yamaji, M. Hoshino, T. Sato, T. Morifuji, M. Sunohara, and M. Bonkohara, "Current Status of Research and Development for 3D Chip Stack Technology," *Jpn. J. Appl. Phys.*, Vol. 40, No. 4B, 2001, pp. 3032–3037.

[17] Sekhar, V., S. Lu, A. Kumar, T. Chai, V. Lee, S. Wang, X. Zhang, C. Premchandran,

V. Kripesh, and J. H. Lau, "Effect of Wafer Back Grinding on the Mechanical Behavior of Multilayered Low-k for 3D-Stack Packaging Applications", *IEEE/ECTC Proceedings*, 2008, pp. 1517-1524..

[18] Reiche, M., and G. Wagner, "Wafer Thinning: Techniques for Ultrathin Wafers," in *Proc. Adv. Packag.*, March 2003.

[19] Yeung, B. H., V. Hause, and T.-Y. Lee, "Assessment of Backside Processes Through Die Strength Evaluation," *IEEE Trans. Adv. Packag.*, Vol. 23, No. 3, August 2000, pp. 582–587.

[20] Chen, S., T.-Y. Kuo, H.-T. Hu, J.-R. Lin, and S.-P. Yu, "The Evaluation of Wafer Thinning and Singulating Processes to Enhance Chip Strength," in *IEEE/ECTC Proceeding*, Vol. 2, May–June 2005, pp. 1526–1530.

[21] Jiun, H. H., I. Ahmad, A. Jalar, and G. Omar, "Effect of Wafer Thinning Methods Toward Fracture Strength and Topography of Silicon Die," *Microelectron. Rel.*, Vol. 46, Nos. 5–6, May–June 2006, pp. 836–845.

[22] Chen, J., and I. De Wolf, "Study of Damage and Stress Induced by Back-Grinding in Si Wafers," *Semicond. Sci. Technol.*, Vol. 18, No. 4, 2003, pp. 261–268.

[23] Blech, A., and D. Dang, "Silicon Wafer Deformation after Backside Grinding," *Solid State Technol.*, Vol. 37, No. 8, 1994, pp. 74–76.

[24] Lee, S., S. Sim, Y. Chung, Y. Jang, and H. Cho, "Fracture Strength Measurement of Silicon Chips," *Jpn. J. Appl. Phys.*, Vol. 36, No. 6A, 1997, pp. 3374–3380.

[25] Gaulhofer, E., and H. Oyer, "Wafer Thinning and Strength Enhancement to Meet Emerging Packaging Requirements," in *Proc. IEMT Eur. Symp.*, Munich, Germany, April 2000, pp. 1–4.

[26] McHatton, C., and C. M. Gumbert, "Eliminating Back-Grind Defects with Wet Chemical Etching," *Solid-State Technol.*, Vol. 41, No. 11, November 1998, pp. 85–90.

[27] DISCO Corporation, Tokyo, Japan [online]. Available at: http://www.disco.co.jp/.

[28] Oliver, W. C., and G. M. Pharr, "An Improved Technique for Determining Hardness and Elastic Modulus using Load and Displacement Sensing Indentation Experiments", *J. Mater. Res.*, Vol. 7, 1992, pp. 1564–1583.

[29] Volinsky, A., D. F. Bahr, M. D. Kriese, N. R. Moody, and W. W. Gerberich, "Nanoindentation Methods in Interfacial Fracture Testing," in *Interfacial and Nanoscale Failure*, Vol. 18, Amsterdam, Elsevier, 2003, Chap. 13.

[30] Campbell, D. S., *Handbook of Thin Film Technology*, McGraw-Hill, New York, 1970, Chap. 12.

[31] Tayebi, N., A. A. Polycarpou, and T. F. Cony, "Effects of Substrate on Determination of Hardness of Thin Films by Nanoscratch and Nanoindentation Techniques," *J. Mater. Res.*, Vol. 19, No. 6, 2004, pp. 1791–1802.

[32] Wang, L., M. Ganor, and S. I. Rokhlin, "Nanoindentation Analysis of Mechanical Properties of Low to Ultralow Dielectric Constant SiOCH Films," *J. Mater. Res.*, Vol. 20, No. 8, 2005, pp. 2080–2093.

第 5 章 薄晶圆拿持技术

5.1 引言

为了获得小尺寸、轻质量、高性能、低功耗以及宽带宽的 3D IC 集成和 3D Si 集成技术产品，芯片/转接板晶圆的厚度必须非常薄[1,2]。对于存储芯片堆叠，其中每个芯片的厚度不超过 $50\mu m$，并且最终要减薄到 $20\mu m$。无论是有源还是无源转接板，其厚度通常都不超过 $200\mu m$ [1,2]。因此对于 3D IC/Si 集成而言，晶圆减薄和薄晶圆拿持是仅次于 TSV 技术的关键技术，本章对晶圆减薄与薄晶圆拿持的关键问题，如芯片/转接板晶圆、载体晶圆、临时键合、减薄、背面制程以及组装等进行讨论，并研究这些问题潜在的解决方案；讨论与薄晶圆拿持相关的先进材料和设备，重点讨论支撑晶圆或载体晶圆技术，并简单介绍无载体晶圆的拿持技术。

5.2 晶圆减薄与薄晶圆拿持

晶圆减薄不是太困难，绝大多数背磨机都可以胜任此项工作，并可将晶圆磨削至 $5\mu m$ 厚。然而，在整个半导体加工和组装过程中，薄晶圆的拿持是个难题。通常在对芯片/转接板晶圆进行背面磨削露出 TSV Cu 之前，需将其临时键合到另一个载体晶圆（支撑晶圆）上，然后再完成后续的半导体制程，如金属化、钝化、凸点下金属层（UBM）制作以及封装工艺。以上过程结束以后，从载体晶圆上移除薄晶圆又是另一个巨大的挑战。

5.3 黏合是关键

黏合材料是薄晶圆拿持的关键材料，如何选择用于薄晶圆临时键合和解键合的黏合材料是许多研究的重点[1~22]。对于黏合材料的一般要求包括：①临时键合后，黏合材料要经得住工艺环境和温度载荷的考验；②解键合时，黏合材料应该能溶解并且容易清理干净；③解键合之后，在薄晶圆上应无任何残留。

5.4 薄晶圆拿持问题与可能的解决方案

中国台湾工业技术研究院（ITRI）已经深入研究了 200mm 和 300mm 薄晶圆的拿持问题，详见参考文献 [3，4]。研究发现，薄晶圆拿持中最关键的步骤是晶圆减薄以及真空腔中 SiO_2 的等离子体增强化学气相沉积（PECVD）。因此，只需要基于 PECVD 和晶圆减薄这两个条件来选取合适的黏合材料。

参考文献 [3] 表明，不同的黏合剂在所有耐化学性实验中均未发生明显改变或分层。因此在临时键合和解键合时，通过向供应商咨询黏合剂的化学特性，负责处理薄晶圆的人员便可以跳过黏合剂的耐化学性实验。研究发现，具有较薄黏结层的临时键合晶圆的总厚度偏差（TTV）要比具有较厚黏结层（$100\mu m$）的好很多[3]，预磨 Si 载体晶圆只能暂时改善初始较差的 TTV。同时，参考文献 [3] 还发现，由于红外设备能真实、直接并且精确地观测到薄晶圆层，基于红外的 ISIS 测量法要比接触测量法更为可信。另外，参考文献 [3] 给出了厚度分别为 $50\mu m$ 和 $20\mu m$ 的 300mm 晶圆的临时键合与解键合工艺过程。

基于参考文献 [3]、[4] 的研究工作，表 5.1 给出了芯片/转接板晶圆和载体晶圆临时键合、减薄、背面制程、解键合和组装[4]等薄晶圆拿持中可能遇到的问题以及可能的解决方案。接下来，给出一些最新的研究结果。

表 5.1　薄晶圆拿持中可能发生的问题及可能的解决方案

检查对象	可能发生的问题	可能的解决方案
芯片/转接板晶圆	磨削后边缘开裂或崩边	边缘修整，修整深度＝目标晶圆厚度＋$100\mu m$，修整宽度＝1mm 采用直径更大的载体晶圆(201mm 或 301mm)
	黏合剂厚度	黏合剂厚度＞凸点高度＋$20\sim50\mu m$
	临时键合排出气体	提高烘烤温度和时间 使用不含溶剂的材料
	排气时形成的孔洞	涂覆黏合材料之前在 150℃ 下预烘烤 30min 在氧化剂中加入 Si_3N_4 以隔绝气体
载体(支撑)晶圆	载体晶圆；Si 基体	进行背面抛光以实现红外对中
	载体晶圆；玻璃基体	需要考虑静电卡盘的承载能力
临时键合	涂覆的黏合材料太厚以至于在键合后溢出	减小键合力 微调边缘修整参数 涂覆后，对边缘进行冲洗
	键合后 TTV 太大	提高涂覆均匀性 增大键合力并升高温度 粗磨载体晶圆以改善初始 TTV
减薄过程	为解决 TSV Cu 胀出进行背面磨削后发生 Cu 污染	利用干法蚀刻解决 Cu 胀出 利用 Si 湿法蚀刻解决 Cu 胀出

续表

检查对象	可能发生的问题	可能的解决方案
背面制程	钝化时对温度有限制	研发热稳定性更好的材料 降低工艺温度,如用聚合物钝化代替 PECVD 沉积 SiO_2
	较大的种子层刻蚀量会丢失凸点	减小种子层厚度以减小刻蚀量
	在背面种子层蚀刻过程中会刻蚀到 Sn	对于 Cu/Sn 微凸点结构,H_2SO_4/H_2O_2 方法会损伤 Sn,因此需要新的抑制剂
解键合	在化学清洗过程中,由于薄膜胶带发生收缩致薄晶圆开裂	采用耐化学性良好的薄膜胶带 在化学清洗过程中注意保护薄膜胶带
	薄膜胶带与器件之间发生分层	在机械解键合时使用黏结强度更高的薄膜胶带
组装	当芯片或转接板从薄膜胶带上移除时,微凸点/UBM/RDL 发生剥落	采用经过 UV 固化且黏结强度较低的薄膜胶带 在背面制程中增大 UBM、微凸点以及 RDL 的黏结强度

5.4.1 200mm 薄晶圆的拿持

开始阶段,为了评价临时键合黏合层的性能,将直径 200mm 晶圆临时键合到一个支撑晶圆上,即一个涂覆 $20\mu m$ 厚专用黏合剂的载体晶圆。采用背面磨削与 CMP 工艺将该晶圆减薄至 $50\mu m$ 厚,然后对其耐化学性和微凸点工艺质量进行测试。

(1) 临时键合后进行晶圆减薄

在薄晶圆拿持的初始阶段,检查临时键合和后续背面磨削工艺的效果是十分必要的。图 5.1 所示为减薄后的临时键合晶圆,分别使用了黏合剂 A 和黏合剂 B_v.1。使用黏合剂 A 的晶圆其整个边缘未出现裂纹或崩边,而使用黏合剂 B_v.1 的晶圆出现了明显的崩边。由此可见,采用强度足够高的黏合剂进行临时键合是十分重要的。提高黏合剂 B 的黏结强度后,制成新的黏合剂 B_v.2,崩边问题就完全解决了,晶圆边缘看起来完好无损,如图 5.1 右下图所示。因此,增大黏合剂的黏结强度无疑是一种有效的方法,可以保证薄晶圆不发生开裂或崩边。

(2) 解键合结果

这部分讨论 200mm 临时键合晶圆的解键合结果。该临时键合晶圆包括一个厚度为 $50\mu m$ 的薄晶圆以及厚度为 $20\mu m$ 的黏合剂 A,主要考察不同胶带类型(A、B、C 和 D)对解键合效果的影响。图 5.2 所示为解键合的流程图,其中第三步是去除残留的黏合剂。实验结果表明,胶带的选用十分重要。如图 5.2 左下图所示,胶带 A 与清洗剂互相作用,导致局部皱褶,在晶圆中产生不均匀应力,在几个小时后最终致使晶圆碎裂。

下面给出了从 A 到 D 四种胶带与化学清洗剂相互作用的结果,如图 5.3 所示。除了上面讲过的 A 型胶带会出现皱褶外,B、C 和 D 型胶带也会出现,但 D 型胶

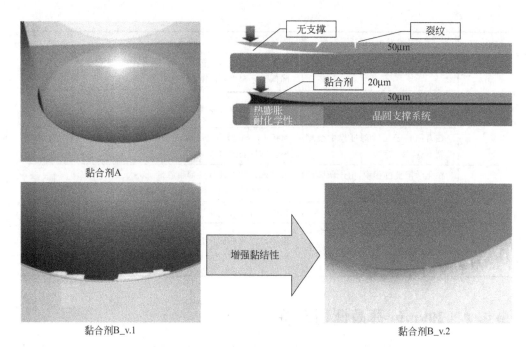

图 5.1　背面磨削后的晶圆，分别采用黏合剂 A、黏合剂 B_v.1 和黏合剂 B_v.2 进行临时键合

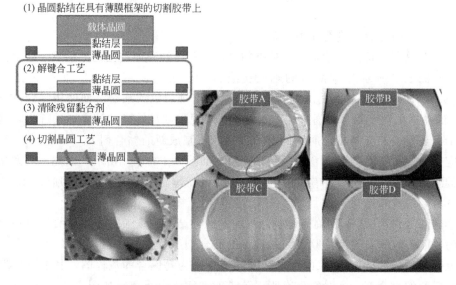

图 5.2　采用薄膜框架进行解键合的流程图

带的皱褶最小，这表明胶带 D 与清洗剂的相互影响最小。

实际上，A、B 和 C 是一般的切割胶带，而 D 是用于晶圆背面磨削的专用胶带。这种磨削胶带通常也用于切割，尽管其价格比普通胶带要高。

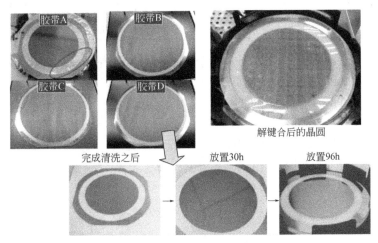

图 5.3 四种胶带与化学清洗方法相互影响的结果，
胶带 D 的表面很平整且晶圆无裂纹产生

因此，选择 D 型胶带用于后续的集成工艺，因为它能保持表面平整以避免晶圆出现裂纹。图 5.3 所示为使用 D 型胶带解键合之后，放置 96h 晶圆未产生裂纹。

此外，选取黏合剂 A 对具有互补金属氧化物半导体（CMOS）电路的 200mm 晶圆进行评估，其厚度为 $50\mu m$。涂覆完黏合剂 A，完成临时键合、减薄以及其他封装工艺之后，要对临时键合晶圆做最后一步处理：解键合。图 5.3 所示为采用带有薄膜框架的胶带 D 对晶圆进行解键合的结果，可以看出除去残留的黏合剂之后，晶圆未发生开裂或崩边，尽管胶带表面不是十分平整。与已报道的解键合实例相比，如对 $70\mu m$ 厚的 200mm 晶圆利用激光烧蚀技术解键合[8]，本文给出了一种不同于高成本玻璃载体晶圆的解键合方法，并在下文评估其用于 300mm 晶圆解键合的效果。

（3）临时键合后的耐化学性测试

将临时键合晶圆切分成四等份做耐化学性测试，测试条件及结果如图 5.4 所示，其中包含了黏合剂 A 和黏合剂 B_v.2。测试的目的是检查封装中采用的化学药剂是否会影响黏合材料。图 5.4 给出了试样初始状态的图像与质量，以及经过显影、去除光刻胶（PR）、蚀刻和化学电镀测试之后的图像与质量。对于黏合剂 A 和 B，完成电镀 Cu 或 Sn 之后，所有试样均未发生明显改变或分层，只是质量有微小的增加，增加幅度小于 5%。因此，两种黏合剂均可用于普通的封装工艺。根据供应商提供的黏合剂耐化学特性，可以直接选用合适的黏合剂进行晶圆的临时键合与解键合。

（4）晶圆背面微凸点工艺的黏结性能评价

除了耐化学性，还需要测试微凸点制程中黏合剂的性能。晶圆微凸点制程如图 5.5～图 5.9 所示，依次包括 PECVD、溅射、版图制作、电镀、移除光刻胶以及回流工艺等。一般来讲，PEVCD 工艺质量随着温度从 200℃增加至 300℃而有所

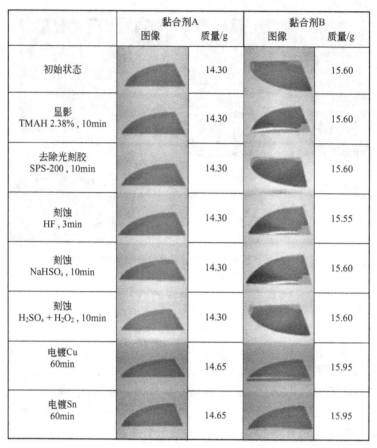

图 5.4 背面微凸点工艺的黏结性能评估，所有方法均未发生分层，且质量改变＜5%

提高[9]，因此，可设置 PECVD 的工艺温度为 250℃用于沉积 1μm 厚的 SiO_2 层，时间为 10min 左右。然后，在 170℃温度下，在晶圆上溅射 0.05μm 厚的 Ti 层以及 0.12μm 厚的 Cu 层。完成掩膜显影之后，再电镀 3μm 厚的 Cu 层和 4μm 厚的 Sn 层。最后，在 240℃温度下进行回流形成直径 20μm、节距 100μm 的微凸点。除了 PECVD、溅射以及回流工艺需在高温下进行以外，其他所有工序都在室温下进行。此外，只有 PECVD 和溅射工艺需在真空腔里进行，如图 5.5 所示。

对于黏合剂 A，经过 PECVD 工艺后，晶圆边缘并未出现崩边或裂纹，如图 5.5 所示。然而，晶圆表面随机出现了一些芯片大小（5mm×5mm）的凹坑，如图 5.5 和图 5.9 所示。一般认为是黏结层中排出的气体导致了凹坑的产生，并且只有在高温（250℃）真空腔里才会形成。根据前面的经验，黏合剂 A 在 PECVD 工艺中可承受 180℃的高温，且不会产生凹坑。黏合剂 B_v.2 的黏结力更强，因此减薄后不会出现崩边或裂纹，但是在 250℃的 PECVD 工艺中晶圆严重损坏，如图 5.5 所示。黏合剂 B_v.3 的黏结强度稍有减小，但仍足以承受背面磨削时的应力。在完成 PECVD 工艺之后，键合晶圆的完好程度要远好于之前的情况，晶圆整个边

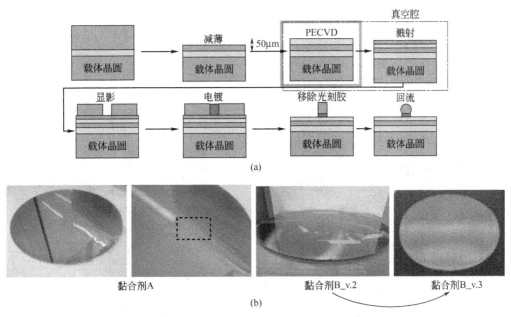

图 5.5 （a）：测试 200mm 键合薄晶圆的背面微凸点成形工艺的流程；
（b）：黏合剂 A、B_v.2 及 B_v.3 所对应的 PECVD 结果，采用黏合剂 A 时产生凹陷（虚线标记处），采用黏合剂 B_v.2 时晶圆碎裂

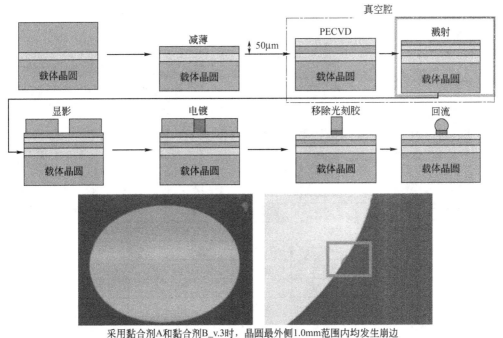

图 5.6 完成溅射后，采用黏合剂 A 和 B_v.3 的晶圆其边缘均发生崩边

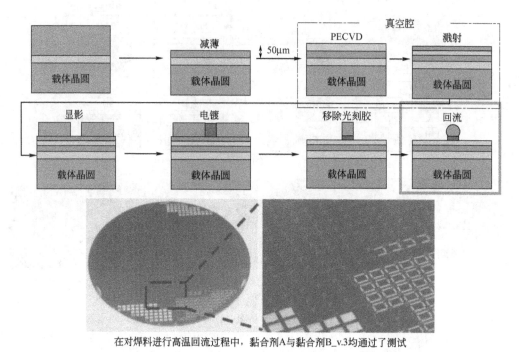

图 5.7　高温回流后（制作微凸点），采用黏合剂 A 和 B_v.3 的晶圆均通过了测试

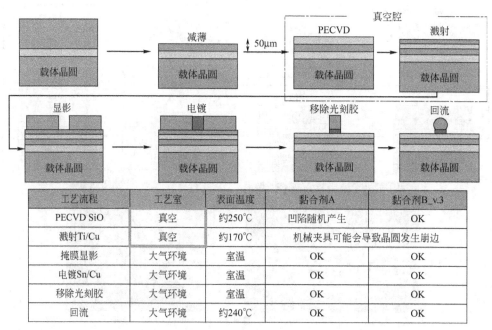

工艺流程	工艺室	表面温度	黏合剂A	黏合剂B_v.3
PECVD SiO	真空	约250℃	凹陷随机产生	OK
溅射Ti/Cu	真空	约170℃	机械夹具可能会导致晶圆发生崩边	
掩膜显影	大气环境	室温	OK	OK
电镀Sn/Cu	大气环境	室温	OK	OK
移除光刻胶	大气环境	室温	OK	OK
回流	大气环境	约240℃	OK	OK

图 5.8　200mm 键合晶圆的背面微凸点制程

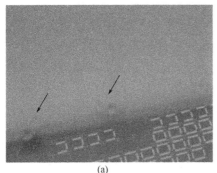 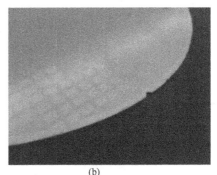

(a) 黏合剂A 晶圆表面随机产生凹坑　　(b) 黏合剂B_v.3 无凹坑产生

图 5.9　晶圆在电镀工艺后的放大图像

缘均未出现崩边或裂纹,如图 5.5 所示。因此,在真空腔中以 250℃进行 PECVD 工艺时可以采用黏合剂 B _ v.3。

在接下来的溅射工艺仍然在真空腔中进行,黏合剂 A 和 B _ v.3 的效果看起来差不多,如图 5.6 所示。溅射均在 170℃温度下进行,此温度是根据粘贴在晶圆表面的温度测量装置测得的。然而,对于两种黏合材料而言,晶圆最外缘 1.0mm 范围内都出现了崩边,如图 5.6 所示。这些微小的崩边可能是加工过程中机械夹具造成的,但是崩边只发生在晶圆外围不重要的区域,并不向晶圆内扩展。因此,对于溅射工艺,基本上可以忽略黏合剂选择问题。

从光刻工艺到最终的微凸点回流成形工艺,所有工艺步都在大气环境中和室温条件下进行。不管是采用黏合剂 A 还是黏合剂 B _ v.3,效果都很好,每一步都没有产生缺陷。图 5.7 所示为完成回流之后的结果,采用的是黏合剂 A。图 5.8 总结了背面微凸点制程的每个工艺步,并给出了简短评价。

黏合剂 A 和黏合剂 B _ v.3 的主要区别在于 PECVD 工艺中晶圆出现凹坑的情况不同,如图 5.9 所示。采用黏合剂 A 时,在临时键合晶圆表面可以观察到随机分布的凹坑,如图 5.9(a) 中的箭头所指的那样。但是对于黏合剂 B _ v.3 则完全不同,如图 5.9(b) 所示。由图 5.9(a) 可以看出,晶圆表面的凹坑几乎同芯片一样大。另一方面,凹坑深度从不到 $1\mu m$ 到超过 $9\mu m$ 不等,最大深度位于凹坑的中心。对于薄黏结层($20\mu m$),凹坑现象十分明显,这会影响薄晶圆的均匀性以及拿持的效果。采用扫描声学显微镜(SAM)可以检测到凹坑[3]。

(5) 200mm 晶圆总结

用于临时键合的黏合材料的化学特性可以从供应商处获得,因此,耐化学性测试可以直接跳过。对于典型的封装工艺,黏合剂的选取只对真空腔中 PECVD 工艺有显著影响。对于其他在大气环境中进行的工艺步,即便是高温(240℃)回流工艺,对黏合剂的要求也要宽松许多,几乎不影响测试结果。黏合剂可能由各种聚合物制造,它们的热稳定性是选择时考虑的一个重要因素,尤其是高温下的热稳定性[9,11]。黏合剂的热稳定性与其抗分解和排气能力有关[12]。据报道,一些黏合剂在 PECVD 过程中可以承受 280℃ 的高温[9]，而另一些黏合剂，如黏合剂 A，则不能承受这样的高温。除了热稳定性，真空环境是另一个关键因素，因为确实发现在大气环境中以 240℃ 的高温进行的微凸点回流工艺没有出现问题。

因此，为了快速选取合适的黏合剂，只需在晶圆减薄后对 PECVD 工艺进行测试。短期内，建议采取聚合物而不是 SiO_2 制作绝缘层。例如，可以采用聚酰亚胺（PI）或聚苯并噁唑（PBO）。聚合物绝缘层可以在大气环境中制作，从而可以避免真空 PEVCD 工艺导致的晶圆凹坑问题。此外，聚酰亚胺由于具有良好的耐高温/耐化学性而广泛用于薄晶圆拿持工艺[7]。长远来讲，需要具有良好热稳定性的黏合剂。

5.4.2　300mm 薄晶圆的拿持

分析 300mm 薄晶圆的拿持仍采用上述两种黏合剂。基于之前 200mm 晶圆的研究结果，建议跳过 300mm 晶圆的耐化学性实验，但薄晶圆拿持材料仍然要经历关键的工艺过程，例如 PECVD 工艺。本研究考虑了 3 种不同结构的 300mm 晶圆：①裸晶圆；②带有 $80\mu m$ 焊锡凸点的晶圆；③带 TSV 和微凸点的晶圆。在临时键合以及减薄工艺之前，沿着宽为 $500\mu m$、深度 $100\mu m$ 的边缘对晶圆进行修边，据报道这样能够提高减薄质量并防止边缘出现裂纹或崩边[10]。

（1）裸晶圆的黏合剂评估

在 3D IC 工艺中，必须要控制减薄工艺后的总厚度偏差（TTV）。TTV 的变化跟许多因素有关，如黏合剂涂层的均匀性、键合工艺以及减薄工艺。由于晶圆尺寸由 200mm 增大至 300mm，所以 TTV 也相应增大。对于薄晶圆黏合材料的分析，需将晶圆减薄至 $50\mu m$ 和 $25\mu m$ 厚。

图 5.10 所示为基于接触测量法测得的晶圆减薄之前的 TTV 数据，包括薄晶圆、薄晶圆的黏合材料以及载体晶圆在内。对于黏合剂 A，厚度为 $25\mu m$ 和 $50\mu m$ 的晶圆其 TTV 均小于 $3\mu m$。对于黏合剂 B_v.3，厚度为 $50\mu m$ 的晶圆其 TTV 约为 $5\mu m$，而 $25\mu m$ 厚的晶圆其 TTV 为 $4.2\mu m$。基于这些测量数据，发现采用黏合剂 A 的晶圆其减薄前的 TTV 更小，这可能与黏合剂特点以及键合工艺有关。黏合剂 A 是在较高温度与加压的情况下进行键合的，而黏合剂 B_v.3 是在室温与极低的键合压力下进行键合的。另外，黏合剂 A 的弹性模量要比黏合剂 B_v.3 大。

接触测量的结果包括晶圆厚度、黏结层厚度以及载体晶圆厚度。对于后续 3D IC 集成中的堆叠工艺来说，测量薄晶圆的厚度更加重要，并且要十分精确。完成晶圆键合与减薄之后，通过红外线（IR）非接触测量法对减薄晶圆厚度进行测量，如图 5.10 所示。测量数据表明，黏合剂 A 的 TTV 要比黏合剂 B_v.3 小很多。

薄晶圆所有的背面制程完成之后，需对其进行解键合。不同的黏合剂，其解键合的方法也不同，采用黏合剂 A 和黏合剂 B_v.3 的晶圆都在室温下完成解键合。解键合后，厚度为 $50\mu m$ 和 $25\mu m$ 的裸晶圆都未出现崩边或裂纹，如图 5.11 所示。晶圆越薄其残余应力越大，应力成倍增大很容易导致薄晶圆损坏[13]。因此，对于

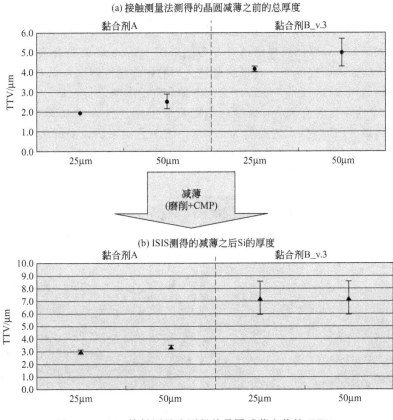

图 5.10 (a) 接触测量法测得的晶圆减薄之前的 TTV；
(b) IR 非接触测量法测得的晶圆减薄之后的 TTV

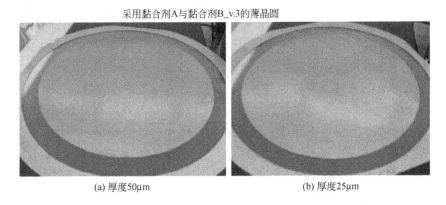

图 5.11 解键合后的薄晶圆

减薄至 50μm 或 25μm 的 300mm 裸晶圆而言，解键合时要实现无裂纹或无崩边是十分困难的。

(2) 含有普通焊锡凸点晶圆的黏合剂评估

在 3D IC 集成工艺中,芯片或转接板晶圆采用的是普通焊锡凸点,其尺寸一般小于 $100\mu m$。本研究利用尺寸 $80\mu m$ 的焊锡凸点进行薄晶圆拿持测试,故黏合剂的厚度必须大于 $80\mu m$ 以保护焊锡凸点,为此采用 $100\mu m$ 厚的黏合剂完全覆盖焊锡凸点。基于前面的研究结果,直接跳过黏合剂的耐化学性测试,只对含有 $80\mu m$ 焊锡凸点晶圆制作 SiO_2 层的 PECVD 工艺中黏合剂的影响进行评估。

图 5.12 所示为含有 $80\mu m$ 焊锡凸点以及采用黏合剂 A 的晶圆其完成 PECVD 工艺后的结果。PECVD 工艺温度为 180℃,该工艺结束后,晶圆表面出现了许多凹坑。这可能是黏合剂 A 中的气体向外排出时引起的,凹坑尺寸与在 200mm 晶圆上观察到的相当。

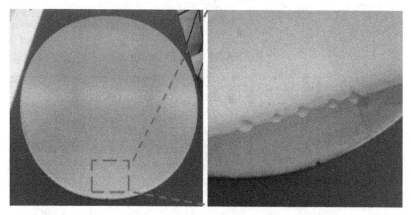

图 5.12 含有 $80\mu m$ 焊锡凸点晶圆的 PECVD 结果,采用厚度 $100\mu m$ 的黏合剂 A。可以看出晶圆边缘处有凹坑产生,可能是有气体排出导致的
(凸点高度,$80\mu m$,黏合剂厚度 $100\mu m$)

图 5.13 所示为在相同的 PECVD 工艺条件下,采用黏合剂 B_v.3 的检测结果。可以看出,晶圆表面外围区域存在较大的孔洞和分层,这些缺陷可能是黏合剂 B_v.3 中的气体向外排出时产生的。根据扫描电子显微镜(SEM)图像,外围区域异常光亮的地方对应孔洞或分层的位置。

不论是黏合剂 A 还是黏合剂 B_v.3,均是在室温下进行解键合。图 5.14 所示为含有 $80\mu m$ 焊锡凸点并采用黏合剂 B_v.3 的 300mm 晶圆解键合结果。可以看出,解键合后未产生崩边或裂纹,并且清洗之后无残留黏合剂。同样,黏合剂 A 也是如此。

(3) 含有 TSV 和焊锡微凸点晶圆的黏合剂评估

直至目前,已经发现 PECVD 工艺中黏合剂 A 会引起凹坑问题,而对于普通焊锡凸点来说,黏合剂 B_v.3 会带来孔洞甚至分层问题。采用黏合剂 A 的晶圆减薄后的 TTV 要好于采用黏合剂 B_v.3 的晶圆。由此可见,两种黏合剂各有优缺点。由于黏合剂 A 能够保证较小的 TTV,而凹坑问题几乎不影响解键合效果,故选取黏合剂 A 对包含 TSV 和焊锡微凸点($25\mu m$ 或更小)的晶圆进行工艺评估。

图 5.13 含有 80μm 焊锡凸点晶圆的 PECVD 结果，采用厚度 100μm 的
黏合剂 B _ v.3。在晶圆表面外围区域观察到较大的孔洞和分层

 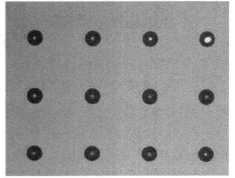

图 5.14 含有 80μm 焊锡凸点晶圆的解键合结果，采用厚度 100μm 的
黏合剂 B _ v.3。解键合并清洗完之后，未发现残留的黏合剂

这里 TSV 直径为 10μm，节距为 40～50μm，深度为 50μm。TSV 上面是直径 20μm 的微凸点，其节距也为 40～50μm。黏结层厚度约为 50μm。

临时键合之后，将晶圆减薄至 50μm。图 5.15 所示为含有 TSV 和微凸点的晶圆减薄后，接触测量法测得的 TTV 数据。TTV 可以控制在 1μm 左右，优于已报道的玻璃键合晶圆的 TTV（2.9μm）[14]。

与之前采用黏合剂 A 的晶圆相比，含有直径 20μm 微凸点的晶圆其 TTV 要好于含有 80μm 焊锡凸点的晶圆。如果采用更大的焊锡凸点，黏结层相应就需要更厚一些。一般来讲，较厚的黏结层其均匀性很难控制。

图 5.15 所示为包含 20μm 微凸点和 TSV 以及采用黏合剂 A 的晶圆，在关键工艺后的检测结果。这里的关键工艺仍然是指 180℃ 下的 PECVD 工艺。完成 PECVD 工艺后发现存在许多凹坑，凹坑区域距离晶圆边缘大概 1～3cm。这可能是临时键合材料中的气体向外排出引起的，也许可以通过调整涂覆或烘烤参数解决。如果涂覆黏合剂之后材料未经过充分烘烤，当工艺温度升高时会导致向外排

ID/位置	1	2	3	4	5	6	7	8	9	TTV	AVG
1	846.0	846.0	845.9	846.2	846.3	846.2	847.0	847.0	846.9	1.1	846.4
2	847.6	847.8	848.0	847.9	847.5	847.6	848.1	848.3	848.0	0.8	847.9
3	848.9	849.3	849.2	848.3	849.1	849.4	848.7	849.0	849.0	1.1	849.0
4	845.5	845.6	845.7	845.6	845.4	845.4	846.0	846.0	845.8	0.6	845.7

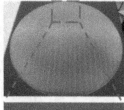

黏合剂 A

微凸点尺寸: 20μm
TSV 直径: 10μm
黏结层厚度: 约50μm

黏合剂 A

图 5.15 采用黏合剂 A 的结果。TSV 直径 10μm，微凸点大小 20μm，黏结层厚度 50μm。TTV 控制良好，但仍有凹坑产生（距离晶圆边缘 1~3cm）

气，这种现象在高温真空腔内尤为明显和严重。

图 5.15 也给出了含微凸点和 TSV 以及采用黏合剂 A 晶圆的解键合结果。300mm 晶圆的厚度约为 50μm，解键合之后未观察到崩边或裂纹。解键合后的晶圆用薄膜框架加以固定，以便拿持和装运。对于包含 20μm 微凸点和 TSV 的晶圆，解键合并清洗后无残留黏合剂[3]。

5.5 切割胶带对含 Cu/Au 焊盘薄晶圆拿持的影响

图 5.16 所示晶圆的厚度为 720μm，其上有若干个 5mm×5mm 芯片，芯片焊盘（30μm×30μm）电镀 Cu 并浸 Au。将一层黏结强度很高的、厚度约为 120μm 的 UV 胶带薄膜黏结在晶圆表面上，然后将晶圆减薄至 50μm。接着切割晶圆，胶带的切割深度为 10~20μm。UV 曝光设备用来溶解胶带并减小芯片与胶带之间的黏结强度，最后去除芯片上的胶带（见图 5.16）[14]。

由图 5.16 可以看出，考虑到切割胶带极高的黏合强度，一些芯片焊盘上的浸 Au 层剥离以后附着在胶带上。这可能是由下面几个原因造成的：①浸 Au 工艺存在问题；②胶带的黏结强度太高；③UV 能量不足以溶解胶带。将 UV 曝光时间延长至原来的 2 倍，结果仍然如此，二次检查浸 Au 池并未发现任何异常，因此结论是切割胶带的黏结强度太高。

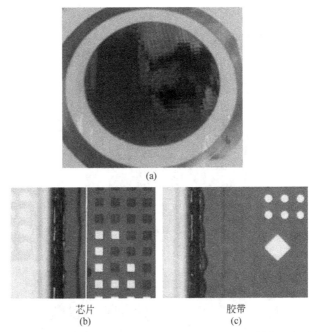

图 5.16 （a）晶圆上有若干 5mm×5mm 芯片，晶圆黏结在切割胶带上；
（b）芯片上浸 Au 层脱落的 Cu 焊盘；（c）附着在胶带上的浸 Au 层

5.6 切割胶带对含有 Cu-Ni-Au 凸点下金属（UBM）薄晶圆拿持的影响

图 5.17 所示晶圆，芯片尺寸为 $5mm \times 5mm \times 720\mu m$，TSV 直径为 $30\mu m$，节距为 $60\mu m$。TSV 的制作工艺如图 5.18 所示，有 6 个关键步骤：①DRIE 工艺制作 TSV 孔；②PECVD 工艺沉积 SiO_2 层；③PVD 工艺沉积阻挡层和种子层；④电镀 Cu 填充 TSV 孔；⑤CMP 工艺去除残余电镀 Cu；⑥TSV Cu 外露。

正面金属化层/UBM 可以制作在含有盲孔 TSV 的厚度为 $720\mu m$ 的晶圆上。然后将此 TSV 晶圆黏结到 $720\mu m$ 厚的载体（支撑）晶圆上，黏合剂选用布鲁尔科技（Brewer Sciences）公司的 9001A 黏合剂。支撑晶圆与盲孔 TSV 晶圆临时键合后，将其减薄至大约 $50\mu m$，以露出 TSV Cu。然后制作背面金属化层以及包括电镀 Cu、电镀 Ni 和浸 Au 的 UBM（见图 5.18）。

将黏合强度很高的 UV 胶带（厚度约为 $120\mu m$ 的薄膜）黏结到含有 Cu-Ni-Au 凸点下金属层的薄晶圆表面，然后将晶圆切割，胶带切割深度为 $10 \sim 20\mu m$。采用 UV 曝光设备溶解胶带，并减小芯片与胶带之间的黏结强度。最后，将胶带从芯片上去除。类似地，可以看到一些浸 Au 层从芯片上剥离附着在 UV 胶带上，因为胶

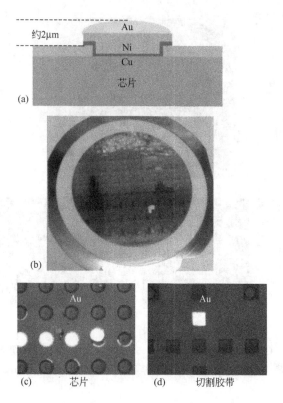

图 5.17 (a) Cu-Ni-Au 凸点下金属层; (b) 含有 Cu/Ni/Au 凸点下金属层和节距为 60μm、直径 30μm TSV 的 CMOS 芯片晶圆; (c) 芯片上浸 Au 层脱落的 Cu 焊盘; (d) 附着在胶带上的浸 Au 层

带的黏结强度太高了。

5.7 切割胶带对含 RDL 和焊锡凸点 TSV 转接板薄晶圆拿持的影响

RDL 分布在 TSV 转接板的两侧，普通焊锡凸点在其底部，如图 5.19 和图 5.20 所示。图 5.18 所示的工艺流程，到第 11 步时制作 TSV 转接板。在解键合之前，含有普通焊锡凸点的 TSV/RDL 转接板与高黏结强度的胶带黏结在一起。完成解键合（机械分离）与切割（图 5.18 中的第 11 步）之后，用 UV 曝光设备溶解胶带并减小芯片和胶带之间的黏结强度。最后，将胶带从芯片上去除，如图 5.21 所示。从图中可以看到一些 RDL 布线甚至普通焊锡凸点都从芯片上剥落下来，说明这种切割胶带的黏结强度对于薄晶圆拿持来说还是太高了。

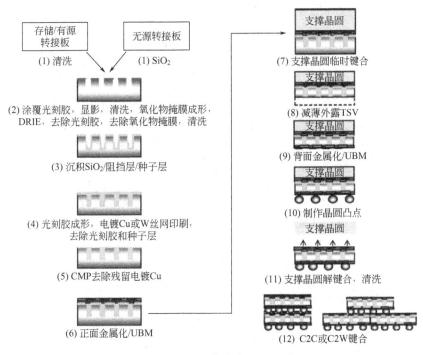

图 5.18 3D IC 集成的 TSV 工艺

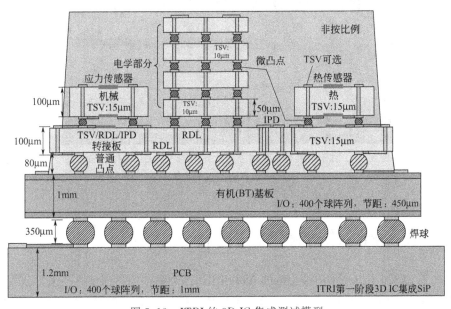

图 5.19 ITRI 的 3D IC 集成测试模型

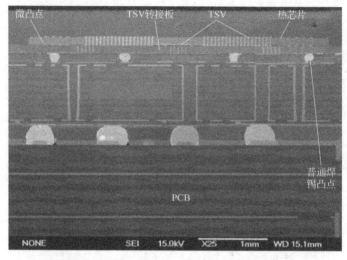

图 5.20 ITRI 的 3D IC 集成测试模型剖面图

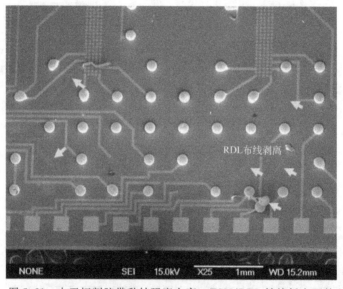

图 5.21 由于切割胶带黏结强度太高，TSV/RDL 转接板底面的布线以及焊锡凸点发生剥离

5.8 薄晶圆拿持的材料与设备

表 5.2 给出了 5 个供应商提供的 6 种用于薄晶圆拿持的黏合材料，这 5 个供应商分别是 Thin-Materials（T-MAT）、布鲁尔科技（BSI）、3M、杜邦（DuPont）以及东京应化（Tokyo Ohka Kogyo，TOK）公司。表 5.2 中第 2 列和第 3 列分别列出了工艺温度限制与可用的黏合剂厚度。可以看到：①绝大部分材料可以承受高达 250℃ 的

工艺温度（除了材料 HT10.10 在 220℃失效以外）；②绝大部分材料（约 100μm 厚）都可以覆盖普通焊锡凸点（除了杜邦公司的材料厚度<20μm 以外）。

载体晶圆对于薄晶圆拿持至关重要。大体上有两种载体晶圆，即硅晶圆与玻璃晶圆。一般来讲，透明的玻璃晶圆用于 UV 固化黏合剂以及利用光热转换（LTHC）激光去除黏合剂时的情况，其成本高于普通硅晶圆。表 5.2 中第 4 列给出了每种黏合材料所需的载体类型。

用于临时键合和解键合的设备是薄晶圆拿持中另一个关键因素，供应商有 Electronic Visions Group（EVG）、SUSS、Tazmo 以及 TOK 公司。一般 TOK 公司的设备只适用于他们自己的黏合剂；SUSS 的设备可以用于 T-MAT、BSI、3M（不包括解键合）与杜邦公司的黏合剂；EVG 的设备可以用于 BSI 的黏合剂；Tazmo 的设备可以用于 3M 的黏合剂，如表中第 5 列所示。

对于芯片/转接板晶圆与支撑晶圆的临时键合，所有供应商提供的黏合剂都必须能使用旋涂机旋涂到芯片/转接板晶圆上，如表 5.2 中第 6 列所示。对于 BSI 公司的 HT10.10、杜邦以及 TOK 公司的黏合剂，载体晶圆无特殊要求。然而对于 T-MAT 的黏合剂，载体必须是弹性体；对于 BSI9001A 黏合剂，载体需要进行区域 2 处理；对于 3M 的黏合剂，需配合使用 LTHC 激光，如表 5.2 中第 7 列所示。

对于解键合工艺，除了 BSI 公司的 HT10.10 黏合剂需要的工艺温度为 180℃以外，所有供应商提供的黏合剂都在室温下进行，如表 5.2 第 9 列所示。然而，解键合方法各不相同有，机械方法（T-MAT）、热方法（BSI-HT10.10）、溶解方法（BSI-9001A、TOK）、激光方法（3M、杜邦），如表 5.2 中第 8 列所示。

不同的材料/设备供应商提供了不同的临时键合和解键合方法，这些方法对于材料、设备、载体晶圆的选择都有很大影响。此外，所选材料的热塑性/热固性毫无疑问会影响工艺的难易程度、温度限制以及最终产品的特性。每种方法都有其优缺点，在研发薄晶圆拿持技术时必须考虑选择合适的黏合剂与设备，以满足产品的需求。

5.9 薄晶圆拿持的黏合剂和工艺指引

5.9.1 黏合剂的选择

① 了解黏合剂的特性以及临时键合、减薄、背面金属化、UBM、RDL 和解键合等工艺对黏合剂的限制极其重要。

② 有两种绝缘层制作方法：PECVD 制作 SiO_2 绝缘层和不采用 PECVD 工艺的聚合物绝缘层，黏合剂的选择取决于所采用的方法。

③ 无论是真空还是大气环境下，键合晶圆都必须有足够的耐化学性并能承受一定的高温。

④ 解键合之后，晶圆不能产生裂纹或崩边。

⑤ 解键合之后进行清洗时，黏合剂应易于去除、无残留，且不能损伤晶圆和胶带。

5.9.2 薄晶圆拿持的工艺指引

表 5.1 和表 5.2 给出了有关薄晶圆拿持技术中临时键合与解键合工艺的一些建议，这些建议对于黏合剂的选择有重要帮助。

表 5.2　世界范围内的薄晶圆拿持工艺系统

材料	工艺温度限制/℃	黏合剂厚度/μm	支撑晶圆	设备(旋涂/键合/解键合)	临时键合		解键合	
					器件	载体	方法	温度
T-MAT	约250	20~200	Si/玻璃	SUSS	旋涂前驱体+等离子体	旋涂弹性体	(1)机械方法 (2)载体上薄膜	室温
BSI(HT10.10)	约220	<100	Si/玻璃	EVG,SUSS	旋涂黏合剂	—	热方法	约180℃
BSI(9001A)(zoneBOND)	约250	<120	Si/玻璃	EVG,SUSS	旋涂黏合剂	区域2处理	(1)机械+溶解方法 (2)器件上薄膜	室温
3M	约250	<125	玻璃	Tazmo,SUSS	旋涂黏合剂	旋涂LTHC键合/UV固化	(1)YAG激光 (2)器件上薄膜	室温
杜邦	>350	2~20	玻璃	SUSS	旋涂黏合剂	—	准分子激光	室温
TOK	约250	<130	多孔支撑板	TOK	旋涂黏合剂	—	溶解方法	室温

5.10　总结与建议

① 在薄晶圆拿持中，晶圆减薄以及在真空腔中采用 PECVD 工艺制作 SiO_2 层是背面制程最关键的步骤，只需要基于 PECVD 和晶圆减薄这两个条件来选取合适的黏合剂。

② 对于不同的黏合剂，在耐化学性测试中均未发生明显变化或分层。因此在临时键合和解键合时，通过向供应商咨询黏合剂的化学特性，负责处理薄晶圆的人员便可以跳过黏合剂的耐化学性实验。

③ 黏结层较薄的键合晶圆其 TTV 要好于黏结层较厚（$100\mu m$）的晶圆。对载体硅晶圆进行粗磨可以暂时弥补初始 TTV，而对于黏结层较厚的晶圆而言，其 TTV 控制还需进一步研究。

④ 由于红外成像（IR）方法可以真实、直接、精确地"看到"薄晶圆层，因

而 ISIS 测量结果要比接触式测量更加有说服力。

⑤ 黏合剂 A 和 B 均可以应用于厚度为 $50\mu m$ 和 $25\mu m$ 的 300mm 晶圆,晶圆解键合后未发现裂纹和崩边。

⑥ 黏合剂 B 可以应用于含 $80\mu m$ 焊锡凸点的 300mm 晶圆,不会发生崩边或硅晶圆开裂,并且清洗后无残留黏合剂。

⑦ 若采用黏合剂 A 作为临时键合材料,$50\mu m$ 厚的 200mm CMOS 薄晶圆解键合之后,未产生裂纹或崩边。此外,同样的方法与结果也适用于含有 $20\mu m$ 焊锡微凸点和 TSV(深度 $50\mu m$)的 300mm 晶圆。

⑧ 在含有 Cu-Au 焊盘、Cu-Ni-Au 凸点下金属层(UBM)以及含有 RDL 和普通焊锡凸点转接板的薄晶圆拿持中,采用黏结强度较高的 UV 薄膜胶带很可能会导致焊盘与 UBM 上的浸 Au 层、RDL 布线以及焊锡凸点剥落。

⑨ 建议选用黏结强度适中的 UV 切割胶带,同时增加 UV 能量以及曝光时间也会有所帮助。

⑩ 本节讨论了薄晶圆拿持中的关键问题,如芯片/转接板晶圆、载体晶圆、临时键合、减薄、背面制程、解键合以及组装等,给出了这些问题可能的解决方案(表 5.1)。

⑪ 表 5.2 给出了薄晶圆拿持中的一些最新材料(Thin Materials、布鲁尔科技、3M、杜邦以及 TOK)和设备(SUSS、EVG、Tazmo 以及 TOK)。

⑫ 表 5.1 和表 5.2 对薄晶圆拿持给出了一些建议,有助于针对特定产品在工艺、材料和设备之间权衡、取舍。

5.11 3M 公司的晶圆支撑系统

图 5.22 和图 5.23 所示为 3M/SUSS 公司最新的晶圆支撑系统,该系统可以帮

图 5.22 3M/SUSS 公司的晶圆支撑系统

图 5.23　3M/SUSS 公司采用玻璃支撑晶圆的支撑系统

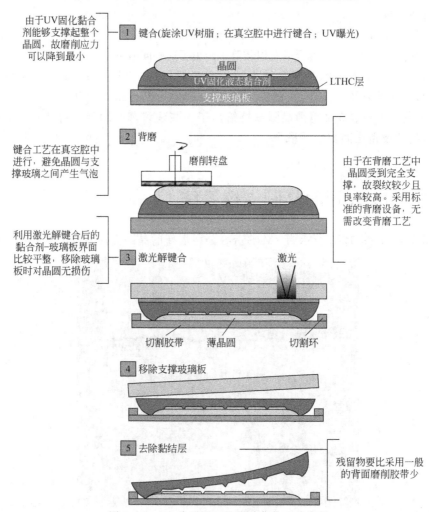

图 5.24　3M 公司用于晶圆减薄的工艺流程

助传统的背磨设备将晶圆减薄至 $20\mu m$。3M 系统的关键在于它能提供一个刚性的、平整的支撑表面,可使移除硅晶圆时的应力降到最小,从而避免裂纹和崩边的发生。整个系统包括用于键合、解键合、去除黏合剂的设备与耗材,如 3M 的 UV 固化液态黏合剂 LC-2201、可循环多次反复使用的玻璃支撑晶圆以及 3M 的光热转换器。

在 3M 系统中,玻璃支撑晶圆(如图 5.23 所示)在背面磨削过程中用于支撑晶圆。采用 UV 固化液态黏合剂将器件晶圆和玻璃支撑晶圆进行键合。经过背面磨削后,如图 5.24 所示,将晶圆转移到切割胶带上,利用 LTHC 激光进行解键合。然后黏合剂便可从晶圆上除去,残留物要比采用一般的背面磨削胶带少。这个系统也可以用于其他半导体和封装工艺,只要器件晶圆与玻璃支撑晶圆之间的热失配程度在允许范围之内。

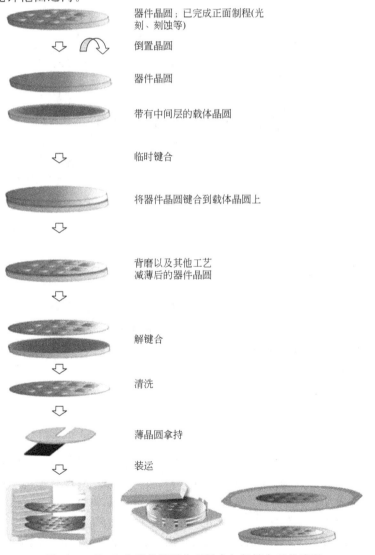

图 5.25　EVG 公司的晶圆临时键合与解键合工艺流程

5.12 EVG 公司的临时键合与解键合系统

EVG 和布鲁尔科技公司给出了一种新的将器件晶圆临时键合到支撑晶圆上的方法，不仅可以进行减薄工艺，还可以完成后续的一系列工艺，包括高温沉积、蚀刻、光刻、介电层制作、电镀以及化学清洗等，如图 5.25 和图 5.26 所示。

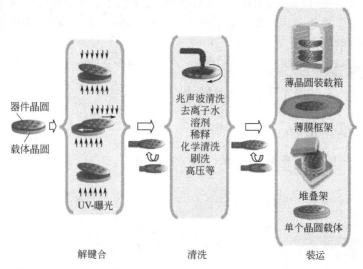

图 5.26 EVG 公司的解键合工艺流程，包括解键合、清洗以及装运

5.12.1 临时键合

器件晶圆与支撑晶圆的正面通过黏合剂 Wafer BOND 黏结起来，这是布鲁尔科技公司的一种 HT 黏合材料，需采用可以旋涂和喷涂的涂覆腔（见图 5.27）进行涂覆。这两种晶圆同时转移到键合腔（EVG850TB），然后小心地对中并在高温下完成真空键合。一旦器件晶圆被临时键合到支撑晶圆上，就可以进行背面制程了，包括背面磨削、蚀刻、金属化以及 TSV 制作等。

5.12.2 解键合

在解键合工艺中，薄晶圆首先通过热激活滑移法从支撑晶圆上移除，然后在清洗室中去除残留的黏合剂，然后以适当的形式进行装运，如薄膜框架载体、专用晶圆装载箱或堆叠架，如图 5.25 和图 5.26 所示。载体晶圆也需要进行清洗，然后用于下一个解键合工艺。近年来，EVG 公司推出了他们最新的 EVG850 "XT 框架" 临时键合和解键合系统，使用的是布鲁尔科技公司（Brewer Science）的低温 Zone BOND 黏合材料。

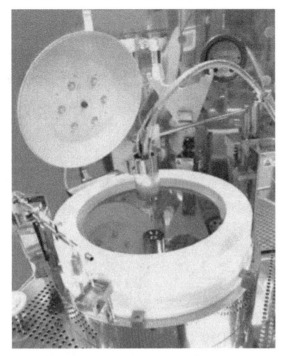

图 5.27　EVG 公司的晶圆涂覆腔，可进行旋涂和喷涂

5.13　无载体的薄晶圆拿持技术

5.13.1　基本思路

直到现在，人们一直在讨论的是将支撑晶圆与 TSV 晶圆键合到一起，通过增强弯曲刚度和强度进行薄晶圆拿持，其中黏合是关键，而且需要先进设备的辅助。最近，一种称为无载体拿持[6,15]的新方法被提出，无需使用黏合剂，成本也大为降低。此方法核心思想就是对晶圆中间部分进行背磨，而对外围边圈不作处理。这个外围边圈起到刚性支撑作用，可以使晶圆具有足够的刚度，从而不需要支撑晶圆，这样可以简化工艺并节省材料和设备成本。

5.13.2　设计与工艺

图 5.28 所示为该方法的工艺步骤，从正面含有器件/TSV 的晶圆（约 700μm 厚）开始。首先将晶圆翻转，在晶圆正面涂覆保护层。然后利用标准切割工具在晶圆背面切出狭缝，其关键参数是狭缝与径向轴的夹角。接着采用迪斯科（Disco）公司的 "Taiko" 系列磨削工具对晶圆进行背磨，但只对晶圆中间部分进行磨削。

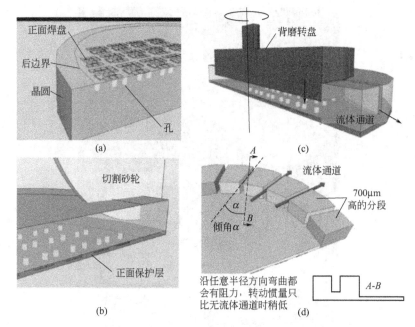

图 5.28 （a）初始状态的 TSV 晶圆；（b）在晶圆背面切出狭缝；
（c）背磨中的晶圆；（d）背磨后的晶圆

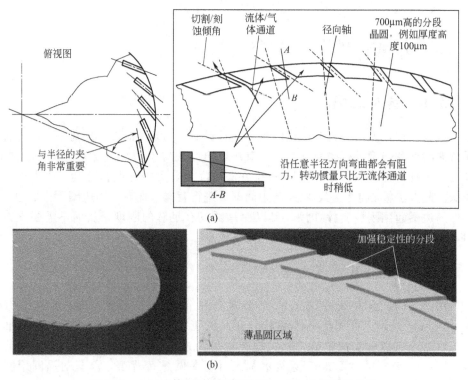

图 5.29 （a）液体/气体通道的倾角；（b）背磨后的晶圆

图 5.29 所示为该设计的核心以及磨削后的晶圆,其弯曲刚度和强度如图 5.30 所示,可以看到晶圆边缘留有一圈未磨削部分。该晶圆设计方法要满足两个条件[6,15]:

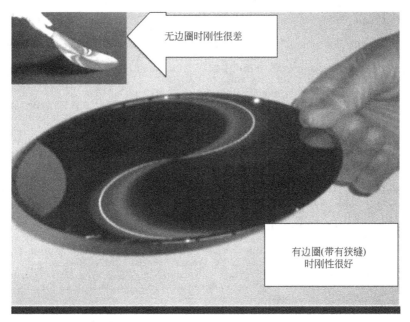

图 5.30 200mm 晶圆的拿持

① 需要在晶圆外圈切出流体通道,以便磨抛液流出;
② 制得晶圆的转动惯量应保持不变,这需要调整狭缝与径向轴之间的夹角。

5.13.3 总结与建议

① 这是一种低成本的薄晶圆拿持工艺,因为不需要黏合剂、临时键合与解键合。
② 该工艺的可靠性也很高,因为避免了黏合剂所带来的问题,如清洗以及残留等。
③ 很容易就可以将该工艺整合进湿法制程中。
④ 对于一些更薄的晶圆,例如用于存储芯片堆叠的 $50\mu m$ 甚至 $20\mu m$ 厚的晶圆,或者是 $100\mu m$ 厚的转接板晶圆,还需要进行更多的研究。

5.14 参考文献

[1] Lau, J. H., *Reliability of RoHS-Compliant 2D and 3D IC Interconnects*, McGraw-Hill, New York, 2011.
[2] Lau, J. H., C. K. Lee, C. S. Premachandran, and A. Yu, *Advanced MEMS*

Packaging, McGraw-Hill, New York, 2010.

[3] Tsai, W., H. H. Chang, C. H. Chien, J. H. Lau, H. C. Fu, C. W. Chiang, T. Y. Kuo, Y. H. Chen, R. Lo, and M. J. Kao, "How to Select Adhesive Materials for Temporary Bonding and De-Bonding of Thin-Wafer Handling in 3D IC Integration?" *IEEE/ECTC Proceedings*, Orlando, FL, June 2011, pp. 989–998.

[4] Chang, H., J. H. Lau, W. Tsai, C. Chien, P. Tzeng, C. Zhan, C. Lee, M. Dai, H. Fu, C. Chiang, T. Kuo, Y. Chen, W. Lo, T. Ku, and M. Kao, "Thin Wafer Handling of 300mm Wafer for 3D IC Integration", *44th International Symposium on Microelectronics*, Long Beach, CA, October 2011, pp. 202–207.

[5] Kettner, P., J. Burggraf, and K. Bioh, "Thin Wafer Handling and Processing: Results Achieved and Upcoming Tasks in the Field of 3D and TSV," *11th IEEE Electronic Packaging Technology Conference*, 2009, pp. 787–789.

[6] Bieck, F., S. Spiller, F. Molina, M. Topper, C. Lopper, I. Kuna, T. Seng, and T. Tabuchi, "Carrierless Design for Handling and Processing of Ultrathin Wafers," *60th IEEE/ECTC Proceedings*, 2010, pp. 316–322.

[7] Itabashi, T, and M. P. Zussman, "High Temperature Resistance Bonding Solutions Enabling Thin Wafer Processing (Characterization of Polyimide Base Temporary Bonding Adhesive for Thinned Wafer Handling)," *60th IEEE/ECTC Proceedings*, 2010, pp. 1877–1880.

[8] Dang, B., P. Andry, C. Tsang, J. Maria, R. Polastre, R. Trzcinski, A. Prabhakar, and J. Knickerbocker, "CMOS Compatible Thin Wafer Processing Using Temporary Mechanical Wafer, Adhesive and Laser Release of Thin Chips/Wafers for 3D Integration," *60th IEEE/ECTC Proceedings*, 2010, pp. 1393–1398.

[9] Tamura, K., K. Nakada, N. Taneichi, P. Andry, J. Knickerbocker, and C. Rosenthal, "Novel Adhesive Development for CMOS-Compatible Thin Wafer Handling," *60th IEEE/ECTC Proceedings*, 2010, pp. 1239–1244.

[10] Justin, W., C. Tai, V. Rao, S. David, D. Fernandez, Y. Li, S. Wen, M. Serene, and J. Lee, "Evaluation of Support Wafer System for Thin Wafer Handling," *12th IEEE/EPTC proceedings*, 2010, pp. 580–584.

[11] Charbonnier, J., S. Cheramy, D. Henry, A. Astier, J. Brun, N. Sillon, A. Jouve, S. Fowler, M. Privett, R. Puligadda, J. Burggraf, and S. Pargfrieder, "Integration of a Temporary Carrier in a TSV Process Flow," *59th IEEE/ECTC Proceedings*, 2009, pp. 865–871.

[12] Hermanowski, J., "Thin Wafer Handling: Study of Temporary Wafer Bonding Materials and Process," *IEEE International Conference on 3D System Integration*, 2009, pp. 1–5.

[13] Zhang, X., A. Kumar, Q. X. Zhang, Y. Y. Ong, S. W. Ho, C. H. Khong, V. Kripesh, J. H. Lau, D.-L. Kwong, V. Sundaram, Rao R. Tummula, and Georg Meyer-Berg, "Application of Piezoresistive Stress Sensors in Ultra Thin Device Handling and Characterization," *Sensors and Actuators A: Physical*, Vol. 156, 2009, pp. 2–7.

[14] Miyazaki, C., H. Shimamoto, T. Uematsu, Y. Abe, K. Kitaichi, T. Morifuji, and S. Yasunaga, "Development of High Accuracy Wafer Thinning and Pickup Technology for Thin Wafer (Die)," *IEEE CPMT Symposium Japan*, August 24–26, 2010, pp. 139–142.

[15] Spiller, S., F. Molina, J. Wolf, J. Grafe, A. Schenke, D. Toennies, M. Hennemeyer, T. Tabuchi, and H. Auer, "Processing of Ultrathin 300-mm Wafers with Carrierless Technology," *IEEE/ECTC Proceedings*, Orlando, FL, 2011, pp. 984–988.

[16] Zhou, S., C. Liu, X. Wang, X. Luo, and S. Liu, "Integrated Process for Silicon Wafer Thinning," *IEEE/ECTC Proceedings*, Orlando, FL, 2011, pp. 1811–1814.

[17] Kwon, W., J. Lee, V. Lee, J. Seetoh, Y. Yeo, Y. Khoo, N. Ranganathan, K. Teo, and S. Gao, "Novel Thinning/Backside Passivation for Substrate Coupling Depression of 3D IC," *IEEE/ECTC Proceedings*, Orlando, FL, 2011, pp. 1395–1399.

[18] Lee, J., V. Lee, J. Seetoh, S. Thew, Y. Yeo, H. Li, K. Teo, and S. Gao, "Advanced Wafer Thinning and Handling for Through Silicon Via Technology," *IEEE/ECTC Proceedings*, Orlando, FL, 2011, pp. 1852–1857.

[19] Halder, S., A. Jourdain, M. Claes, I. Wolf, Y. Travaly, E. Beyne, B. Swinnen, V. Pepper, P. Guittet, G. Savage, and L. Markwort, "Metrology and Inspection for Process Control During Bonding and Thinning of Stacked Wafers for Manufacturing 3D SICs," *IEEE/ECTC Proceedings*, Orlando, FL, 2011,

pp. 999–1002.
[20] Jourdain, A., T. Buisson, A. Phommahaxay, A. Redolfi, S. Thangaraju, Y. Travaly, E. Beyne, and B. Swinnen, "Integration of TSVs, Wafer Thinning and Backside Passivation on Full 300-mm CMOS Wafers for 3D Applications," *IEEE/ECTC Proceedings*, Orlando, FL, 2011, pp. 1122–1125.
[21] Knickerbocker, J. U., P. Andry, B. Dang, R. Horton, C. Patel, R. Polastre, K. Sakuma, E. Sprogis, C. Tsang, B. Webb, and S. Wright, "3D Silicon Integration," *58th IEEE/ECTC Proceedings*, 2008, pp. 538–543.
[22] Zhang, X., A. Kumar, Q. X. Zhang, Y. Y. Ong, S. W. Ho, C. H. Khong, V. Kripesh, J. H. Lau, D.-L. Kwong, V. Sundaram, Rao R. Tummula, and Georg Meyer-Berg, "Application of Piezoresistive Stress Sensors in Ultra Thin Device Handling and Characterization," *Sensors and Actuators A: Physical*, Vol. 156, 2009, pp. 2–7.

第 6 章　微凸点制作、组装与可靠性

6.1　引言

前面章节已经提到，3D IC 集成通常由带有硅通孔（TSV）的堆叠薄芯片和焊锡微凸点组成。微凸点技术是 3D IC 集成的第三项重要技术，仅次于 TSV 技术与薄晶圆拿持技术。需要注意的是，3D Si 集成只采用 TSV 将晶圆/芯片进行互连，而不采用焊锡微凸点。

焊锡材料是用于组装的最主要互连材料，可用于：①将芯片与光电子器件组装到基板上；②将电子与光电子封装/组件/模块组装到印刷电路板上；③集成 3D IC SiP（系统级封装）器件。焊锡材料为电子和光电子器件/组件/模块的互连提供了极大的灵活性和便利性。焊锡材料的特性促进了组装技术的多样化，加快了半导体组装技术的发展。3D IC 封装以及 3D IC 集成就是这样的例子。对于这些组装技术而言，焊锡材料起到了电信号互连与机械"粘接"作用，也正因为此，焊点互连组装及其可靠性成为这些技术发展的关键问题。

许多国家已经立法禁止使用有害物质，RoHS 禁止在封装材料中使用铅元素（Pb），提倡使用绿色环保材料。为了遵从该法案，过去的 10 年中电子产业已经花费了数十亿美元，致力于减少铅的使用。例如，研发新型无铅材料，研究高质量的无铅焊接工艺以及研究无铅焊点的表征和可靠性评估方法等。本章中，主要针对无铅焊锡（Sn）进行研究。

正如前面提到的，3D IC 集成[1,2]定义为将摩尔芯片在垂直方向（z 向）借助 TSV、薄芯片/转接板和微凸点进行堆叠，以实现高性能、低功耗、宽带宽、小尺寸和低成本的目标。因此，焊锡微凸点成为 3D IC 集成的重要支撑技术之一。倒装芯片采用的普通焊球（直径约 $100\mu m$）体积过大，无法用于 3D IC SiP 集成[3,4]。3D IC SiP 集成需要更小的焊球（直径$\leqslant 25\mu m$），这样的小焊球称为微凸点。

晶圆上制作普通焊锡凸点最成熟、最常用的方法是电镀工艺[3,4]。那么是否可以采用同样的工艺和工艺参数制作微凸点呢？这是本章 A 部分所讨论的重点（6.2～6.7 节）。本章 B 部分（6.8～6.11 节）则重点讨论晶圆超细节距微凸点的制作、组装及其可靠性问题。

A 部分：晶圆微凸点制作工艺

6.2 内容概述

本节研究用于 3D IC 集成的 300mm 晶圆细节距无铅焊锡微凸点的制作及表征方法，重点探讨共形和自下而上两种镀 Cu 工艺。测量微凸点中 Cu 与 Sn 的体积含量，并通过剪切实验、微凸点时效试验以及测试前后微凸点剖面的 SEM 图像等表征微凸点。最后，讨论利用普通焊锡凸点的传统电镀工艺制作微凸点的工艺。

6.3 普通焊锡凸点制作的电镀方法

可控塌陷芯片（C4）中使用的普通焊锡凸点及其类似的焊锡凸点，制作的电镀工艺如图 6.1 所示。一般情况下，Al/Cu 衬垫的大小约为 $100\mu m$，目标凸点的高度为 $100\mu m$。在钝化层开口之后（一般不需要），先将晶圆整个表面溅射一层 $0.1\sim0.2\mu m$ 厚的 Ti 或 TiW，然后再溅射 $0.3\sim0.8\mu m$ 厚的 Cu、Ti-Cu 或 TiW-Cu，这些金属层称为凸点下金属层（UBM）。为了得到高度为 $100\mu m$ 的凸点，先在 Ti-Cu 或 TiW-Cu 上覆盖一层 $40\mu m$ 厚的光刻胶，接着利用焊锡凸点掩膜（紫外线 UV 曝光）确定焊区图形。光刻胶层的开口比钝化层中衬垫的开口宽 $7\sim10\mu m$。然后在 UBM 上电镀 $6\sim25\mu m$ 厚的 Cu 层之后，再电镀无铅焊料。电镀时在电镀池中通以稳态或脉冲电流，并以晶圆作为阴极。为了使焊锡凸点高度最终能够达到 $100\mu m$，电镀焊料需高出光刻胶层约 $15\mu m$，并呈蘑菇状。然后移除光刻胶，利用过氧化氢刻蚀去除多余的 Ti-Cu 或 TiW-Cu。最后进行回流，焊料融化时由于表面张力的作用形成光滑、截平的球状焊锡凸点。

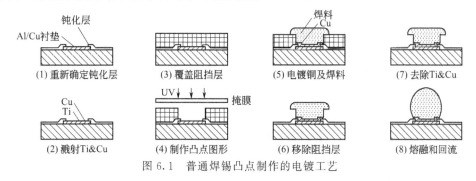

图 6.1 普通焊锡凸点制作的电镀工艺

这种方法的一个缺点是制得的焊锡凸点高度不一致。一般在电镀过程中，由于电流施加在晶圆边界上，电流密度的变化导致靠近晶圆边界的凸点要高于位于晶圆中心区域的凸点，详见参考文献 [3, 4]。

本研究的重点是考察能否采用与普通焊锡凸点相同的制作工艺和参数来制作 3D IC 集成中的焊锡微凸点。

6.4　3D IC 集成 SiP 的组装工艺

3D IC 集成 SiP 的组装工艺至少有 3 个重要步骤：①在摩尔晶圆上制作微凸点；②制作 TSV/RDL/IPD 转接板，正面有 UBM、背面有 UBM 或焊锡凸点的晶圆；③摩尔芯片到转接板、芯片到晶圆（C2W）键合。图 6.2 为部分工艺的示意图，TSV/RDL/IPD 转接板和晶圆微凸点的制作工艺详见参考文献 [5～16]，C2W 键合工艺详见参考文献 [17～20]。除了改变一些工艺参数外，文献 [6] 重点研究采用与普通焊锡凸点相同的电镀工艺、相同的 UBM（Ti-Cu）和焊料以及相同的工艺流程制作 300mm 晶圆微凸点的方法。

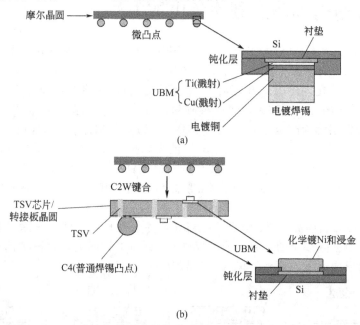

图 6.2　(a) 在摩尔晶圆上制作微凸点；(b) 单个摩尔芯片与 TSV 转接板或者与正面有 UBM 层、背面有 UBM 层或 C4 凸点的晶圆键合（C2W）

6.5　晶圆微凸点制作的电镀方法

6.5.1　测试模型

研究 300mm 晶圆上制作无铅焊锡微凸点的方法，并与两种不同的芯片进行键合。如图 6.3 所示，两种芯片的尺寸参数为：①芯片 1 大小为 18mm×22mm，衬垫数目多于7300个，节距为170μm和340μm；②芯片2大小为10mm×10mm，衬垫数目多于2800个，节距为50μm和150μm。采用辣根过氧化物酶（HRP）显影以及光学显微镜（OM）图像测得两个芯片的钝化层开口大小为30～35μm，如图6.4所示。为了测试焊锡微凸点/焊点的性能及可靠性，在将带有微凸点的芯片与TSV芯片或者无源转接板互连时，需将两个芯片的引脚交错互连以形成菊花链。

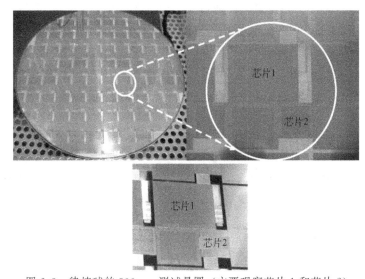

图6.3 待植球的300mm测试晶圆（主要观察芯片1和芯片2）

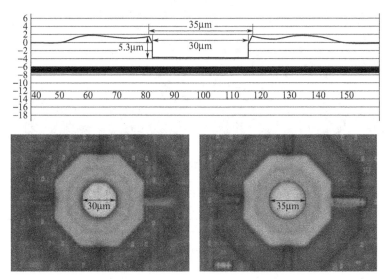

图6.4 采用HRP显影及OM成像得到的芯片钝化层开口轮廓

6.5.2 采用共形Cu电镀和Sn电镀制作晶圆微凸点

微凸点的几何尺寸如图6.5所示。微凸点的最大尺寸不超过30μm，因而必须对原始钝化层［第一层聚酰亚胺（PI）］的开口进行调整，为此需在原始钝化层上覆盖一层厚度为 $4\mu m$ 的新 PI 层，然后利用开口为 $20\mu m$ 的掩膜在新钝化层上制得 $24\sim25\mu m$ 的开口，如图 6.6 所示。接着通过物理气相沉积（PVD），在新钝化层以及硅晶圆的衬垫上依次溅射 100nm 厚的 Ti 阻挡层、300nm 厚的 Cu 种子层，从而形成一层极薄的 UBM。然后在晶圆上旋涂一层 $9.5\mu m$ 厚的光刻胶，固化后厚度大约为 $8.5\mu m$。利用开口为 $25\mu m$ 的掩膜确定光刻胶的开口，以便后续电镀 $11\mu m$ 厚的共形 Cu 以及 $5\mu m$ 厚的 Sn，如图 6.7 所示。电镀完成后，去除多余的光刻胶并进行湿法刻蚀，得到的微凸点如图 6.8～图 6.10 所示。

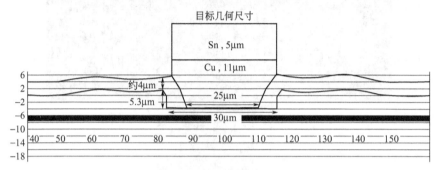

图 6.5　采用 HRP 显影及 OM 成像得到的芯片钝化层开口轮廓，通过重新定义钝化层开口控制微凸点的最终形状

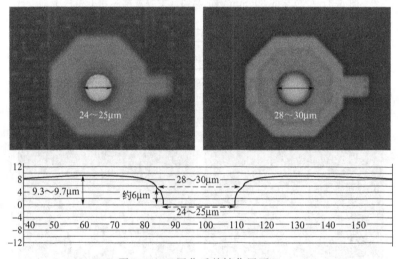

图 6.6　PI 固化后的钝化层开口

由于电镀共形 Cu 时受到钝化层与光刻胶开口的限制，由图 6.8 和图 6.9 可以看出，沿着钝化层和光刻胶侧壁的 Cu 高达 $18.23\mu m$，高出新钝化层约 $10.83\mu m$，远超过了目标高度 $11\mu m$。因此导致电镀 Sn 的高度远小于目标高度 $5\mu m$，如图 6.9 和图 6.10 所示。

图 6.11 所示为电镀共形 Cu 以及焊锡微凸点的剪切测试结果。测试时，剪切

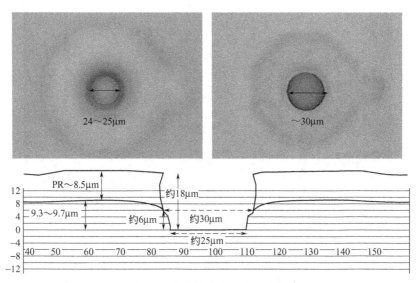

图 6.7 钝化层上的光刻胶开口

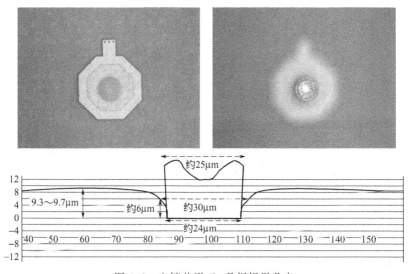

图 6.8 电镀共形 Cu 及焊锡微凸点

工具在两个不同的位置施加剪力：一个位于第二层 PI 表面上方 $2\mu m$ 处，一个位于第二层 PI 表面上方 $6\mu m$ 处。在 $2\mu m$ 位置测得平均剪力为 $16gf/$微凸点，远大于 $6\mu m$ 位置的平均剪力（$4gf/$微凸点）。两种剪切位置对应的微凸点失效模式如图 6.12 所示。在 $2\mu m$ 处施加剪力时，主要是 Cu 柱体受剪切作用；在 $6\mu m$ 处施加剪力时，Cu 柱和焊锡均受到剪切作用。这些微凸点是不能接受的，因为它们可能会导致 C2W 键合以及焊锡微凸点的可靠性问题。

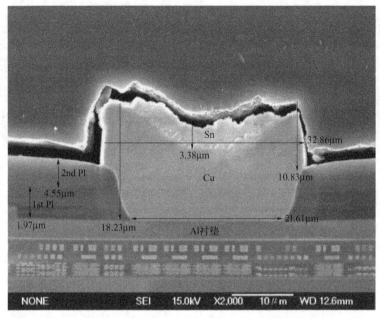

图 6.9 共形 Cu 及焊锡微凸点的 SEM 图像（微凸点从环氧灌封胶处开始分离）

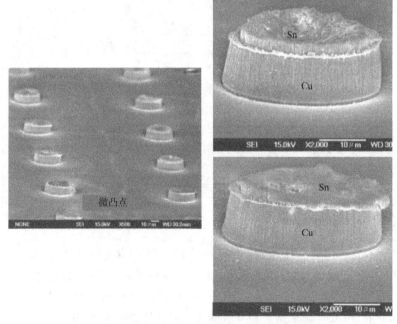

图 6.10 共形 Cu 及焊锡微凸点的 3D 图像

第 6 章 微凸点制作、组装与可靠性

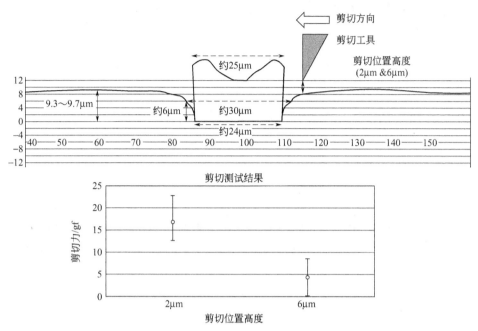

图 6.11 电镀共形 Cu 及焊锡微凸点的剪切测试结果

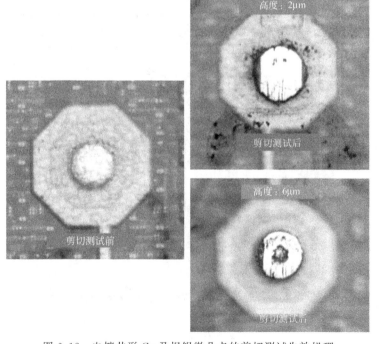

图 6.12 电镀共形 Cu 及焊锡微凸点的剪切测试失效机理

199

6.5.3 采用非共形 Cu 电镀和 Sn 电镀制作晶圆微凸点

如图 6.13 所示的无铅微凸点，Cu 层厚度为 $9\mu m$，Sn 层厚度为 $10\mu m$。这次采用非共形 Cu 电镀，即自下而上镀 Cu，而掩膜、钝化层、UBM、光刻胶、湿法刻蚀等工艺均与 6.5.2 节相同。

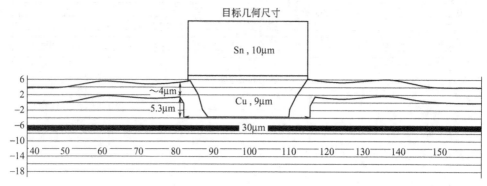

图 6.13 采用非共形 Cu 电镀、Sn 电镀的微凸点几何尺寸（Cu 较少、Sn 较多）

图 6.14 和图 6.15 所示为采取自下而上的方法制作晶圆微凸点。可以看出，焊锡（Sn）的体积比采用共形 Cu 电镀方法时的体积大很多，也比 Cu 的体积大得多。

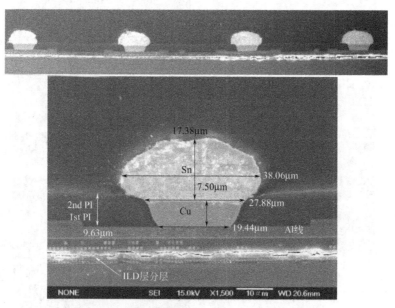

图 6.14 未经回流处理的非共形电镀微凸点的 SEM 图（图中 ILD 分层是在制作试样过程中产生的）

剪切实验的结果如图 6.16 所示。在 $2\mu m$ 处施加剪力时，测得平均剪切力为

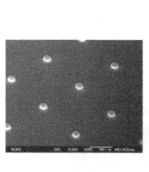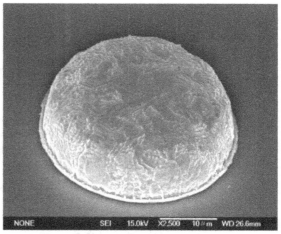

图 6.15 未经回流处理的非共形电镀微凸点的 3D 图像

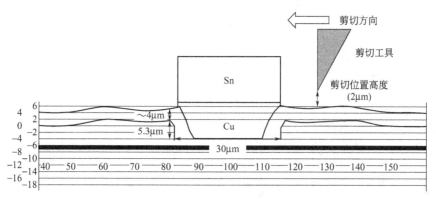

凸点	剪切力/gf	凸点	剪切力/gf	凸点	剪切力/gf	凸点	剪切力/gf	凸点	剪切力/gf
1	2.62	2	3.50	3	3.11	4	3.54	5	3.65
6	3.78	7	3.70	8	3.32	9	3.31	10	4.11
11	2.98	12	3.35	13	3.49	14	3.60	15	3.45
16	3.34	17	3.97	18	3.43	19	2.92	20	3.26
21	3.65	22	3.42	23	3.31	24	3.66	25	3.75
26	3.90	27	3.67	28	3.88	29	3.45	30	2.92

凸点平均剪切力：3.47gf/凸点

图 6.16 非共形电镀微凸点的剪切测试结果

3.5gf/微凸点。由于主要是对焊锡部分施加剪切作用，所以剪切力大小要比 6.5.2 节中得到的值小，失效模式如图 6.17 所示。

无铅微凸点的时效效应如图 6.18 所示。可以看出，微凸点在 150℃ 温度下时效 72h 之后，金属间化合物（IMC）Cu_6Sn_5 已经十分明显。时效处理后得到的平均剪切力为 2.1gf/微凸点，小于未经时效处理时的剪切力（3.5gf/微凸点）。

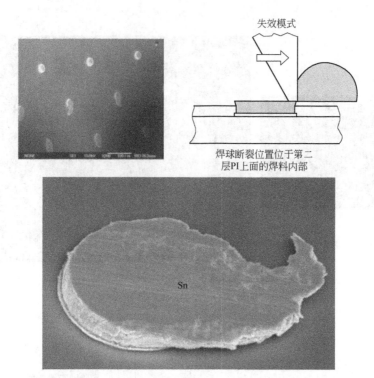

图 6.17 非共形电镀微凸点的典型失效模式

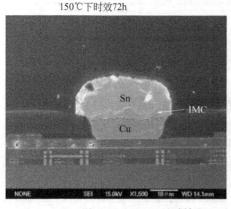
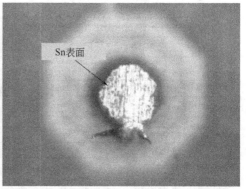

图 6.18 150℃下时效 72h 后,非共形电镀微凸点的失效模式

6.6 制作晶圆微凸点的电镀工艺参数

不能将制作普通焊锡凸点的电镀工艺参数用于制作焊锡微凸点,原因在于微凸点与普通焊锡凸点有较大不同。两者最大的不同在于体积,微凸点的体积比普通焊锡凸点小了 20 多倍。相比于普通焊锡凸点,微凸点中 IMC、电迁移以及无法用 X 射线观察到的[1]柯肯达尔(Kirkendall)孔洞的影响更为重要。因此,不同于普通

焊锡凸点，为了减少 IMC 生长、电迁移效应和柯肯达尔孔洞，焊锡微凸点在钎焊之前不做回流处理。此外，一旦微凸点中产生 X 射线能够观察到的较大孔洞，其危害要比在普通焊锡凸点中同样大小的孔洞大得多。因此，与普通焊锡凸点的组装工艺不同，微凸点组装工艺中不使用助焊剂，以避免在回流过程中形成孔洞。微凸点的可靠性要比普通焊点差，因此填充下填料对于微凸点而言十分重要。为了能够完全填充微凸点间的间隙，应该选用较小粒径的下填料。

尽管微凸点与普通焊点的 UBM 均采用 Ti-Cu 材料，但是两者的 UBM 大小并不相同，厚度至少相差 10 倍。在晶圆上沉积 Ti-Cu 需要使用半导体设备，如 PVD。与普通焊点相比，利用湿法刻蚀去除种子层金属对于微凸点来说更为关键。由于微凸点咬边的裕量更小，所以微凸点的制程窗口也相对更小。

普通焊锡凸点与微凸点所采用的光刻胶材料以及厚度都不同。通常情况下，微凸点的光刻胶厚度比普通焊锡凸点小 4～5 倍。

普通焊锡凸点的衬垫和钝化层开口比微凸点都大，因此，无论采用共形 Cu 电镀还是自下而上电镀，效果差不多。但是，对于微凸点而言，两者区别较大，下文将进一步讨论。

6.7　总结与建议

研究了 3D IC 集成中，300mm 晶圆上细节距微凸点的制程工艺，对比分析了两种不同电镀 Cu 工艺（共形 Cu 电镀与自下而上电镀）所制得的晶圆微凸点。测试了 Cu 和 Sn 体积对微凸点质量的影响。最后，通过剪切实验、时效处理和测试前后微凸点横截面的 SEM 分析对微凸点进行表征。重要结论和建议总结如下：

① 指出并讨论了微凸点与普通焊锡凸点在电镀方法上的主要不同点；
② 由于微凸点的 UBM 尺寸较小，所以共形 Cu 电镀时，得到的 Cu 柱较大而 Sn 较少；
③ 采用非共形 Cu 电镀（自下而上）时，制得微凸点的 Cu-Sn 体积比更为合理；
④ 通过剪切实验测得了微凸点的力学性能参数，平均剪切力为 3.5gf/微凸点；
⑤ 通过分析剪切实验后微凸点横截面的 SEM 图像发现，微凸点的失效模式为 Sn 部分发生剪切断裂；
⑥ 时效处理对微凸点的影响主要表现为 IMC 生长以及抗剪切能力下降；
⑦ 为了更好地推广 3D IC 集成技术，还需要更多的可靠性测试。

B 部分：超细节距晶圆微凸点的制作、组装与可靠性评估

这一部分重点关注节距为 20μm、直径为 10μm 衬垫上微凸点的设计与制作。

芯片尺寸为 5mm×5mm，微凸点数量多达上千个。为了便于测量，采用一种菊花链结构。凸点图案成型后，采用电镀工艺制作微凸点。为了减小湿法刻蚀的咬边量，同时保证电镀的均匀性，需要设计合适的阻挡层及种子层厚度。此外，采用与标准实验不同的剪切实验方法测量微凸点的强度。将无铅 Cu-Sn 微凸点芯片与硅晶圆键合在一起（C2W），并在芯片与晶圆之间的空隙填充专用下填料。测量了无下填料键合芯片的抗剪强度，通过开路/短路测量、扫描声学层析成像（SAT）分析以及横截面的 SEM 分析，评估键合质量与下填料填充的完整性，采用热循环（−55～125℃）实验测试堆叠 IC 的可靠性。最后，探讨了超细节距（节距为 10μm 的 5μm 衬垫）无铅焊锡微凸点的制程工艺[17]。

6.8　细节距无铅焊锡微凸点

6.8.1　测试模型

细节距无铅焊锡微凸点的制作及组装的测试模型如下：①芯片大小为 5mm×5mm，节距为 20μm、直径为 10μm 的衬垫 1600 个；②10mm×10mm 的芯片与 200mm 的晶圆键合在一起。采用电镀工艺制作晶圆微凸点，其结构包括 Cu RDL、Cu 垫以及焊锡。电镀完成后，焊锡经过回流在 Cu 垫上形成微凸点。

为了研究微凸点微结构的演变对键合强度的影响，先将微凸点分别以 100℃、125℃、150℃三个温度进行回流以及 15min、30min、1h、10h 和 30h 的时效处理，然后利用 Dage4000 型试验机对微凸点进行剪切测试（剪切速率为 100μm/s）。最后，通过场发射扫描电子显微镜（FESEM）观察试样截面的形貌以及 IMC 的生长情况，利用能谱仪（EDS）确定 IMC 的组分。

6.8.2　微凸点制作

微凸点的制作涉及各种不同的技术问题。采用 Semitool Raider-M 电镀机制作微凸点，利用光刻机完成整个光刻工艺。与微凸点制作工艺相关的问题中，UBM 的刻蚀最为重要[21~24]。本研究中，采用传统的湿法刻蚀来完成种子层的刻蚀。

微凸点的电镀工艺流程如图 6.19 所示。首先利用 PVD 方法在整个测试晶圆表面蒸镀一层极薄的 Ti/Cu 种子层，为电镀工艺提供导电路径。种子层包括 50nm Ti 和 120nm Cu，其中 Ti 的作用是使得 Cu 更好附着在氧化层上。然后旋涂一层光刻胶并曝光形成所需的图形，接着采用 PECVD 制得氧化物钝化层，并由离子反应刻蚀（RIE）确定其开口形状。经过种子层溅射、光刻、电镀以及种子层刻蚀等一系列工艺之后，得到键合衬垫。制得的晶圆包含节距为 20μm、直径为 10μm 的衬垫，每个衬垫上镀有 3μm 厚的 Cu 和 3μm 厚的 Sn。

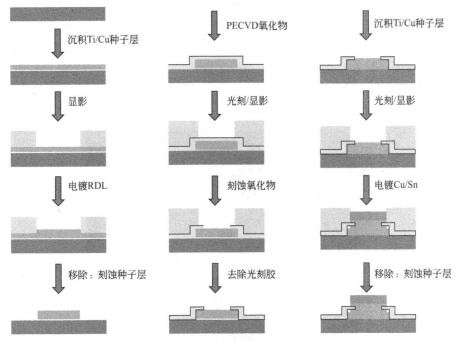

图 6.19 具有 RDL 晶圆微凸点的工艺流程

6.8.3 微凸点表征

(1) 微凸点的均匀性

图 6.20 和图 6.21 分别表示的是电镀完 Cu、Sn 之后，晶圆的 OM 与 SEM 图像。本研究顺利完成了 RDL 和微凸点的工艺集成，所制作微凸点衬垫的节距为 20μm、

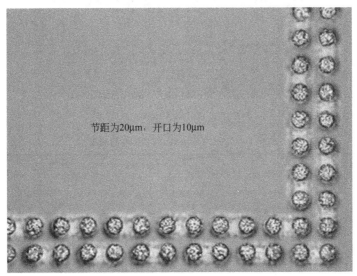

图 6.20 节距为 20μm 的 CuSn 凸点的 OM 图像

直径为 $10\mu m$，最大对中误差不超过 $1\mu m$，如图 6.21 所示。电镀完 Cu、Sn 之后，需对微凸点进行表征测量。利用轮廓仪测量微凸点的高度，分析电镀微凸点的均匀性。为了减小湿法刻蚀的咬边量，同时保证电镀的均匀性，需要设计合理的阻挡层及种子层厚度（Ti=50nm，Cu=120nm）。如表 6.1 所示，包括 CuSn 凸点高度以及 RDL 厚度在内，微凸点总高度的变化小于 10%。该微凸点的均匀性已达到标准，可以进行后续的键合工艺。

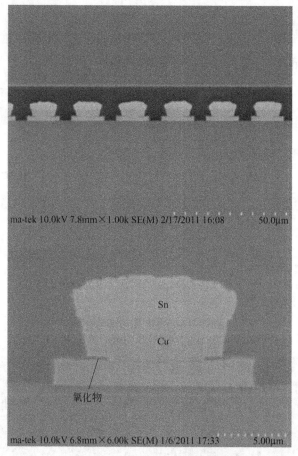

图 6.21　节距为 $20\mu m$ 的 CuSn 凸点的 SEM 图像

表 6.1　节距为 $20\mu m$ 微凸点的高度变化

项目	CuSn 凸点高度（$20\mu m$）；节距为 $20\mu m$
底部	8.66
中部	9.44
左侧	8.49
右侧	8.50
顶部	7.76
均匀度	9.76%

不同回流时间下微凸点的微观结构如图 6.22 所示,可以看出,经过 1 次回流之后(即在回流态再经历 1 次回流),回流态微凸点中 IMC 由扇贝状转变为平坦状。Cu_6Sn_5 的厚度定义为 Cu 柱顶端到 IMC 层顶部的距离,测量不同条件下 Cu_6Sn_5 的厚度,每次测得 15 个值然后平均。以 1 次回流为例,回流后 Cu_6Sn_5 的平均厚度从 2.24μm 增加至 2.38μm。采用 EDS 测得 IMC 的化学组分为 53.6Cu-44.4Sn(%),即 Cu_6Sn_5,焊料中 Cu-Sn 反应已达到稳定[25]。Lee 等人的研究表明[26],回流温度为 240℃时,在纯焊料合金与 Cu 柱的界面处会形成扇贝状的 Cu_6Sn_5,但是当焊料合金中存在 Cu 原子时,Cu_6Sn_5 会由扇贝状转变为平坦状。因此,在回流的初始阶段,金属间化合物 Cu_6Sn_5 以扇贝状出现,这是由于焊锡中无 Cu 原子存在,在后续的回流过程中 Cu_6Sn_5 转变为平坦状。

图 6.22 微凸点形貌图:(a) 回流态;(b) 经历 1 次回流;
(c) 经历 2 次回流;(d) 经历 3 次回流

经历 2 次、3 次回流之后,微凸点中 Cu_6Sn_5 的平均厚度分别增长至 2.42μm、2.44μm。微凸点中 Cu_6Sn_5 的厚度与回流次数的关系如图 6.23 所示。可以看出,在图中阶段 1,Cu_6Sn_5 的生长主要由 Cu 和 Sn 反应机制控制。而在阶段 2、阶段 3,由于 Cu_6Sn_5 周围的 Sn 原子过少,Cu_6Sn_5 生长减缓,Cu_6Sn_5 的生长受扩散机制控制[27,28]。

(2) 微凸点剪切强度

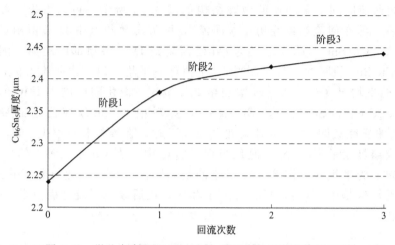

图 6.23 微凸点中 IMC(Cu_6Sn_5) 的厚度与回流次数的关系

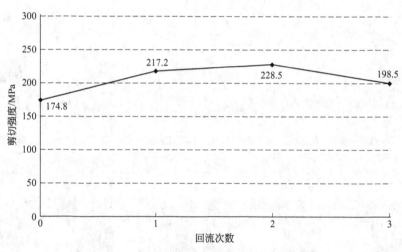

图 6.24 微凸点的剪切强度与回流次数的关系

分别取不同条件下 15 个微凸点进行剪切测试确定其剪切强度。如图 6.24 所示,经过 1 次、2 次回流之后,回流态微凸点的剪切强度由 174.8MPa 分别提高至 217.2MPa、228.5MPa,原因在于微凸点中 Cu_6Sn_5 在不断生长,其剪切模量比 Sn 高[29,30]。然而,由于 Cu_6Sn_5 与 Cu 柱界面上形成了孔洞,导致第 3 次回流之后,微凸点的剪切强度下降到 198.5MPa,如图 6.25(a) 所示。当给微凸点施加剪切力时,在 Cu_6Sn_5 与 Cu 柱界面上会产生裂纹,并沿着界面扩展直至微凸点与 Cu 柱完全分离,如图 6.25(b) 所示。孔洞的形成是由于 Cu 原子的过度消耗。Hon 等人的研究发现[31],350℃温度下焊接 Sn9Zn3.5Ag 20s 时,在 Sn9Zn3.5Ag 焊料合金与 Cu 柱的界面上也会发生相同的现象。尽管所有试样的失效机制都相同,但是经过 3 次回流之后界面上形成的孔洞,加快了微凸点的失效,这表明预处理会削弱微凸点的剪切强度。

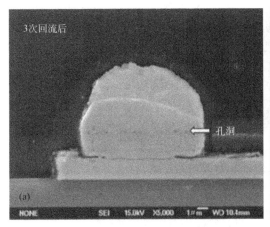
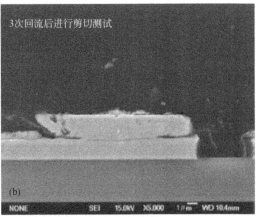

图 6.25　微凸点形貌图：(a) 3 次回流后；(b) 剪切测试后

（3）时效效应

图 6.26 所示为微凸点分别在 100℃、125℃、150℃温度下进行时效处理，所得到剪切强度曲线。从图中可以看出，时效 15min 后试件的剪切强度有所提高，尤其温度为 125℃、150℃时更为明显。这是由于这一阶段 Cu 柱与焊锡的界面上有 Cu_6Sn_5 不断生长，具有较高硬度的 Cu_6Sn_5 的体积分数增大提高了微凸点的剪切强度。

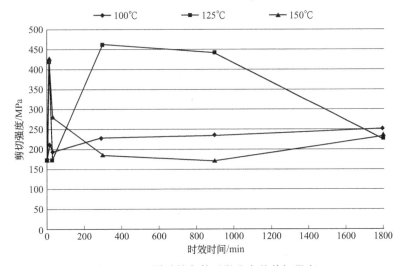

图 6.26　不同时效条件下微凸点的剪切强度

当时效时间增至 30min 时，三种温度下试样的剪切强度分别下降了 8.8%、89% 和 34.7%，这是因为 Cu_6Sn_5 过度生长使得微凸点的体积收缩，从而导致缺陷的产生[32]。由于微凸点时效处理后的失效机制与回流后微凸点中 Cu_6Sn_5 和 Cu 柱界面的断裂机理比较相似（如图 6.27 所示），这表明相比于倒装芯片和球栅阵列封装中的普通焊锡凸点，微凸点中的相变会更快地引发缺陷的产生。

图 6.27 100℃下时效 15min 时，微凸点的失效模式

从图 6.23 中可以看出，第一次回流后 Cu_6Sn_5 的生长受到抑制，其原因是 Cu_6Sn_5 周围的 Sn 原子数量明显减少，同时由于焊料体积的减小，使得 Cu 原子扩散至 Sn 基体的通道变窄。此外，这一阶段开始进行另一个反应——$Cu_6Sn_5+9Cu\rightarrow 5Cu_3Sn$[32]。$Cu_3Sn$ 的生成会产生两方面的影响：①使焊料的体积增加了 56.33%，平衡了 Cu_6Sn_5 生长导致的 6.39% 体积收缩；②在 100℃、125℃温度下时效 5h 后，微凸点的剪切强度分别增大到 226.9MPa、461.9MPa。然而在相同条件下，由于更高的时效温度加快了 Cu_3Sn 的生长和柯肯达尔孔洞的产生，所以微凸点的剪切强度减小至 184.8MPa。因此，剪切强度的降低是由 Cu 柱与焊锡之间的缺陷扩散引起的。在 125℃温度下时效 30min，微凸点会产生同样的变化，剪切强度下降至 229.7MPa。然而，当时效温度为 100℃时剪切强度下降并不明显，原因是 100℃下 Cu_3Sn 的生长以及柯肯达尔孔洞的形成速率都很低。

6.9 C2C 互连细节距无铅焊锡微凸点的组装

6.9.1 组装方法、表征方法与可靠性评估方法

采用两种方法进行组装，即热压键合（TCB）组装与自然回流焊。表 6.2 列出了进行可靠性评估的测试项目和条件。所有的试样首先都需要进行预测试（采用 JESD22-A113D，LV3），筛选出早期失效的试样，然后挑选数十个合格试样对微

凸点进行可靠性评估。可靠性测试采取 $-55\sim125$℃的温度循环试验（JESD22-A104B），保温时间为 5min，升降温速率为 15℃/min。测试过程中定时测量试样的电阻，直到完成 3000 次温度循环。判断试样失效的标准为：其电阻值变化等于或超过初始电阻值的 20%。完成可靠性测试后，采用扫描声学显微镜（SAM）观测下填料中生成的孔洞，并以此对试样进行评估。

表 6.2 可靠性评估的测试条件

测试项目	测试条件
预处理	烘烤(125℃,24h)→浸润(30℃/60%RH,192h)→回流(260℃,3 次)
温度循环试验(TCT)	$-55\sim125$℃，温度循环 3000 次，保温时间为 5min，变温速率为 15℃/min

最后，将失效的试样置于环氧树脂中进行镶样，然后固定在 SiC 板上，采用 Al_2O_3 磨砂纸进行抛光。接着利用 SEM（JEOL，Japan）观察试样剖面，研究微凸点的形貌、IMC 厚度以及芯片、下填料、转接板之间界面的形貌。与此同时，利用 EDS（Oxford，UK）分析金属间化合物的化学成分。

6.9.2 C2C 自然回流焊组装工艺

Lee 和 Kim[33]通过射频（RF）等离子体去除了 Au20Sn 共晶合金表面的氧化层，开发了无助焊剂焊接技术。本研究中，首先利用有机溶剂（MX2302；Kyzen，US）去除芯片与转接板间残余的助焊剂，然后采用以 N_2 为反应物的等离子体刻蚀设备（Creating Nano Technologies，Taiwan）去除化学方法无法清理的残留物，最后借助 SAM（Hitachi；Japan）频率为 75MHz 的探针确定下填料中孔洞的生成情况。

6.9.3 C2C 自然回流焊组装工艺效果的表征

图 6.28 所示为采用传统回流焊工艺制得的微凸点，微凸点的互连准确，界面上形成扇贝状的金属间化合物 Cu_6Sn_5，其平均厚度为 2.23μm，与图 6.22(a)中的金属间化合物相似。

为了研究微凸点的剪切强度，利用 Dage4000 型试验机以 100μm/s 的速率对上层芯片施加剪切力，测得微凸点的平均剪切强度为 138.1MPa，比回流态微凸点的剪切强度低 21%，原因是一些微凸点的高度较低，回流过程中未形成互连，如图 6.29 所示。

筛选出微凸点互连较好的试样进行可靠性评估。测试之前，利用混合清理方法去除上下芯片之间的残余助焊剂[34]。填充完下填料之后，采用 SAT 检查微间隙的填充效果，如图 6.30 所示。可以看出，利用有机溶剂湿法去除结合 N_2 等离子体干法去除方法已经将残余的助焊剂清除干净，并且下填料中未发现气孔。

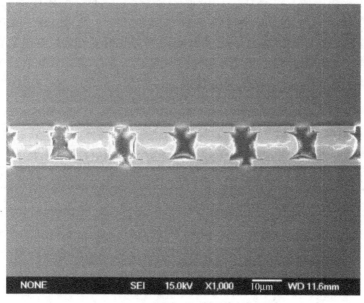

图 6.28 采用传统回流焊工艺得到的微凸点

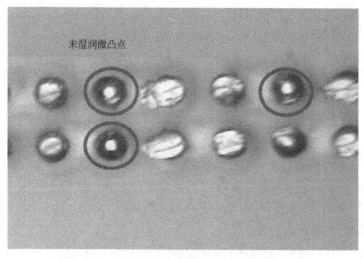

图 6.29 传统回流焊工艺制得的未润湿微凸点

6.9.4 C2C 热压键合 (TCB) 组装工艺

首先对微凸点进行预处理,即利用等离子体去除微凸点表面的氧化物及污染物。然后采用 SÜSS FC-150 热压键合机将上层芯片与底层芯片进行互连,TCB 过程的温度加载曲线如图 6.31 所示。接着选用纳美仕(Namics)公司生产的下填料对芯片与基板之间的微间隙进行填充,填充结束后,在 150℃ 温度下固化 30min,所选下填料的材料参数见表 6.3。

图 6.30 利用溶剂与等离子体清理微间隙并填充完下填料后的 SAT 图像

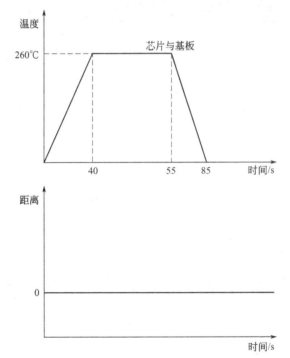

图 6.31 TCB 过程的温度加载曲线

表 6.3 下填料的材料参数

参　数	下填料
T_g/℃	135
黏度/Pa·s	9

续表

参　数	下填料
CTE/(ppm/℃)	42
填料颗粒含量/%	50
填料颗粒尺寸/μm	0.3
弹性模量/GPa	6.5

6.9.5　C2C 热压键合（TCB）组装工艺效果的表征

采用 TCB 组装工艺得到的微凸点截面形貌如图 6.32 所示，可以看出除了挤压出去的焊锡，剩余的都与 Cu 充分反应，在界面处形成 Cu_6Sn_5。测得单个焊点的平均剪切强度为 292.4MPa，高于传统回流焊工艺制得的焊点剪切强度。之后采用相同的下填料填充芯片间的微间隙，根据已有的结果[35]，预处理后并未发现下填料有分层现象发生。

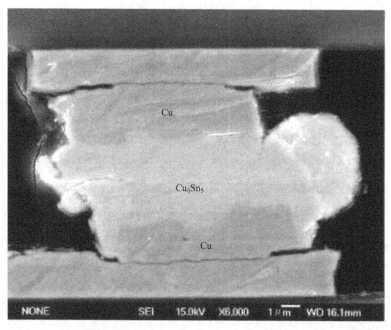

图 6.32　TCB 工艺制得微凸点的截面形貌

6.9.6　组装可靠性评估

（1）热循环测试条件与测试结果

热循环测试条件见表 6.2，温度从 -55℃ 升至 125℃，再降至 -55℃，保温时间为 5min，升降温速率为 15℃/min。当电阻变化超过 20% 时判为失效。试样总数为 34 个，共完成 3000 次热循环，有 16 个试样失效。图 6.33 以及表 6.4 所示为测试结果的 Weibull 分布,可以看到,中位秩(50%)对应的 Weibull 斜率为 3.31,试样的特征寿命(失效概率为 63.2%)为 3349 个热循环。采用 TCB 工艺得到的 CuSn 微凸点其平均寿命定义为平均失效时间:(MTTF)=3349Γ(1+1/3.31)=3005 个循环,其中 Γ 为伽马函数。计算得到 $F(3005)=1-\exp[-(3005/3349)^{3.31}]=0.51$,表明当器件达到了平均寿命时,失效概率达到 51%。

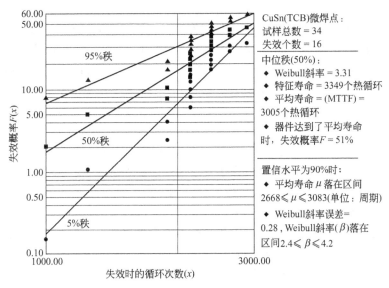

图 6.33 CuSn 微凸点(TCB)热循环测试结果的 Weibull 分布曲线,置信水平为 90%

表 6.4 试样总数、失效个数、特征寿命、平均寿命以及温度循环测试得到的 CuSn(TCB)、Sn2.5Ag(回流焊)、Sn2.5Ag(TCB)微凸点达到平均寿命时的失效概率

微凸点	失效数/样本数	威布尔斜率	特征寿命(循环次数)	平均寿命(循环次数)	达到平均寿命时的失效概率/%
CuSn(TCB)	16/34	3.31	3349	3005	51
Sn2.5Ag(回流焊)	45/47	1.02	980	972	63
Sn2.5Ag(TCB)	27/72	1.15	5006	4765	61

(2)置信度为 90% 的测试结果

考虑置信度为 90% 的情况,确定 90% CuSn(TCB)微凸点测试试样的 Weibull 斜率和平均寿命的分布区间。为此,利用如下方程确定秩为 5% 和 95% 的情况:

$$1-(1-z)^n-nz(1-z)^{n-1}-\frac{n(n-1)}{2!}z^2(1-z)^{n-2}-\cdots$$

$$-\frac{n(n-1)\cdots(n-j+1)}{(j-1)!}z^{j-1}(1-z)^{n-j+1}=G$$

式中，G 为已确定的秩；z 为失效概率；n 为样本总数；j 为失效序数。对于 $G=0.05$，$n=34$，$j=1$ 时 $z=0.15\%$；$j=2$ 时 $z=1.06\%$；$j=3$ 时 $z=2.45\%$；$j=4$ 时 $z=4.12\%$；$j=5$ 时 $z=5.98\%$；$j=6$ 时 $z=7.98\%$；$j=7$ 时 $z=10.08\%$；$j=8$ 时 $z=12.28\%$；$j=9$ 时 $z=14.56\%$；$j=10$ 时 $z=16.9\%$；$j=11$ 时 $z=19.3\%$；$j=12$ 时 $z=21.8\%$；$j=13$ 时 $z=24.3\%$；$j=14$ 时 $z=26.9\%$；$j=15$ 时 $z=29.51\%$；$j=16$ 时 $z=32.18\%$。当 $G=0.95$，$n=34$，$j=1$ 时 $z=8.43\%$；$j=2$ 时 $z=13.21\%$；$j=3$ 时 $z=17.34\%$；$j=4$ 时 $z=21.25\%$；$j=5$ 时 $z=24.93\%$；$j=6$ 时 $z=28.46\%$；$j=7$ 时 $z=31.89\%$；$j=8$ 时 $z=35.22\%$；$j=9$ 时 $z=38.47\%$；$j=10$ 时 $z=41.65\%$；$j=11$ 时 $z=44.76\%$；$j=12$ 时 $z=47.82\%$；$j=13$ 时 $z=50.82\%$；$j=14$ 时 $z=53.78\%$；$j=15$ 时 $z=56.68\%$；$j=16$ 时 $z=59.5\%$。秩为 5% 和 95% 的曲线如图 6.33 所示。因此当置信度为 90% 时，CuSn 微凸点（TCB）的疲劳寿命落在 5%～95% 的置信区间内。例如，CuSn 微凸点（TCB）的实际疲劳寿命 μ 落在 $2668\leqslant\mu\leqslant3083$（次循环）区间内，这意味着 100 个试样中有 90 个的平均寿命大于 2668 个热循环，小于 3083 个热循环，其余试样的平均寿命情况并不可知。该寿命预测方法十分有效，可以确定 CuSn 微凸点（TCB）的最低寿命（即 2668 个热循环），并且能够与微凸点的标准寿命进行对比（这取决于测试条件及用途）。本测试的不确定度为 10%，为了通过提高置信水平来降低不确定度，平均寿命区间必须更宽，其下限值更小。

(3) 实际 Weibull 斜率和实际特征寿命

由于只有 16 个试样失效，因此 Weibull 斜率的误差对结果的影响十分重要，可以利用如下公式对 Weibull 斜率进行计算：

$$\frac{1}{\sqrt{2\pi}}\int_{-\infty}^{E\sqrt{2N}}e^{-t^2/2}dt=\frac{(1+C)}{2}$$

式中，E 为 Weibull 斜率误差；N 为失效样品数；C 为目标置信度。本测试中 $C=90\%$，$N=16$，算得 $E=28\%$。因此，CuSn 微凸点（TCB）的 Weibull 斜率 β 落在 $2.4\leqslant\beta\leqslant4.2$ 区间内，如图 6.33 和表 6.5 所示。

表 6.5 置信度为 90%CuSn（TCB）微凸点的实际 Weibull 斜率及的实际平均寿命

待测目标		CuSn 微凸点
平均寿命(MTTF)	试样	3005 个循环
	达到平均寿命时的失效概率	51%
	置信度为 90% 的实际平均值 μ	$2668\leqslant\mu\leqslant3083$ 个循环
Weibull 斜率	试样	3.31
	置信度为 90% 的威布尔斜率误差	28%
	置信度为 90% 的真实威布尔斜率	$2.4\leqslant\beta\leqslant4.2$

图 6.34 所示为采用 TCB 工艺制得的 CuSn（TCB）微凸点、采用传统回流焊工艺制得的 Sn2.5Ag 微凸点以及采用 TCB 工艺制得的 Sn2.5Ag 焊点的 Weibull 曲线，三种情况所采取的热循环测试条件与失效准则都相同。Sn2.5Ag（回流焊）试样总数为 47 个，Sn2.5Ag（TCB）试样总数为 72 个，Sn2.5Ag（回流焊）失效试样有 45 个，Sn2.5Ag（TCB）失效试样有 27 个。Sn2.5Ag（回流焊）试样的 Weibull 斜率、特征寿命、平均寿命分别为 1.02、980 个循环、972 个循环。Sn2.5Ag（TCB）试样的 Weibull 斜率、特征寿命、平均寿命分别为 1.15、5006 个循环、4765 个循环。

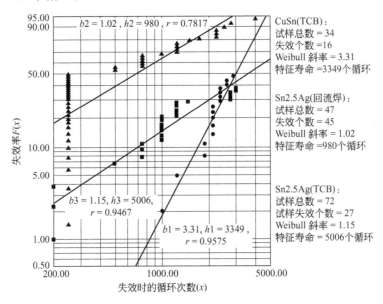

图 6.34 温度循环测试得到的 CuSn（TCB）、Sn2.5Ag（回流焊）以及 Sn2.5Ag（TCB）微凸点的 Weibull 分布图

（4）不同微凸点的平均寿命对比

许多情况下，都是根据有限的实验数据来比较两种产品的质量和一致性。寿命预测的难点之一就是要利用小样本数来估计总体寿命，而根据有限的实验数据对比两种产品总体的寿命尤为困难。如果通过测试确定一种产品优于另一种，那么如何说明样本的置信度 P 就是该产品总体的置信度（此处的置信度 P 不同于之前所讨论置信度 C）呢？这里提供了一个简单的方法来判断一种产品的质量（平均寿命 MTTF）是否优于另一种产品，无需考虑样本与总体之间的差异[1]。具体公式如下：

$$P = \frac{1}{1 + \frac{\lg 1/q}{\lg 1/(1/q)}}$$

式中，
$$q = 1 - \frac{1}{\left[1 + \left(\frac{t+4.05}{6.12}\right)^5\right]^{40/7}}$$

$$t = \frac{\sqrt{1+\sqrt{T}}\,(\rho-1)}{\rho\Omega_2 + \Omega_1}$$

$$T = (r_A - 1)(r_B - 1)$$

$$\Omega_1 = \sqrt{\frac{\Gamma(1+2/\beta_B)}{\Gamma^2(1+1/\beta_B)} - 1}$$

$$\Omega_2 = \sqrt{\frac{\Gamma(1+2/\beta_A)}{\Gamma^2(1+1/\beta_A)} - 1}$$

$$\rho = \frac{M_A}{M_B}$$

式中，M_A 为样本 $A(S_A)$ 的平均寿命；M_B 为样本 $B(S_B)$ 的平均寿命；r_A 为样本 S_A 的失效数；r_B 为样本 S_B 的失效数；β_A 为样本 S_A 的 Weibull 斜率；β_B 为样本 S_B 的 Weibull 斜率。可以看出，该置信度 P（所求值）与之前的置信度 C（已知量）不同。

表 6.6 所示为 CuSn 微凸点（TCB）、Sn2.5Ag 微凸点（回流焊）、Sn2.5Ag 微凸点（TCB）的平均寿命以及置信度 P 的对比结果。可以看出，CuSn 微凸点（TCB）的质量（平均寿命）优于 Sn2.5Ag 微凸点（回流焊），其置信度为 96%。这表明对比 100 次，其中有 96 次得到的结果均为 CuSn 微凸点（TCB）的质量优于 Sn2.5Ag 微凸点（回流焊）。

表 6.6 CuSn（TCB）微凸点、Sn2.5Ag（回流焊）微凸点、Sn2.5Ag（TCB）微凸点质量（平均寿命）以及置信度 P 的对比

不同微凸点的平均寿命对比	MTTF(平均寿命)比值	置信度 P
Sn2.5Ag(TCB)与 Sn2.5Ag(回流)	4765/972	92%
Sn2.5Ag(TCB)与 CuSn(TCB)	7765/3005	51%
CuSn(TCB)与 Sn2.5Ag(回流)	3005/972	96%

从表中还可以看出，Sn2.5Ag[❶] 微凸点（TCB）的质量"优于"CuSn 微凸点（TCB），但是该结论的置信度仅为 51%，即对比 100 次只有 51 次是如此。由于置信度过低，该结论从统计学上讲可以忽略。实质上，Sn2.5Ag 微凸点（TCB）与 CuSn 微凸点（TCB）的平均寿命比较相近。

最后，通过对比 Sn2.5Ag 微凸点（TCB）与 Sn2.5Ag 微凸点（回流焊）的平均寿命发现（如表 6.6 所示），Sn2.5Ag 微凸点（TCB）的质量（平均寿命）优于 Sn2.5Ag 微凸点（回流焊），其置信度为 92%。这表明对比 100 次，其中有 92 次得到的结果均为 Sn2.5Ag 微凸点（TCB）的质量优于 Sn2.5Ag 微凸点（回流焊）。

---

❶ 此处 Sn2.5Ag 表示 2.5%（质量分数）的 Ag 与 Sn 的合金，下同。

6.10 超细节距晶圆无铅焊锡微凸点的制作

6.10.1 测试模型

测试模型包含 3200 个节距为 10μm、直径为 5μm 的衬垫。芯片与基板的尺寸分别为 5mm×5mm 和 10mm×10mm，置于 200mm 的晶圆上。节距 10μm、直径 5μm 衬垫上的微凸点结构与节距 20μm、直径 10μm 衬垫上的微凸点结构相同。

6.10.2 微凸点制作

节距 10μm 的微凸点与节距 20μm 的微凸点制作工艺相同，一些初步的结论总结如下。

6.10.3 超细节距微凸点的表征

如图 6.35 所示为节距 10μm、直径 5μm 的衬垫上微凸点的 OM 图像，SEM 图像如图 6.36 所示。可以看出，相比于节距 20μm 的微凸点，节距 10μm 的微凸点上焊锡的晶粒尺寸更大，并且其体积一致性也较差。不过，采用厚度为 50nm 的 Ti 阻挡层以及厚度为 120nm 的 Cu 种子层有效地减少了咬边量，使得 Ti 层的咬边量仅为 0.67μm，如图 6.37 所示。

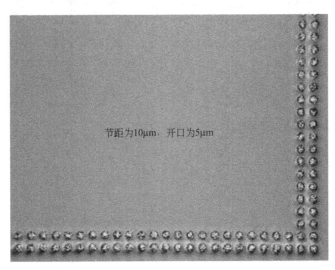

图 6.35 节距 10μm 微凸点的 OM 图像

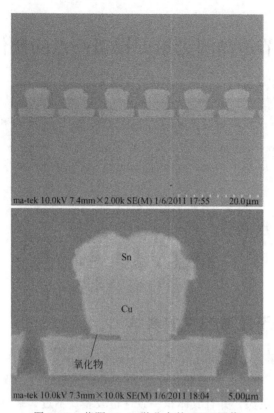

图 6.36 节距 10μm 微凸点的 SEM 图像

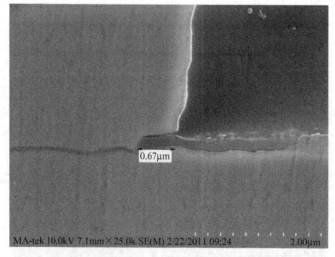

图 6.37 节距 10μm 微凸点上 Ti 层的咬边情况（咬边量为 0.67μm）

6.11 总结与建议

针对 3D IC 集成，本章主要研究了无铅细节距焊锡微凸点的制作、组装及其可靠性，并且给出了微凸点的一些特征参数和可靠性测试结果。此外，还研究了节距为 $10\mu m$、直径为 $5\mu m$ 的衬垫上超细节距微凸点的制作工艺。一些重要的结论以及建议总结如下[17]。

① 针对无铅焊锡微凸点的制作，提出了将 RDL 与直径 $10\mu m$、节距 $20\mu m$ 衬垫上的微凸点集成在一起的工艺路线。

② 在回流过程中，焊锡与 Cu 柱界面上的金属间化合物 Cu_6Sn_5 由扇贝状逐渐转变为平坦状。

③ 3 次回流之后，在 Cu_6Sn_5 与 Cu 柱的界面上会产生一些微小孔洞，从而降低了微凸点的剪切强度。

④ 采用 TCB 工艺制得的微凸点其剪切强度要高于传统回流焊工艺制得的微凸点，原因是 TCB 工艺制得的微凸点中焊锡全部与 Cu 反应生成 Cu_6Sn_5。此外，传统回流焊工艺制得的微凸点不一致，因而其剪切强度较低。

⑤ 给定置信度为 90%，采用 TCB 工艺制得的微凸点，其实际平均寿命落在 [2668，3083] 次循环区间内。对比寿命下限值（2668 个循环）和标准寿命值（根据应用条件确定），可以粗略估计 TCB 工艺制得的微凸点的可靠性。

⑥ CuSn 微凸点（TCB）的平均寿命高于 Sn2.5Ag 微凸点（回流焊），其置信度为 96%，即对比 100 次，有 96 次结果为真，其余 4 次结果未知。

⑦ Sn2.5Ag 微凸点（TCB）的平均寿命高于 Sn2.5Ag 微凸点（回流焊），置信度为 92%，即对比 100 次，有 92 次为真，其余 8 次结果未知。

⑧ Sn2.5Ag 微凸点（TCB）的平均寿命高于 CuSn 微凸点（TCB），置信度为 51%。由于置信度过低，说明实质上两者的平均寿命比较相近。

⑨ 已经证明了一种用于制作超细节距微凸点的新设计和工艺的可行性，该方法能够有效减小咬边量。当然，还需要更多的特征参数和可靠性测试结果进行验证。

6.12 参考文献

[1] Lau, J. H., *Reliability of RoHS-Compliant 2D and 3D IC Interconnects*, McGraw-Hill, New York, 2011.
[2] Lau, J. H., C. K. Lee, C. S. Premachandran, and A. Yu, *Advanced MEMS Packaging*, McGraw-Hill, New York, 2010.
[3] Lau, J. H., *Flip-Chip Technology*, McGraw-Hill, New York, 1995.
[4] Lau, J. H., *Low-Cost Flip-Chip Technology*, McGraw-Hill, New York, 2000.
[5] Hsin, Y. C., C. Chen, J. H. Lau, P. Tzeng, S. Shen, Y. Hsu, S. Chen, C. Wn, J.

Chen, T. Ku, and M. Kao, "Effects of Etch Rate on Scallop of Through-Silicon Vias (TSVs) in 200-mm and 300-mm Wafers," *IEEE/ECTC Proceedings*, Orlando, FL, June 2011, pp. 1130–1135.

[6] Lau, J. H., P-J Tzeng, C-K Lee, C-J Zhan, M-J Dai, Li Li, C-T Ko, S-W Chen, H. Fu, Y. Lee, Z. Hsiao, J. Huang, W. Tsai, P. Chang, S. Chung, Y. Hsu, S-C Chen, Y-H Chen, T-H Chen, W-C Lo, T-K Ku, and M-J Kao, J. Xue, and M. Brillhart,, "Wafer Bumping and Characterizations of Fine-Pitch Lead-Free Solder Microbumps on 12" (300mm) wafer for 3D IC Integration", *Proceedings of IMAPS International Conference*, Long Beach, CA, October 2011, pp. 650-656.

[7] Chen, J. C., P. J. Tzeng, S. C. Chen, C. Y. Wu, J. H. Lau, C. C. Chen, C. H. Lin, Y. C. Hsin, T. K. Ku, and M. J. Kao, "Impact of Slurry in Cu CMP (Chemical Mechanical Polishing) on Cu Topography of Through-Silicon Vias (TSVs), Re-distributed Layers, and Cu Exposure," *IEEE/ECTC Proceedings*, Orlando, FL, June 2011, pp. 1389–1394.

[8] Tsai, W., H. H. Chang, C. H. Chien, J. H. Lau, H. C. Fu, C. W. Chiang, T. Y. Kuo, Y. H. Chen, R. Lo, and M. J. Kao, "How to Select Adhesive Materials for Temporary Bonding and De-Bonding of Thin-Wafer Handling in 3D IC Integration?" *IEEE/ECTC Proceedings*, Orlando, FL, June 2011, pp. 989–998.

[9] Dorsey, P., "Xilinx Stacked Silicon Interconnect Technology Delivers Breakthrough FPGA Capacity, Bandwidth, and Power Efficiency," Xilinx white paper: Virtex-7 FPGAs, WP380, October 27, 2010, pp. 1–10.

[10] Banijamali, B., S. Ramalingam, K. Nagarajan, and R. Chaware, "Advanced Reliability Study of TSV Interposers and Interconnects for the 28nm Technology FPGA," *IEEE/ECTC Proceedings*, Orlando, FL, June 2011, pp. 285–290.

[11] Banijamali, B., S. Ramalingam, N. Kim, and R. Wyland, "Ceramics vs. Low CTE Organic Packaging of TSV Silicon Interposers," *IEEE/ECTC Proceedings*, Orlando, FL, June 2011, pp. 573–576.

[12] Kim, N., D. Wu, D. Kim, A. Rahman, and P. Wu, "Interposer Design Optimization for High Frequency Signal Transmission in Passive and Active Interposer using Through Silicon Via (TSV)," *IEEE/ECTC Proceedings*, Orlando, FL, June 2011, pp. 1160–1167.

[13] Chai, T., X. Zhang, J. H. Lau, C. Selvanayagam, K. Biswas, S. Liu, D. Pinjala, G. Tang, Y. Ong, S. Vempati, E. Wai, H. Li, B. Liao, N. Ranganathan, V. Kripesh, J. Sun, J. Doricko, and C. Vath, "Development of Large Die Fine-Pitch Cu/Low-k FCBGA Package with Through Silicon Via (TSV) Interposer," *IEEE Transactions on Components, Packaging and Manufacturing Technology*, Vol. 1, No. 5, 2011, pp. 660–672.

[14] Selvanayagam, C., J. H. Lau, X. Zhang, S. Seah, K. Vaidyanathan, and T. Chai, "Nonlinear Thermal Stress/Strain Analysis of Copper Filled TSV (Through Silicon Via) and Their Flip-Chip Microbumps," *IEEE/ECTC Proceedings*, Orlando, FL, May 27–30, 2008, pp. 1073–1081. Also, *IEEE Transactions on Advanced Packaging*, Vol. 32, No. 4, November 2009, pp. 720–728.

[15] Yu, A., N. Khan, G. Archit, D. Pinjalal, K. Toh, V. Kripesh, S. Yoon, and J. H. Lau, "Development of Silicon Carriers with Embedded Thermal Solutions for High-Power 3D Package," *IEEE Transactions on Components, Packaging and Manufacturing Technology*, Vol. 32, No. 3, September 2009, pp. 566–571.

[16] Zhang, X., A. Kumar, Q. X. Zhang, Y. Y. Ong, S. W. Ho, C. H. Khong, V. Kripesh, J. H. Lau, D.-L. Kwong, V. Sundaram, Rao R. Tummula, and Georg Meyer-Berg, "Application of Piezoresistive Stress Sensors in Ultrathin Device Handling and Characterization," *Journal of Sensors & Actuators: A. Physical*, Vol. 156, November 2009, pp. 2–7.

[17] Lee, C. K., T. C. Chang, Y. Huang, H. Fu, J. H. Huang, Z. Hsiao, J. H. Lau, C. T. Ko, R. Cheng, K. Kao, Y. Lu, R. Lo, and M. J. Kao, "Characterization and Reliability Assessment of Solder Microbumps and Assembly for 3D IC Integration," *IEEE/ECTC Proceedings*, Orlando, FL, June 2011, pp. 1468–1474.

[18] Zhan, C., J. Juang, Y. Lin, Y. Huang, K. Kao, T. Yang, S. Lu, J. H. Lau, T. Chen, R. Lo, and M. J. Kao, "Development of Fluxless Chip-on-Wafer Bonding Process for 3D Chip Stacking with 30-μm Pitch Lead-Free Solder Micro Bump Interconnection and Reliability Characterization," *IEEE/ECTC Proceedings*, Orlando, FL, June 2011, pp. 14–21.

[19] Huang, S., T. Chang, R. Cheng, J. Chang, C. Fan, C. Zhan, J. H. Lau, T. Chen, R. Lo, and M. Kao, "Failure Mechanism of 20-μm-Pitch Micro Joint Within a

Chip Stacking Architecture," *IEEE/ECTC Proceedings*, Orlando, FL, June 2011, pp. 886–892.

[20] Lin, Y., C. Zhan, J. Juang, J. H. Lau, T. Chen, R. Lo, M. Kao, T. Tian, and K. N. Tu, "Electromigration in Ni/Sn Intermetallic Micro Bump Joint for 3D IC Chip Stacking," *IEEE/ECTC Proceedings*, Orlando, FL, June 2011, pp. 351–357.

[21] Yu, A., A. Kumar, S. W. Ho, H. W. Yin, J. H. Lau, K. C. Houel, Sharon L. P. Siang, X. Zhang, D.-Q. Yu, N. Su, M. C. Bi-Rong, J. M. Ching, T. T. Chun, V. Kripesh, C. Lee, J. P. Huang, J. Chiang, S. Chen, C.-H. Chiu, C.-Y. Chan, C.-H. Chang, C.-M. Huang, and C.-H. Hsiao, "Development of Fine Pitch Solder Microbumps for 3D Chip Stacking," *IEEE/ECTC Proceedings*, December 2008, pp. 387–392.

[22] Yu, A., J. H. Lau, S. W. Ho, A. Kumar, W. Y. Hnin, D.-Q. Yu, M. C. Jong, V. Kripesh, D. Pinjala, and D.-L. Kwong, "Study of 15-μm-Pitch Solder Microbumps for 3D IC Integration," *IEEE/ECTC Proceedings*, Orlando, FL, June 2009, pp. 6–10.

[23] Reed, J. D., M. Lueck, C. Gregory, A. Huffman, and J. M. Lannon, Jr., "High Density Interconnect at 10-μm Pitch with Mechanically Keyed Cu/Sn-Cu and Cu-Cu Bonding for 3D Integration," *IEEE/ECTC Proceedings*, Orlando, FL, June 2010, pp. 846–852.

[24] Zhang, W., P. Limaye, Y. Civale, R. Labie, and P. Soussan, "Fine Pitch Cu/Sn Solid State Diffusion Bonding for Making High Yield Bump Interconnections and Its Application in 3D Integration," *IEEE Proceedings of Electronics System Integration Technology Conference*, Berlin, Germany, 2010, pp. 1–4.

[25] Tu, K. N., F. Ku, and T.Y. Lee, "Morphological Stability of Solder Reaction Products in Flip Chip Technology," *J. Electronic Materials*, Vol. 30, No. 9, 2001, pp. 1129–1132.

[26] Lee, T. Y., W. J. Choi, K. N. Tu, J. W. Jang, S. M. Kuo, J. K. Lin, D. R. Frear, K. Zeng, and J. K. Kivilahti, "Morphology, Kinetics, and Thermodynamics of Solid-State Aging of Eutectic SnPb and Pb-Free Solders (Sn3.5Ag, Sn3.8Ag0.7Cu, and Sn0.7Cu) on Cu," *Journal of Materials Research*, Vol. 17, No. 2, 2002, pp. 291–301.

[27] Shen, J., Y. C. Chan, and S. Y. Liu, "Growth Mechanism of Ni_3Sn_4 in a Sn/Ni Liquid/Solid Interfacial Reaction," *Acta Materialia*, Vol. 57, 2009, pp. 5196–5206.

[28] Dybkov, V. I., "Reaction Diffusion and Solid State Chemical Kinetics," Chapter 1, IPMS Publications, New York, 2002.

[29] Frear, D. R., S. N. Burchett, H. S. Morgan, and J. H. Lau, eds., *The Mechanics of Solder Alloy Interconnects*, Van Nostrand Reinhold, New York, 1994, p. 60.

[30] Schwartz, Mel, "Tin and Alloys: Properties," in *Encyclopedia of Materials, Parts and Finishes*, 2nd ed., CRC Press, Boca Raton, FL, 2002.

[31] Hon, M. H., T. C. Chang, and M. C. Wang, "Phase Transformation and Morphology of the Intermetallic Compounds Formed at the Sn9Zn3.5Ag/Cu Interface in Aging," *Journal of Alloys and Compounds*, Vol. 458, 2008, pp. 189–199.

[32] Bader, S., W. Gust, and H. Hieber, "Rapid Formation of Intermetallic Compounds by Interdiffusion in the Cu-Sn and Ni-Sn Systems," *Acta Metall. Mater.*, Vol. 43, No. 1, 1995, pp. 329–337.

[33] Kim, J., and C. C. Lee, "Fluxless Wafer Bonding with Sn-Rich SnAu Dual-Layer Structure," *Materials Science and Engineering A*, Vol. 417, Nos. 1–2, 2006, pp. 143–148.

[34] Chang, T. C., R. Cheng, P. Chang, Y. Hung, J. Chang, T. Yang, and S. Huang., "Reliability Characterization of 20-μm-Pitch Microjoints Assembled by a Conventional Reflow Technique," *IEEE Proceedings of International Conference on Electronics Packaging*, Nara, Japan, 2011, pp. 221–226.

[35] Chang, J. Y., S. Huang, R. Cheng, F. Leu, C. Zhan, and T. Chang, "High Throughput Chip on Wafer Assembly Technology and Metallurgical Reactions of Pb-Free Micro-Joints Within a 3D IC Package," *IEEE Proceedings of International Conference on Electronics Packaging*, Hokkaido, Japan, 2010, pp. 159–164.

第7章 微凸点的电迁移

7.1 引言

前面章节中已经提到过,对于3D IC集成来说普通焊锡凸点(直径80~100μm)的体积过大,需采用体积更小的微凸点(直径≤25μm)。相比于大体积的焊点,微凸点会导致焊点中的电流密度大幅增加。例如,当直径为20μm的微凸点中通入0.05A的电流时,电流密度高达约10^4A/cm^2,该值远大于直径为100μm的焊点中通入相同电流所产生的电流密度。

其次,随着微凸点体积的减小,由电迁移(EM)引发的电流聚集效应更加明显。再次,相关文献中的实验数据表明,超细节距CuSn微凸点在电迁移过程中其微结构会发生变化,并伴随大量的柯肯达尔孔洞产生。分析结果表明,电迁移会加剧微凸点中焊料的相变,导致其微结构显著退化。与大体积的焊点相比,这种情况在超细节距微凸点中更为明显。最后,由于焊料体积较小,金属间化合物(IMC)在微凸点的电迁移过程中具有关键作用。对于倒装芯片焊点,电流聚集引发的电迁移失效方面的问题已经得到了广泛研究,并且有不少文献发表,如参考文献[1~42]。本章研究电流引起的焊点失效,首先研究大节距、大体积焊点[43],然后研究细节距、小体积焊点[44]。

7.2 大节距大体积微焊锡接点

7.2.1 测试模型与测试方法

(1) 焊锡凸点

图7.1所示为用于测试的焊锡凸点,Cu柱上是Sn2.5Ag焊料。完成电镀后,微凸点的尺寸为:20μm厚的Sn2.5Ag焊料,20μm厚的Cu柱。衬垫直径约为25μm,硅芯片尺寸为10mm×8mm,上面布有13413个节距为50μm或100μm的微凸点。载芯片上的凸点下金属层由5μm厚的化学镀Ni与0.1μm厚的浸Au组成,其直径约为20μm。硅芯片和载芯片上的Al线路层厚度为1.5μm,宽度为25μm。

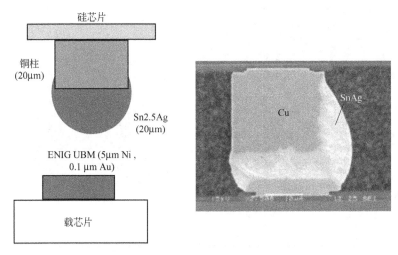

图 7.1 由铜柱、SnAg 焊料、化学镀 Ni 以及浸 Au（ENIG）UBM 组成的待测试样

（2）测试方法

图 7.2 所示为组装好的待测试样，图 7.3 所示为用于测量单个焊锡凸点连续电阻的开尔文（Kelvin）结构示意图。可以看出：①开尔文电路中设有四个测量点；②有两个与被测凸点相邻的备选凸点；③待测试样倒装在 20mm×20mm 的载芯片上；④载芯片采用芯片胶与 PCB 板粘接，并采用 Au 线与 PCB 互连，如图 7.2 所示；⑤分别使用两组探针测量电流方向以及电压压降。为了避免引线键合界面发生

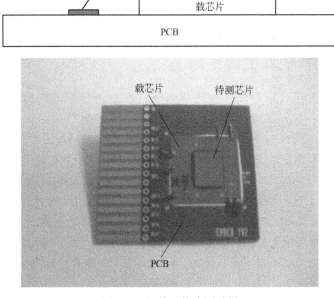

图 7.2 组装后的待测试样

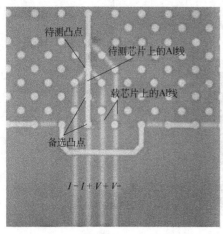
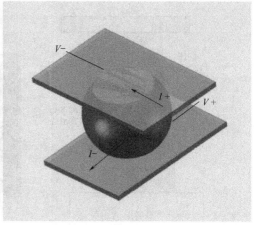

图 7.3　四点开尔文结构及电流方向和电压测量方法

电流引起的失效，在一对焊盘上焊接四条 Au 线。测试时，电流从芯片边缘处的焊锡凸点通入。

7.2.2　测试步骤

采用集成模块可靠性分析系统（MIRA，QualiTau）进行电迁移（EM）测试，温度为 140℃。根据焊锡凸点直径的大小分别通入密度为 $2.04A/cm^2$ 和 $4.08A/cm^2$ 的电流。电流从化学镀 Ni 和浸 Au（ENIG）一端通向 Cu 柱一端，每种电流都测试 6 个试样。在进行电迁移测试之前，通过测量焊锡凸点的电阻温度系数（TCR）考虑焦耳（Joule）热对测试结果的影响。该测试方法将焊锡凸点本身作为温度计，根据两种电流密度所对应的测量结果，测得芯片的实际温度分别为 148℃和 165℃。

7.2.3　测试前试样的微结构

图 7.4 所示为组装后形成互连的两个凸点在电迁移测试前的微结构，主要由 Cu 柱、SnAg 焊料以及 ENIG UBM 组成[43]。根据 X 射线能谱仪（EDX）的分析结果发现，凸点中铜柱一侧和 ENIG 一侧均形成了金属间化合物 $(Cu,Ni)_6Sn_5$。相关文献研究表明，当 SnAg 焊料被 Ni 润湿时会生成金属间化合物（IMC）$NiSn$[36]。同时，SnCu 以及 SnAgCu 焊料中 Cu 含量的变化会导致 IMC 由 $(Cu,Ni)_3Sn_4$ 转变为 $(Cu,Ni)_6Sn_5$。此外，微量的 Ni 会使 Cu 在液体 Sn 中的溶解度减小，从而加快 Cu UBM 上 Cu_6Sn_5 的生长速度[36]。考虑到微凸点中 Cu UBM 与 ENIG 焊盘间的距离很近，Cu 可以在很短的时间内扩散至 ENIG 焊盘中，因而 Cu 的含量能够达到在 ENIG 上形成 $(Cu,Ni)_6Sn_5$ 的临界值。SnAg 焊点中最常见的金属间化合物为 Ag_3Sn，如图 7.4 所示。

Cu 柱与焊锡凸点的互连电阻在 10～40mΩ 之间，电阻值的变化主要是因为回

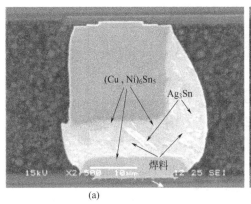 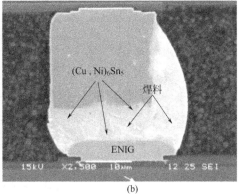

图 7.4 由 Cu 柱和 SnAg 焊料构成的焊锡凸点

流焊后凸点微结构与初始的不同。对于微凸点互连而言，初始界面的微结构可能与回流焊后的略有不同，即使同一芯片上相邻的两个微凸点也会如此。如图 7.4(a) 所示，ENIG 一侧的 IMC 层较薄，UBM 一侧的 IMC 层与 ENIG 一侧的 IMC 层被剩余的焊料隔开。通过对比发现，图 7.4(b) 中 UBM 与 ENIG 上都形成了较厚的 IMC 层，并且在某些区域发生融合。由于 IMC 层的电阻值高于焊料，故图 7.4(b) 中焊点的电阻值要高于图 7.4(a) 中焊点的电阻值。

7.2.4　140℃、低电流密度条件下测试后的试样

(1) 140℃、电流密度为 $2.04\times10^4 \text{A/cm}^2$ 条件下的测试试样

将三个互连好的凸点通入密度为 $2.04\times10^4 \text{A/cm}^2$ 的电流，如图 7.5(a) 所示。可以看出，直至 1000h，电阻值缓慢增大，但未发现凸点失效。图 7.5(b) 为电阻相对退化率随电流加载时间的变化曲线，比电阻值的变化率更易观察。相对退化率定义为测得的电阻值与初始电阻值的比值。如图 7.5(b) 所示的三条曲线，电阻值增加到初始值的 125%～150%。通过观察微凸点互连的横截面发现，在电流作用下其微结构发生了变化。

(2) 低电流密度下试样的微结构

通以密度为 $2.04\times10^4 \text{A/cm}^2$ 的电流 1000h 后，焊锡凸点的典型微结构如图 7.6 所示。显然，$(Cu,Ni)_6Sn_5$ 变薄了，界面上剩余的焊料减少了，靠近 Cu 的一侧形成了 Cu_3Sn。在该测试条件下焊料与 Cu 柱界面发生如下反应：

$$Cu+Sn+Ni \longrightarrow (Cu,Ni)_6Sn_5 \tag{7.1}$$

$$Cu+Cu_6Sn_5 \longrightarrow Cu_3Sn \tag{7.2}$$

从式 (7.1) 可以看出，消耗焊锡的同时电阻也在升高，原因在于生成的 IMC 其电阻率较高。由式 (7.2) 可以看出，由于 Cu_3Sn 比 Cu_6Sn_5 的电阻低，Cu_3Sn 的形成和生长会减缓"电阻值的升高"。因此，电阻值变化与材料的扩散、IMC 的生长密切相关。Cu、SnAg 焊料、Cu_3Sn 以及 Cu_6Sn_5 的电阻率分别为 $1.7\mu\Omega\cdot\text{cm}$、

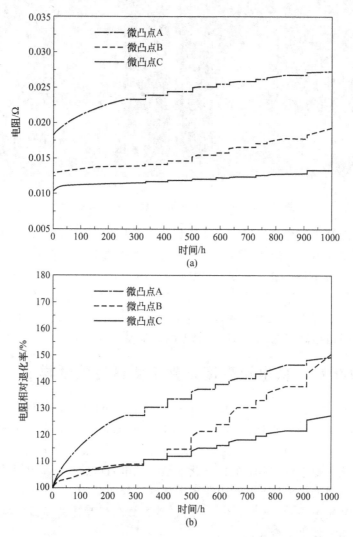

图 7.5 （a）电阻值与电流加载时间的关系曲线；
（b）电阻相对退化率随电流加载时间的关系曲线（电流密度为 $2.04×10^4 A/cm^2$）

$11.5\mu\Omega \cdot cm$、$8.3\mu\Omega \cdot cm$ 和 $17.5\mu\Omega \cdot cm$[45]。由此可见，Cu 与焊锡的消耗、IMC 的生长都会使得焊点的电阻值升高。

在较薄的 Cu_3Sn 层中发现了少量的柯肯达尔孔洞，形成柯肯达尔孔洞的原因是 Cu 从 Cu_3Sn 到 $(Cu,Ni)_6Sn_5$ 的扩散速率高于 Sn 从 $(Cu,Ni)_6Sn_5$ 到 Cu_3Sn 的扩散速率[38]。随着 Cu_3Sn 的生长，结果会形成更多的柯肯达尔孔洞。Xu 等人[31]研究了温度为 135℃、电流密度为 $1.6×10^4 A/cm^2$ 的条件下，Cu 柱上厚度为 $90\mu m$ 的 SnAgCu 焊料中的电迁移问题。由于焊料与 Cu 的体积比更大，因而电流加载 1290h 后，IMC 的主要成分为 $(Cu,Ni)_6Sn_5$，未发现有柯肯达尔孔洞以及裂纹产生[31]。

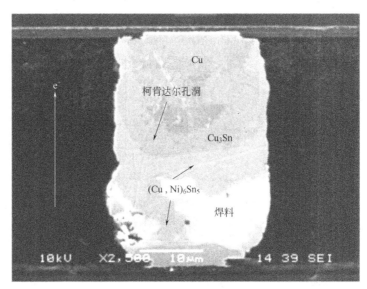

图 7.6 低电流密度（$2.04 \times 10^4 \mathrm{A/cm^2}$）条件下焊点的典型微结构图

7.2.5 140℃、高电流密度条件下测试后的试样

（1）140℃、电流密度为 $4.08 \times 10^4 \mathrm{A/cm^2}$ 条件下的测试试样

当采用较高电流密度（$4.08 \times 10^4 \mathrm{A/cm^2}$）进行测试时，试样发生失效。3 个试样的电阻变化曲线如图 7.7(a) 所示，焊点均由 Cu 柱、SnAg 焊料以及 ENIG UBM 组成。当电阻达到一定值时，几乎保持不变直到电阻值突然陡增，从而说明此时焊点互连发生开路。如图 7.7(a) 所示，三条曲线中有两条在达到稳定值之前，电阻值在持续增大，而另外一个试样的电阻值前 100h 内不断升高，然后再降至稳定阶段。显然，由于电阻值的升高发生在初始阶段[5]，故很难将焊点电阻值的异常变化归因于大体积焊点中发生的焊料基体相变。此外，从图 7.7(a) 还可以看出，试样在高温高电流密度下测试 100h 后，焊料几乎被全部消耗。

（2）高电流密度测试后试样的微结构

正如 7.2.4 节所讨论的，IMC 的生长对电阻值的变化有重要影响。如图 7.7(a) 所示，微凸点 D 的初始电阻值为 0.01Ω，与其他微凸点相比电阻值略小。根据 7.2.3 节的分析可以推断，微凸点 D 在初始阶段形成了很少的 IMC。因此，在 EM 测试中，随着 IMC $(Cu, Ni)_6Sn_5$ 的生长，电阻值将增大得更快，如图 7.7(b) 所示，其电阻值会超过初始值的 3 倍左右。当 Cu_3Sn 的生长占主导时，电阻值逐渐下降并达到稳定状态。对于微凸点 E 和 F，回流焊之后形成更厚的 $(Cu, Ni)_6Sn_5$ 层。在进行 EM 测试时，微凸点 E 和 F 中 $(Cu, Ni)_6Sn_5$ 的生长速度较慢，并且 Cu_3Sn 比微凸点 D 中更早生成。因此，Cu_3Sn 的生成并无反向效应，高电阻率 $(Cu, Ni)_6Sn_5$ 的生长弥补了由于 Cu_3Sn 的生长引起的电阻减小。

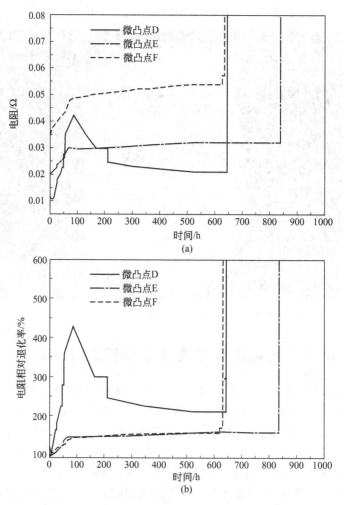

图 7.7 (a) 电阻值与电流加载时间的关系曲线；
(b) 电阻相对退化率与电流加载时间的关系曲线（电流密度为 $4.08×10^4 A/cm^2$）

通以密度为 $4.08×10^4 A/cm^2$ 的电流 132h 后，微凸点的微结构如图 7.8(a) 所示。可以看出：①由于电子力以及焦耳热的影响，加快了材料的扩散速率；②凸点内无残余焊料；③（Cu，Ni）$_6$Sn$_5$ 和 Cu$_3$Sn 变薄；④Cu$_3$Sn 中产生大量柯肯达尔孔洞。当焊料消耗完全且 IMC 互扩散达到平衡时，电阻值将不再变化，这就解释了为什么图 7.7 中的电阻曲线最后达到稳定状态。从图 7.8(a) 中可以看出，焊点内部出现了柯肯达尔孔洞，并且 ENIG 衬垫附近也萌生了小的水平裂纹。

图 7.8(b) 所示为高电流密度下测试 640h 后的失效凸点的截面图。可以看出，柯肯达尔孔洞连成垂直裂纹贯穿了（Cu，Ni）$_6$Sn$_5$ 层。沿着 ENIG 衬垫的（Cu，Ni）$_6$Sn$_5$ 中形成了一条连续微裂纹，导致了焊点开路。ENIG 一侧的微结构细节如图 7.8(c) 所示。这种微裂纹与通常倒装芯片焊点中由电流引起的宽裂纹不同[26,27]，原因是，在本测试条件下，裂纹是由机械应力引起的，然后沿着晶界或

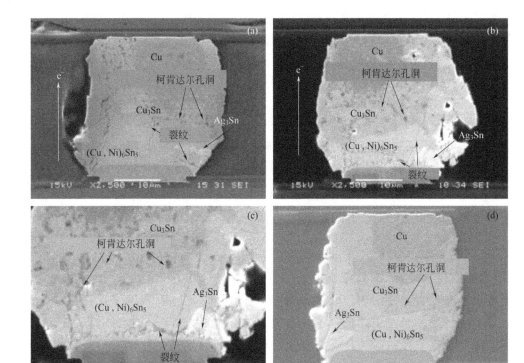

图 7.8 SnAg 凸点在高电流密度条件下（$4.08 \times 10^4 \mathrm{A/cm^2}$）的微结构：
(a) 通电 132h 后的截面图；(b) 通电 640h 后的截面图；
(c) 通电 640h 的截面图（放大图）；(d) 未通电时的截面图

晶相边界扩展。测试结束后发现一个有意思的现象：一些失效的凸点又会呈现较小的电阻，但是加热到测试温度时，焊点再次出现开路。

图 7.8(d) 是与被测凸点相邻且未通电流的凸点截面图。可以看出：该凸点中产生的柯肯达尔孔洞和裂纹较少；界面处未发现明显的裂纹；Cu_3Sn 层厚度较小，并且柯肯达尔孔洞的尺寸较小。这表明加载电流可以促进或加快柯肯达尔孔洞的形成[5]。

7.2.6 焊锡接点的失效机理

高电流密度下，电迁移会在零通量边界处，如两种材料的界面处引发失效[42]。置于高电流密度下的微凸点，有以下特征：①当焊料全部转化成 IMC 时，可能形成的零通量界面为 Cu/Cu_3Sn、$Cu_3Sn/(Cu, Ni)_6Sn_5$、$(Cu, Ni)_6Sn_5/Ag_3Sn$ 和 $(Cu, Ni)_6Sn_5/ENIG$；②在多种 IMC 焊点中，加载电流促进了柯肯达尔孔洞在 Cu/Cu_3Sn 界面处的产生以及在 $(Cu, Ni)_6Sn_5/ENIG$ 界面处的生长；③水平裂纹以及垂直裂纹的形成与柯肯达尔孔洞有关。

两种力学性能不同的材料，在它们界面处存在界面应力，界面应力容易引起沿界面的断裂。Cu_3Sn 的热膨胀系数（CTE）为 $16.3\times10^{-6}/℃$，而 $(Cu，Ni)_6Sn_5$ 的热膨胀系数为 $19\times10^{-6}/℃$，两者不匹配会引起焊点内部产生应力。此外，由于电流作用在芯片边缘处的焊点上，芯片与载芯片之间的 CTE 不匹配会导致两侧边缘处焊点在测试温度下产生应力。Cu_3Sn 和 Cu_6Sn_5 的断裂韧性分别为 $5.72MPa/m^{1/2}$ 和 $2.8MPa/m^{1/2}$[39]，因此可以推断，当焊点在高电流密度和机械应力的共同作用下，与 Cu_3Sn 相比裂纹更容易在 $(Cu，Ni)_6Sn_5$ 中扩展。这就解释了为什么焊点在靠近 $(Cu，Ni)_6Sn_5/ENIG$ 界面处发生开路，而不是在形成大量柯肯达尔孔洞的 Cu_3Sn 处。另一方面，当焊点置于低电流密度时并未失效，原因在于：①电子流较小；②引起机械应力的焦耳热效应并不明显；③反应剩余的焊料可以释放焊点内部的机械应力。

7.2.7 总结与建议

对 Cu 柱、SnAg 焊料以及 ENIG UBM 组成的微凸点进行了电迁移测试。发现当微凸点置于 140℃、电流密度为 $2.04\times10^4A/cm^2$ 条件下，经历 1000h 后未发生失效。当在相同温度下，电流密度升高至 $4.08\times10^4A/cm^2$ 时，由于焦耳热以及电子力的影响，焊料很快被消耗，同时生成多种 IMC，即 Cu_3Sn 和 $(Cu，Ni)_6Sn_5$。在微凸点中，除了 Cu_3Sn 中有大量柯肯达尔孔洞形成外，还发现在 $(Cu，Ni)_6Sn_5/ENIG$ 界面处也有孔洞形成，而裂纹在这些孔洞处形成和扩展，从而导致微凸点失效。研究表明，当微凸点置于高温、高电流密度条件下，多种 IMC 的形成是凸点电迁移失效的重要因素。一些重要结果和建议如下。

① 微凸点置于高温（140℃）、高电流密度（$2.04\times10^4A/cm^2$，$4.08\times10^4A/cm^2$）条件下时，电阻升高主要原因是 IMC 的形成。

② Cu 柱、SnAg 焊料、ENIG UBM 组成的微凸点在 140℃、电流密度为 $4.08\times10^4A/cm^2$ 的条件下，发现经过一段时间之后（50~200h）电阻值达到稳定，此时 Cu、Ni、Sn 等不同材料间的扩散达到平衡，IMC 的生长也达到稳定状态。高温、高电流密度条件下，在 Cu_3Sn 中形成大量柯肯达尔孔洞。由于 $(Cu，Ni)_6Sn_5$ 属于脆性材料，在 $(Cu，Ni)_6Sn_5$ 晶界处存在较多的裂纹，这些裂纹会引起微凸点的突然失效。

③ Cu 柱、SnAg 焊料、ENIG UBM 组成的微凸点在 140℃、电流密度为 $2.04\times10^4A/cm^2$ 的条件下，由于焦耳热影响较小，通电 1000h 后电阻值依然未达到稳定。电子流方向对材料的扩散有一定影响，当电子流从 Cu UBM 通向 ENIG 时，电阻值增大较快。

④ 测试结果表明，微凸点的电迁移失效模式和机制与普通焊点完全不同，原因在于微凸点电阻值的升高是由 IMC 的生长以及相变引起的，并非由扁平裂纹的形成引起的。

7.3 小节距小体积微焊锡接点

本节中,给出了节距 30μm SnAg 焊锡接点的电迁移测试过程及结果[44]。焊点结构包括 SnAg 焊料和 5μm Cu/3μm Ni UBM,用于同 Cu UBM 微凸点进行对比。由于微凸点中电阻变化值极小,故采用开尔文结构来测量电阻值的微小变化,同时利用菊花链结构测量电迁移过程中焦耳热对细节距微凸点互连的影响。研究结果表明,与倒装芯片焊点相比,该微凸点具有更好的抗电迁移能力。这种特性主要与电迁移过程中焊点微结构的演变有关,即大部分剩余的 Sn 转化为 IMC。

7.3.1 测试模型与方法

(1) 测试模型

测试模型为芯片上芯片(COC)结构,包括上层芯片和下层芯片,尺寸分别为 5.1mm×5.1mm、15mm×15mm。两个芯片的厚度和微凸点的结构均相同,微凸点数目多于 3000 个,节距为 30μm。微凸点直径为 8μm,高度为 30μm。微凸点的结构如图 7.9 所示,Al 衬垫上的 UMB 由 5μm 厚的 Cu、3μm 厚的 Ni 以及 5μm 厚 Sn2.5Ag 组成。Sn2.5Ag 材料由电镀、回流两道工序制得,最后在 UBM 上形成一层 Sn2.5Ag。测试模型的详细说明见图 7.9。

测试模型的详细说明

参数	上层芯片	下层芯片
芯片尺寸 (X, Y)	(5.1mm, 5.1mm)	(15mm, 15mm)
芯片厚度	750μm	
凸点节距	30μm	
凸点尺寸	直径:8μm	
凸点高度 (Sn2.5Ag/Ni/Cu)	5μm/3μm/5μm	

外观图

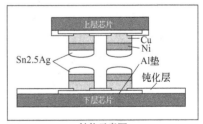

结构示意图

图 7.9 COC 测试模型结构简图

(2) 测试模型的组装

组装流程如图 7.10 所示,首先采用等离子体清洗芯片的上下表面,清除微凸点上的污染物以及氧化层。然后将上下芯片进行对中互连。本研究采用的是 Toray

FC3000WS 键合机，借助于间隙控制热键合工艺，该方法可以很好地控制键合过程中上下芯片之间的距离。在 300℃下保温 15s，对上下芯片进行倒装键合，理想的微凸点（退火前）如图 7.11(a) 所示。

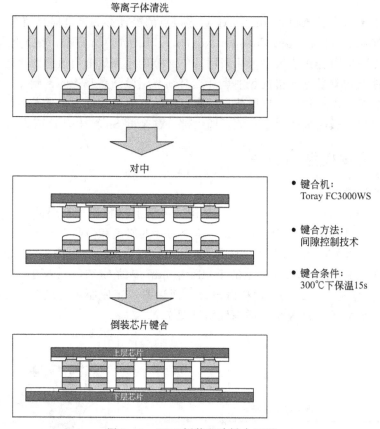

图 7.10 COC 倒装芯片键合过程

（3）退火处理

为了研究 SnAgCuNi 微凸点的电迁移行为，选择两种微凸点进行研究：类型 I 为普通微凸点，类型 II 为退火后的微凸点。相比于 I 型微凸点，II 型为经过了 300℃下保温 15s 的退火处理，微凸点中形成了 IMC。这两种微凸点的区别如图 7.11 所示，退火处理后 IMC 厚度增加了。

（4）电迁移测试与数值分析

为了精确测定单个微凸点的电阻值，采用四点开尔文结构进行电迁移测试。对 COC 试样通以高电流，环境温度为 150℃，现场测量并记录测试过程中的电压和电阻值，根据电阻值的变化监测微凸点互连的失效时间。

采用 ANSYS 有限元软件建立 3D 有限元模型计算微凸点内的电流密度分布，发现最大电流密度与微凸点的 EM 寿命相关。3D 有限元模型如图 7.12 所示。

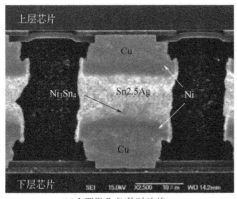
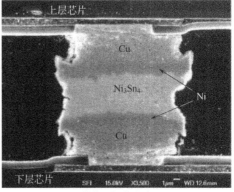

(a) Ⅰ型微凸点(热时效前)　　　　　　(b) Ⅱ型微凸点(热时效后)

图 7.11　(a) Ⅰ型微凸点（退火前）的截面图；(b) Ⅱ型焊锡接点（退火后）的截面图

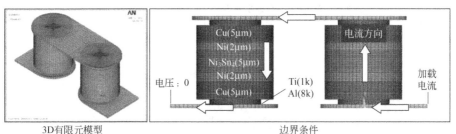

3D有限元模型　　　　　　边界条件

材料参数

	Cu	Ni	Al线	Ni_3Sn_4	Ti
电阻率 /mΩ·cm	1.7	6.8	2.7	28.5	43.1

图 7.12　3D 有限元模型

7.3.2　结果与讨论

（1）电性能

准备好Ⅰ型试样之后，采用四点电阻测量法精确测定单个微凸点的电阻值。利用四点开尔文结构测量试样通以－10～10mA 电流时的电压值，单个微凸点的电阻值可以根据电压-电流关系计算得到，结果如图 7.13 所示。可以看出，当电流值超过 5mA 时，电阻值接近 15mΩ；当电流低于±5mA 时，电阻值变化迅速。

对经过退火处理的Ⅱ型 COC 试样进行电性能测试，两种微凸点的电阻值对比情况如图 7.14 所示。当所加电流超过±5mA 时，Ⅱ型微凸点的电阻值趋于稳定，接近 20mΩ；当所加电流低于±5mA 时，Ⅱ型微凸点的平均电阻值比Ⅰ型微凸点更加稳定。原因是Ⅱ型微凸点中 IMC 的转化十分彻底，使得电阻值更高且更为稳定，尤其在电流极小的情况下。

（2）数值分析结果

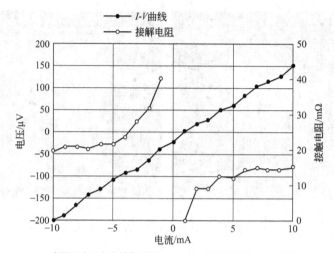

- 利用四点开尔文结构测量−10~10mA电流所对应的电压值；
- 单个微凸点的电阻值可以根据电压-电流关系计算得到；
- 当电流值≥5mA时，电阻值接近15mΩ；
- 当电流值<±5mA时，电阻值的发散程度呈增大趋势。

图7.13　四点电阻测量结果（Ⅰ型）

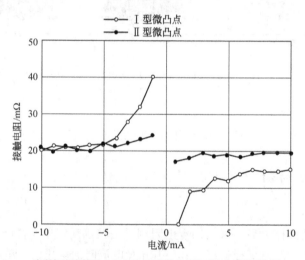

- 当所加电流低于±5mA时，Ⅱ型微凸点的平均电阻值比Ⅰ型微焊点更加稳定；
- Ⅱ型微凸点中的IMC转化更加完全，使其电阻值更高且更为稳定，尤其在电流极小的情况下。

图7.14　Ⅰ型与Ⅱ型微凸点四点电阻测量结果的对比

图7.15(a) 所示为加载电流为0.13A情况下，微凸点中的电流密度分布。分析结果表明，最大电流密度发生在靠近Cu UBM的Al线拐角处，尤其当电流从拐角流入并流出时，该现象更为明显。采用有限元方法分析了不同电流加载条件下各材料中的最大电流密度，当电流为0.13A时，Al线、UBM以及凸点上的最大电

流密度分别为 $1.25 \times 10^6 \mathrm{A/cm^2}$、$6.94 \times 10^5 \mathrm{A/cm^2}$、$5.96 \times 10^4 \mathrm{A/cm^2}$。此外，最大电流密度 J_{max} 与加载电流的关系曲线如图 7.15(b) 所示。从图中可以看出：①加载电流越大，J_{max} 越大；②Al 线中的 J_{max} 高于 UBM 和凸点中的 J_{max}；③凸点中的 J_{max} 最小；④J_{max} 与加载电流呈线性关系。

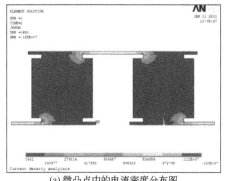

(a) 微凸点中的电流密度分布图
(b) 最大电流密度与加载电流的关系曲线

- 加载电流越大，J_{max} 越大；
- Al 线中的 J_{max} 高于 UBM 和凸点中的 J_{max}；
- 凸点中的 J_{max} 最小；
- J_{max} 与加载电流呈线性关系；
- 即使凸点中 J_{max} 小于 $10^4 \mathrm{A/cm^2}$，Al 线中的 J_{max} 也可能大于 $10^6 \mathrm{A/cm^2}$。

图 7.15 数值分析结果

（3）实验结果

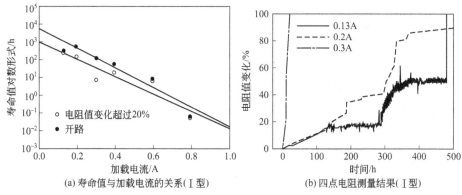

(a) 寿命值与加载电流的关系（Ⅰ型）
(b) 四点电阻测量结果（Ⅰ型）

- 加载高电流时，根据电阻值变化 20% 测得的试样寿命值小于电路开路时测得的寿命值，但两者很接近；
- 随着加载电流减小，电路开路时测得寿命值远高于其他方法测得的寿命值。

图 7.16 Ⅰ型微凸点

采用四点开尔文结构进行 EM 测试的实验结果如图 7.16 和图 7.17 所示，给出了两种寿命时间：一种是当试样完全开路时测得的寿命；另一种是当试样电阻值的变化超过 20% 时测得的寿命。在加载电流分别为 0.13A、0.2A、0.3A、0.4A、

0.6A 和 0.8A 的情况下，所有的 I 型微凸点在 600h 内均发生开路，但在达到开路失效时间的一半时，其电阻值变化已经超过 20%。图 7.16 中的曲线描绘了各种加载电流下 I 型微凸点的寿命，其中 Y 轴取寿命值的对数形式。从图中可以看出，加载高电流时，根据电阻值变化 20% 测得的试样寿命值小于电路开路时测得的寿命值，但两者很接近。随着加载电流减小，电路开路时测得寿命值远高于其他方法测得的寿命值。如图 7.16 中所示，当加载电流高于 0.3A 时，电阻值急剧变化，并且试样 100h 内发生失效。加载 0.13A 的电流 292h 后，试样的电阻值变化超过 20%，尽管试样在加载 480h 后仍未失效。

当电流高于 0.3A 时，Al 线由于焦耳热的作用发生融化，这是导致 I 型微凸点互连失效的主要原因，焊点电阻值变化很快直至失效。当加载电流低于 0.2A 时，焦耳热的影响减小，并且焊点电阻值得变化也较为稳定。当电流为 0.2A 时，如图 7.16(a) 所示的两种方法测得的寿命值相差很大。失效发生在 Al 线、UBM 以及凸点处。

由于退火处理过程中 II 型微凸点内部有 IMC 形成，所以抵抗 EM 能力较强，测得的结果如图 7.17 所示。电流密度相同的情况下，未经过退火处理的 I 型微凸点其寿命值明显低于 II 型微凸点。当加载电流较大时，根据电阻值变化 20% 测得的试样寿命值小于电路开路时测得的寿命值。而当加载电流较小时，根据电路开路时测得寿命值高于其他方法测得的寿命值，但结果相差不大。当电流为 0.2A 和 0.13A 时，试样均未发生失效，电阻值变化低于 10%，故图 7.17 中未给出其失效曲线。

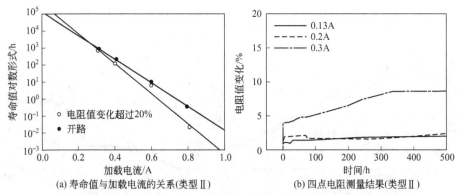

(a) 寿命值与加载电流的关系(类型 II)　　(b) 四点电阻测量结果(类型 II)

● 电流密度相同的情况下，未经过热退火处理的 I 型微凸点其寿命值明显低于 II 型微凸点；
● 当加载电流较大时，根据电阻值变化 20% 测得的试样寿命值小于电路开路时测得的寿命值。

图 7.17　II 型微凸点

如图 7.17 所示，加载 0.3A 的电流 926h 后，II 型微凸点发生失效；加载 0.2A 的电流 1470h 后，II 型微凸点仍未失效。加载过程中，焊点电阻值的变化均在 10% 以内。当加载电流为 0.13A 时，电阻值变化仍低于 10%。由于抵抗 EM 的能力较强，所以 II 型微凸点的电阻值能够长时间保持稳定，可以推测其寿命将超过 1000h。

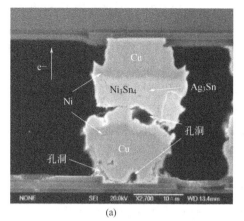 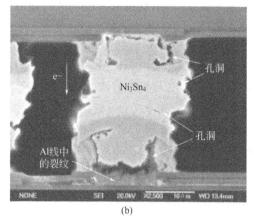

(a)　　　　　　　　　　　　　　　(b)

- 试样失效，失效部位于Al线、UBM以及凸点处；
- 图(a)中微凸点中部和阴极一侧的UBM中出现孔洞；
- 图(b)中微凸点阴极和阳极两侧的UBM中均有裂纹产生，并且阴极一侧的Al线发生融化现象。

图 7.18　0.13A 的电流加载 484h 后，Ⅰ型微凸点的失效模式

（4）失效模式

加载电流为 0.13A 时，Ⅰ型微凸点截面的 SEM 图像如图 7.18 和图 7.19 所示，电子流方向为箭头所指方向。当电流密度较高时，由于电流聚集导致的热时效使得微凸点中剩余的焊料逐渐转变为 IMC。根据这两张 SEM 图像可以发现，电迁

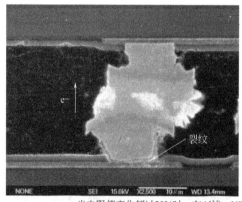 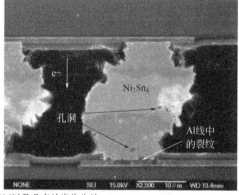

当电阻值变化超过20%时，在Al线、UBM以及凸点处发生失效

图 7.19　Ⅰ型微凸点的失效模式（加载电流＝0.13A，电阻变化＞20%）

移测试后结构中存在三种缺陷：焊料中的孔洞、UBM 中的孔洞以及 Al 融化导致的裂纹。图 7.18(a) 中微凸点自下而上通入 0.13A 的电流，发现在微凸点的中部和阴极一侧的 UBM 中出现孔洞。

图 7.18(b) 中微凸点自上而下通入电子流，阴极和阳极两侧的 UBM 中均有裂纹产生，并且阴极一侧的 Al 线发生融化现象。文献［41］给出的实验数据表明，当 Al 互连中的电流密度达到 10^6 A/cm^2 时，Al 互连的阴极和阳极均会发生电迁移破坏。本研究中，Al 线中的电流密度达到了 1.25×10^6 A/cm^2，电流聚集产生

的焦耳热使得 Al 线温度升高,从而加剧了 Al 线中的电迁移,同时加快了 Ni 和 Cu 的扩散。

当加载电流为 0.13A 时,Ⅰ型微凸点电阻值的变化超过 20%,其截面的 SEM 图像如图 7.19 所示。可以看到微凸点阴极和阳极处均有孔洞产生,20% 的微凸点中都发现了 Al 线的电迁移损伤。加载电流为 0.3A 时Ⅱ型微凸点的失效机制与上述相似。这表明在 COC 封装结构中,Al 线中的电流密度依然可以达到很高的水平。通过对比发现,较大焊点中电迁移造成的孔洞主要在阴极附近,但微凸点不遵循这一规律。

(5) 菊花链测试

除了对试样进行电迁移测试以外,还对菊花链进行了初步的电迁移测试。与单对微凸点的电迁移测试不同,由 Al 线产生的焦耳热使得测试温度大大升高。当炉温为 56℃ 时,将四条菊花链通以 0.12A 的电流,单个试样的表面温度能够达到 150℃。因此,在相同的电迁移测试条件下,菊花链的寿命比单对微凸点的寿命更短,而且菊花链的失效机制也与微凸点不同。为了研究电流引起 Ni 消耗带来的极化效应,将一条连接Ⅰ型微凸点的菊花链在 150℃ 下通以 0.12A 的电流,通电时间为 1 天,然后抛光并与相邻的菊花链进行对比。由于 Si 芯片具有较高的热导率,可以假设相邻菊花链承受相同的温度。

如图 7.20 和图 7.21 所示的四个微凸点,凸点 A 和 B 位于通电菊花链上,凸点 C 和 D 位于相邻的未通电菊花链上。与凸点 C 和 D 相比,通电凸点阴极处的 Ni 层更薄且短,趋于消失。

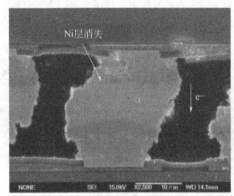

(a) 凸点A   (b) 凸点B

- 将一条连接Ⅰ型微凸点的菊花链在150℃下通以0.12A的电流,通电时间为1天,然后进行抛光;
- 凸点阴极处的Ni层变得更薄且短,趋于消失;
- 在高温条件下,材料相变更容易导致微凸点失效。

图 7.20 通以电流的菊花链结构

正如人们所知的,COC 试样微凸点中的电流聚集效应也许没有倒装芯片上的焊锡接点严重。然而,在电迁移测试中仍然存在两个比较重要的可靠性问题:①高电流密度下 Al 线中的电迁移问题;②高温条件下材料相变引起的焊锡凸点失效

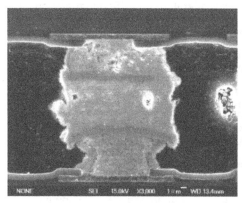
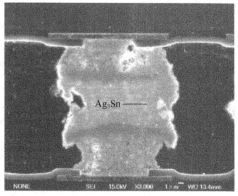

(a) 凸点 C　　　　　　　　　　　　(b) 凸点 D

不通电流时，微凸点中未发现材料相变引起的失效

图 7.21　未通电流的菊花链结构

问题。

7.3.3　总结与建议

本研究中，对节距 30μm 的 COC 微凸点进行了电迁移测试，制作了 I 型（未退火）和 II 型（退火）两种微凸点。重要的结果及建议总结如下：

① 相比于未经退火处理的微凸点，退火处理的微凸点具有更稳定、更高的抵抗电迁移的能力。

② 电迁移测试中，当电流密度超过 $10^4 A/cm^2$ 时，I 型微凸点值电阻迅速增大，600h 后失效。此外，II 型微凸点具有较高的抵抗电迁移的能力，比 I 型微凸点寿命更高。

③ 电迁移失效发生在 Al 线以及 UBM 处，说明微凸点互连中 Al 线的电流密度较高，失效主要是电流聚集导致的焦耳热引起的。Al 线中的电流密度过高可能是微凸点互连的关键可靠性问题之一。

④ 研究发现，焊点中 Ni 层引起的失效情况与菊花链中电子流的方向有关。材料相变引起的失效是另一个重要的细节距微凸点互连可靠性问题。

⑤ 在进一步的电迁移测试中，Al 线中的电流密度应小于 $10^6 A/cm^2$，可使电迁移测试持续的时间更长。此外，增加加载通道有助于提高效率。

7.4　参考文献

[1] Tu, K. U., "Recent Advances on Electromigration in Very Large-Scale-Integration of Interconnects," *J. Appl. Phys.*, Vol. 94, No. 9, 2003, pp. 5451–5473.

[2] Lin, K. L., and G. P. Lin, "The Electromigration Investigation on the Newly Developed Pb-Free Sn-Zn-Ag-Al-Ga Solder Ball Interconnect," *IEEE/ECTC Proceedings*, Reno, NV, May 29–June 1, 2007, pp. 1467–1472.

[3] Lin, K. L., and S. M. Kuo, "The Electromigration and Thermomigration Behaviors of Pb-Free, Flip-Chip Sn-3Ag-0.5Cu Solder Bumps." In *IEEE/ECTC Proceedings*, San Diego, CA, May 30–June 2, 2006, pp. 667–672.

[4] Nah, J. W., J. O. Suh, K. N. Tu, S. W. Yoon, C. T. Chong, V. Kripesh, B. R. Su, and C. Chen, "Electromigration in Pb-Free Solder Bumps with Cu Column as Flip-Chip Joints," *IEEE/ECTC Proceedings*, San Diego, CA, May 30–June 2, 2006, pp. 657–662.

[5] Ebersberger, B., R. Bauer, and L. Alexa, "Reliability of Lead-Free SnAg Solder Bumps: Influence of Electromigration and Temperature," *IEEE/ECTC Proceedings*, Lake Buena Vista, FL, May 31–June 3, 2005, pp. 1407–1415.

[6] Rinne, G., "Electromigration in SnPb and Pb-Free Solder Bumps," *IEEE/ECTC Proceedings*, Las Vegas, NV, June 1–4, 2004, pp. 974–978.

[7] Jang, S. Y., J. Wolf, W. S. Kwon, and K. W. Paik, UBM (Under Bump Metallization) Study for Pb-Free Electroplating Bumping: Interface Reaction and Electromigration," *IEEE/ECTC Proceedings*, San Diego, CA, May 28–31, 2002, pp. 1213–1220.

[8] Gan, H., and K. Tu, "Effect of Electromigration on Intermetallic Compound Formation in Pb-free Solder-Cu Interfaces." In *IEEE Proceedings of Electronic & Components Technology Conference*, San Diego, May 28–31, 2002, pp. 1206–1212.

[9] Huang, Z., Chatterjee, R., Justison, P., Hernandez, R., Pozder, S., Jain, A., Acosta, E., Gajewski, D. A., Mathew, V., and Jones, R. E. "Electromigration of Cu-Sn-Cu Micropads in 3D Interconnect," *IEEE/ECTC Proceedings*, Lake Buena Vista, FL, May 27–30, 2008, pp. 12–17.

[10] Rinne, G., "Emerging Issues in the Physics and Control of Electromigration in SnPb and Pb-Free Solder Bumps," *IEEE/ECTC Proceedings*, Singapore, December 10–12, 2003, pp. 72–76.

[11] Su, P., T. Uehling, D. Wontor, M. Ding, and P. S. Ho, "An Evaluation of Electromigration Performance of SnPb and Pb-free Flip Chip Solder Joints," *IEEE/ECTC Proceedings*, Lake Buena Vista, FL, May 31–June 3, 2005, pp. 1431–1436.

[12] Lu, M., D. Y. Shih, P. Lauro, S. Kang, C. Goldsmith, and S. K. Seo, "The Effects of Ag, Cu Compositions and Zn Doping on the Electromigration Performance of Pb-Free Solders," *IEEE/ECTC Proceedings*, San Diego, CA, May 25–29, 2009, pp. 922–929.

[13] Nicholls, L., R. Darveaux, A. Syed, S. Loo, T. Y. Tee, T. A. Wassick, and B. Batchelor, "Comparative Electromigration Performance of Pb-Free Flip Chip Joints with Varying Board Surface Condition," *IEEE/ECTC Proceedings*, San Diego, CA, May 25–29, 2009, pp. 914–921.

[14] Su, P., L. Li, Y. S. Lai, Y. T. Chiu, and C. L. Kao, "A Comparison Study of Electromigration Performance of Pb-Free, Flip-Chip Solder Bumps," *IEEE/ECTC Proceedings*, San Diego, CA, May 25–29, 2009, pp. 903–908.

[15] Lee, J. H., G. T. Lim, Y. B. Park, S. T. Yang, M. S. Suh, Q. H. Chung, and K. Y. Byun, "Size Effect on Electromigration Reliability of Pb-Free, Flip-Chip Solder Bump," *IEEE/ECTC Proceedings*, Lake Buena Vista, FL, May 27–30, 2008, pp. 2030–2034.

[16] Lu, M., P. Lauro, D. Y. Shih, R. Polastre, C. Goldsmith, D. W. Henderson, H. Zhang, and M. G. Cho, "Comparison of Electromigration Performance for Pb-Free Solders and Surface Finishes with Ni UBM," *IEEE/ECTC Proceedings*, Lake Buena Vista, FL, May 27–30, 2008, pp. 360–365.

[17] Chae, S. H., B. Chao, X. Zhang, J. Im, and P. S. Ho, "Investigation of Intermetallic Compound Growth Enhanced by Electromigration in Pb-Free Solder Joints," *IEEE/ECTC Proceedings*, Reno, NV, May 29–June 1, 2007, pp. 1442–1449.

[18] Chae, S. H., J. Im, P. S. Ho, and T. Uehling, "Effects of UBM Thickness, Contact Trace Structure and Solder Joint Scaling on Electromigration Reliability of Pb-Free Solder Joints," *IEEE/ECTC Proceedings*, Lake Buena Vista, FL, May 27–30, 2008, pp. 354–359.

[19] Chae, S. H., X. Zhang, H. L. Chao, K. H. Lu, P. S. Ho, M. Ding, P. Su, T. Uehling, and L. N. Ramanathan, "Electromigration Lifetime Statistics for Pb-Free Solder Joints with Cu and Ni UBM in Plastic Flip-Chip Packages," *IEEE/ECTC Proceedings*, San Diego, CA, May 30–June 2, 2006, pp. 650–656.

[20] Kwon, Y. M., and K. W. Paik, "Electromigration of Pb-Free Solder Flip-Chip Using Electroless Ni-P/Au UBM," *IEEE/ECTC Proceedings*, Reno, NV, May 29–June 1, 2007, pp. 1472–1477.

[21] Ebersberger, B., R. Bauer, and L. Alexa, "Reliability of Lead-Free SnAg Solder Bumps: Influence of Electromigration and Temperature," *IEEE/ECTC Proceedings*, Lake Buena Vista, FL, May 31–June 3, 2005, pp. 1407–1415.

[22] Iwasaki, T., M. Watanabe, S. Baba, Y. Hatanaka, S. Idaka, Y. Yoloyama, and M. Kimura, "Development of 30-Micron Pitch Bump Interconnections for COC-FCBGA," *Proceedings of the 56th Electronic Components and Technology Conference*, 2006, pp. 1216–1222.

[23] Gan, H., S. L. Wright, R. Polastre, L. P. Buchwalter, R. Horton, R. Horton, P. S. Andry, C. Patel, C. Tsang, J. Knickerbocker, E. Sprogis, A. Pavlova, S. K. Kang, and K. W. Lee, "Pb-Free Micro-Joints (50-μm Pitch) for the Next Generation Micro-Systems: The Fabrication, Assembly and Characterization," *Proceedings of the 56th Electronic Components and Technology Conference*, 2006, pp. 1210–1215.

[24] Sunohara, M., T. Tokayunaga, T. Kurihara, and M. Higashi, "Silicon Interposer with TSVs (Through-Silicon Vias) and Fine Multilayer Wiring," *IEEE/ECTC Proceedings*, Lake Buena Vista, FL, May 27–30, 2008, pp. 847–852.

[25] Ren, F., X. Zhang, J. W. Nah, K. N. Tu, L. Xu, and J. H. L. Pang, "In-Situ Study of the Effect of Electromigration on Strain Evolution and Mechanical Property Change in Lead-Free Solder Joints," *IEEE/ECTC Proceedings* San Diego, CA, May 30–June 2, 2006, pp. 1160–1163.

[26] Yeh, E., W. J. Choi, K. N. Tu, P. Elenius, and H. Balkan, "Current-Crowding-Induced Electromigration Failure in Flip Chip Solder Joints," *Appl. Phys. Lett.*, Vol. 80, 2002, pp. 580–582.

[27] Zhang, L., S. Ou, J. Huang, K. N. Tu, S. Gee, and L. Nguyen, "Effect of Current Crowding on Void Propagation at the Interface Between Intermetallic Compound and Solder in Flip Chip Solder Joints," *Appl. Phys. Lett.*, Vol. 88, 2006, pp. 012106–1.

[28] Ding, M., G. Wang, B. Chao, P. S. Ho, P. Su, and T. Uehling, "Effect of Contact Metallization on Electromigration Reliability of Pb-Free Solder Joints," *J. Appl. Phys.*, Vol. 99, 2006, pp. 094906–094906-6.

[29] Nah, J., J. O. Suh, K. N. Tu, S. W. Yoon, V. S. Rao, V. Kripesh, and F. Hua, "Electromigration in Flip Chip Solder Joints Having a Thick Cu Column Bump *and a Shallow* Solder Interconnect," *J. Appl. Phys.*, Vol. 100, 2006, pp. 123513–123513-5.

[30] Lai, Y., Y. T. Chia, C. W. Lee, Y. H. Shao, and J. Chen, "Electromigration Reliability and Morphologies of Cu Pillar Flip-Chip Solder Joints," *IEEE/ECTC Proceedings*, Lake Buena Vista, FL, May 27–30, 2008, pp. 330–335.

[31] Xu, L., J. K. Han, J. J. Liang, K. N. Tu, and Y. S. Lai, "Electromigration Induced High Fraction of Compound Formation in SnAgCu Flip Chip Joints with Copper Column," *Appl. Phys. Lett.*, Vol. 92, 2008, pp. 262104–262104-3.

[32] Chang, Y., T. H. Chiang, and C. Chen, "Effect of Voids Propagation on Bump Resistance Due to Electromigration in Flip-Chip Solder Joints Using Kelvin Structure," *Appl. Phys. Lett.*, Vol. 91, 2007, pp. 132113–132113-3.

[33] Yeo, A., B. Ebersberger, and C. Lee, " Consideration of Temperature and Current Stress Testing on Flip Chip Solder Interconnects," *Miroelectron Reliab.*, Vol. 48, 2008, pp. 1847–1856.

[34] Lu, M., D. Y. Shih, P. Lauro, C. Goldsmith, and D. W. Henderson, "Effect of Sn Grain Orientation on Electromigration Mechanism in High Sn-Based Pb-Free Solders," *Appl. Phys. Lett.*, Vol. 92, 2008, pp. 211909–211909-3.

[35] Yang, P., C. C. Kuo, and C. Chen, "The Effect of Pre-Aging on the Electromigration of Flip-Chip Sn-Ag Solder Joints," *J Mater.*, 2008, pp. 77–80.

[36] Laurila, T., V. Vuorinen, and J. K. Kivilahti, "Interfacial Reactions Between Lead-Free Solders and Common Base Materials," *Mater. Sci. Eng.*, Vol. R49, 2005, pp. 1–60.

[37] Wei, C., C. F. Chen, P. C. Liu, and C. Chen, "Electromigration in Sn-Cu Intermetallic Compounds," *J. Appl. Phys.*, Vol. 105, 2009, pp. 023715–023715-41.

[38] Zeng, K., R. Stierman, T. Chiu, D. Edwards, K. Ano, K. N. Tu, "Kirkendall Voids Formation in Eutectic SnPb Solder Joints on Bare Cu and Its Effects on Joint Reliability," *J. Appl. Phys.*, Vol. 97, 2005, pp. 024508–024508-8.

[39] Lee, C., P. Wang, and J. Kim, "Are Intermetallics in Solder Joints Really Brittle?" *IEEE/ECTC Proceedings*, Reno, NV, May 29–June 1, 2007, pp. 648–652.

[40] Labie, R., P. Limaye, K. W. Lee, C. J. Berry, E. Beyne, and I. D. Wolf, "Reliability Testing of Cu-Sn Intermetallic Micro-Bump Interconnections for 3D-Device Stacking," *International Interconnect Technology Conference*, 2008, pp. 19–21.

[41] Ouyang, F.-Y., K. N. Tu, C.-L. Kao, and Y.-S. Lai, "Effect of Electromigration in the Anodic Al Interconnect on Melting of Flip Chip Solder Joints," *Applied Physics Letters*, Vol. 90, No. 2, 2007, pp. 294–296.

[42] Liu, C., J. Chen, Y. Chuang, L. Ke, and S. Wang, "Electromigration-induced Kirkendall voids at the Cu/Cu3Sn Interface in Flip-Chip Cu/Sn/Cu Joints," *Appl. Phys. Lett.*, Vol. 90, Issue 11, 2007, pp. 112114–112114-3.

[43] Yu, D., T. Chai, M. Thew, Y. Ong, V. Rao, L. Wai, and J. H. Lau, "Electromigration Study of 50 μm Pitch Micro Solder Bumps Using Four-Point Kelvin Structure," *IEEE/ECTC Proceedings*, San Diego, CA, May 25–29, 2009, pp. 930–935.

[44] Lin, Y., C. Zhan, J. Juang, J. H. Lau, T. Chen, R. Lo, M. Kao, T. Tian, and K. N. Tu, "Electromigration in Ni/Sn Intermetallic Micro Bump Joint for 3D IC Chip Stacking", *IEEE/ECTC Proceedings*, Orlando, FL, May 2011, pp. 351–357.

[45] Frear, D., H. Morgan, S. Burchett, and J. H. Lau, *The Mechanics of Solder Alloy Interconnects*, Van Nostrand Reinhold, New York, 1994.

第 8 章 芯片到芯片、芯片到晶圆、晶圆到晶圆键合

8.1 引言

对于 3D IC 集成，一般采用芯片到芯片（C2C）或者芯片到晶圆（C2W）键合技术。由于芯片良率与芯片尺寸的不同，除了应用于 MEMS 之外，很少采用晶圆到晶圆（W2W）键合技术。大多数现有的键合技术其温度都超过 300℃[1~3]，如此高的温度需要额外的冷却工序，致使产率较低。因此，需要通过低温键合技术[4~39]缩短生产时间。另外，低温键合可以减小由于键合结构热失配引起的 3D IC 集成器件的损伤。

硅（Si）的 C2C、C2W 以及 W2W 键合温度可以低至室温[13]，例如表面活化键合工艺。但是这种键合技术要求键合表面足够地平整和洁净，使得大规模生产受到限制。本章主要讨论低熔点焊料（如 InAg、InCu、InSn、InNi 以及 InSnCu 等）的键合技术，其键合温度低于 200℃。

8.2 低温焊料键合基本原理

低温键合的基本流程如图 8.1 所示。低温键合是比较完备的工艺，例如瞬态液相（TLP）键合、固液互扩散（SLID）键合、金属间化合物（IMC）键合。将低熔点焊料，如 InAg、InCu、InSn、InNi、InAu、InSnAu、InCuNi 以及 InSnCu 等，覆盖在特制的 UBM 上，完成对中与低温键合之后，所有焊料将与芯片/晶圆上的 UBM 反应生成金属间化合物（IMC），其熔点比焊料熔点高出几百摄氏度。UBM 一般采用 Cu、Cu/Au、Ti/Au、Ni/Au、Ti/Cu 以及 Cu/Ti/Au 等。

IMC 的高熔点特性正是 3D IC 芯片堆叠封装以及 3D MEMS 封装所需要的。例如，将两个芯片低温键合之后，焊料转变成熔点更高的 IMC，当第三个芯片与这两个芯片再次低温键合时，之前形成的互连不会发生融化或移动。又比如，利用低熔点焊料将 MEMS 器件键合到专用 IC（ASIC）晶圆上之后，所键合区域都会形成高熔点的 IMC，当另一晶圆与已经形成互连的 MEMS 器件和 ASIC 晶圆进行键合时，MEMS 器件和 ASIC 晶圆之间的互连不会再融化或移动。此外，采用表面贴装工艺（SMT）和无铅焊料将整个 3D IC 芯片堆叠封装或者 3D MEMS 封装

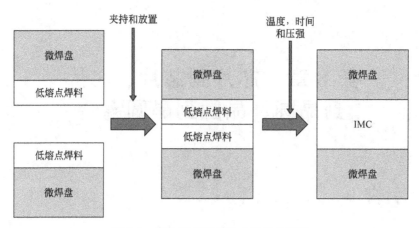

图 8.1 瞬态液相键合技术的基本原理

组装到印刷电路板（PCB）上时，最高回流温度为 260℃，此过程中 3D 堆叠芯片封装中的芯片与芯片之间以及 3D MEMS 中的 ASIC 与 MEMS 之间的 IMC 也不会发生融化或移动。

8.3 低温 C2C 键合 [$(SiO_2/Si_3N_4/Ti/Cu)$ 到 $(SiO_2/Si_3N_4/Ti/Cu/In/Sn/Au)$]

8.3.1 测试模型

图 8.2 和图 8.3 所示为气密性 C2C 和面对面低温键合的示意图。采用 C2W 键合机进行 C2C 键合（如图 8.4 所示），基芯片上键合环的尺寸为 8mm×8mm，厚度为 300μm，种子层（SiO_2、Si_3N_4、Ti、Cu）以及焊料层（In、Sn、Au）结构如图 8.3 所示[15,16]。首先采用连续蒸镀工艺从硅晶圆表面沿着键合环制作种子层，然后电镀 Cu，最后利用电子束连续蒸镀工艺制作焊料层。Cu 层的厚度为 3μm，In 层厚度为 1.6μm，Sn 层厚度为 1.4μm，Au 层厚度为 50nm。为了防止

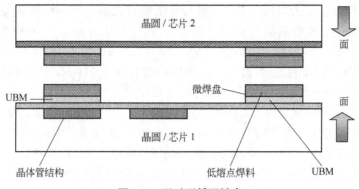

图 8.2 面对面低温键合

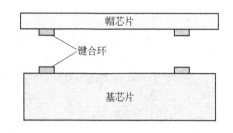
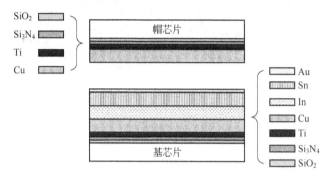

图 8.3 含有键合环的帽芯片与基芯片（上部）
以及键合环的多层结构示意图

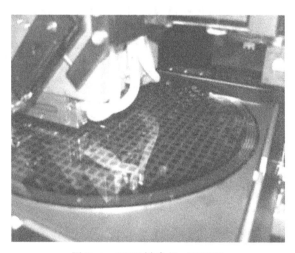

图 8.4 C2W 键合机（SUSS）

试样从真空室取出暴露在空气中时有氧气渗入，需要在 Sn 层上沉积一层极薄的 Au 层。基芯片和帽芯片尺寸均为 10mm×10mm，但帽芯片上没有 In、Sn、Au 层。

图 8.5 所示为蒸镀/电镀后键合环的俯视图及剖面图。从图中可以看出，键合环表面十分粗糙，呈颗粒状结构，其晶粒尺寸为 2~3μm。对这种粗糙的表面进行键合时，需利用熔融焊料层以及加高压的方法填充键合面间的间隙。

图 8.5 蒸镀涂层的键合环表面（左图）及剖面（右图）的 SEM 图像

8.3.2 拉力测试结果

设定键合压强为 1.5MPa，通过拉力测试测得的不同键合温度、键合时间下的键合强度如图 8.6 所示。从图中可以看出，键合温度越高、键合时间越长，则键合强度越高。图 8.7 所示为键合压强为 1.5MPa、键合温度为 180℃、键合时间为 20min 制得的键合对的剖面图。可以看出，互连厚度沿整个键合区域都较为均匀，为 $(6.2\pm0.5)\mu m$；键合对的互连情况较好，未观察到分界线，这说明焊料中间层在键合过程中完全融化并填补了颗粒间隙。

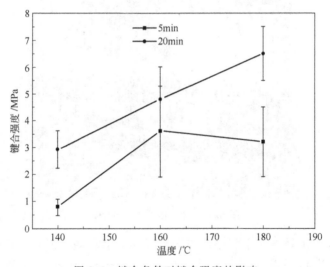

图 8.6 键合条件对键合强度的影响

键合压强为 1.5MPa、键合温度为 180℃、键合时间为 20min 制得的芯片键合界面，其声像图如图 8.8 所示。整个键合环为灰色，说明键合对接触紧密，无孔洞或分层产生。氦气泄漏率测试结果表明，键合对的密封性较好，5 个试样的平均泄漏率为 5.8×10^{-9} $(atm \cdot mL)/s$。

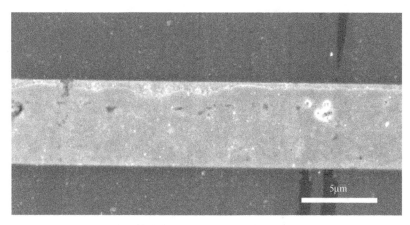

图 8.7　键合压强 1.5MPa、键合温度 180℃、
键合时间 20min 制得的键合对剖面图

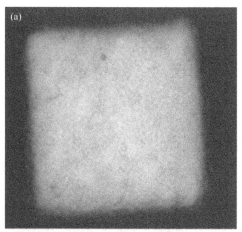

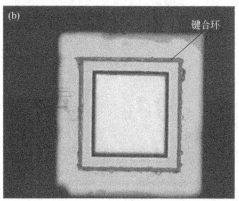

图 8.8　键合效果较好的平整芯片声像图：(a) 键合效果较好的
平芯片；(b) 键合效果较好的键合环（键合压强为 1.5MPa、
键合温度为 180℃、键合时间为 20min）

8.3.3　X射线衍射与透射电镜观察结果

利用X射线衍射（XRD）观察键合对在拉力测试后的断面形貌，检测键合反应区中IMC的成分。结果发现断面生成了$AuIn_2$、Cu_6Sn_5以及$Cu_{11}In_9$相，但未发现Sn和In元素的峰值，如图8.9所示。

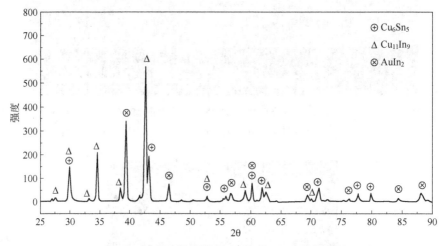

图8.9　键合对拉伸测试后界面的XRD分析结果

由于IMC层相对较薄，所以采用透射电镜（TEM）对其进行局部观察。图8.10为基底硅附近区域的TEM图，图8.11为Cu与焊料中间层之间界面反应区的TEM图。如图8.10所示，基底硅上第一层为SiO_2，然后依次为Si_3N_4、Ti和

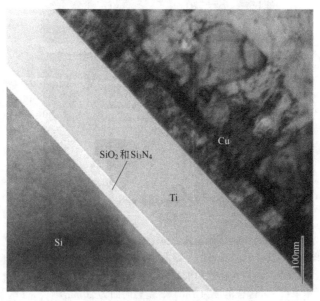

图8.10　基底硅附近区域的TEM图像

Cu。各界面层都无孔洞且厚度均匀,SiO_2 和 Si_3N_4 层在基底硅与后续金属层之间起到了很好的黏结作用。

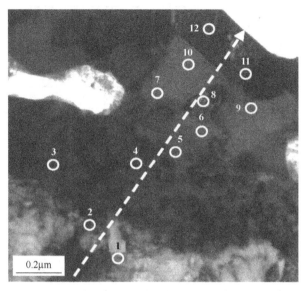

图 8.11 界面区的 TEM 图像

图 8.11 所示为 Cu 与 SnIn 焊料中间层之间的反应区,可以清楚地看到小晶粒结构。测试点的元素组成见表 8.1,主反应区内存在两种 IMC,即 $Cu_6(Sn,In)_5$ 和 $Cu_{11}(In,Sn)_9$,与 XRD 分析结果一致,证明在键合界面处的确存在两种 IMC。从近 Cu 区域到近 In 区域,Cu 的含量从 56% 逐渐减少到 48%。EDX 分析结果表明,仍然存在未参与反应的 Cu。在键合界面边缘处,Cu 的含量急剧下降,以 In 和 Au 为主要成分,如表 8.1 所示。点 11 和点 12 处的成分分别为 $Cu_{5.9}Au_{34}In_{59}Sn_{1.1}$ 和 $Cu_{5.1}Au_{33}In_{60.2}Sn_{1.7}$,对应于平衡相 $AuIn_2$。

表 8.1 图 8.11 中沿界面上各点处元素组成的 EDX 分析结果

测试点	组成/%				相
	Cu	In	Sn	Au	
1	99.5	0.3	0.2	0	Cu
2	99	0.3	0.2	0.5	Cu
3	56	28.5	15.5	0	$Cu_{11}(In,Sn)_9$
4	54.6	31.3	13.7	0.4	$Cu_{11}(In,Sn)_9$
5	52.1	30.4	17.0	0.5	$Cu_{11}(In,Sn)_9$
6	51.4	28.3	19.6	0.7	$Cu_{11}(In,Sn)_9$
7	50.2	31.2	18.3	0.3	$Cu_{11}(In,Sn)_9$

续表

测试点	组成/%				相
	Cu	In	Sn	Au	
8	48.1	21.0	30.4	0.5	η-$Cu_6(SnIn)_5$
9	49.8	23.4	26.6	0.2	η-$Cu_6(SnIn)_5$
10	50.3	20.7	29.0	0	η-$Cu_6(SnIn)_5$
11	5.9	59.0	1.1	34	$Au(In,Sn)_2$
12	5.1	60.2	1.7	33.0	$Au(In,Sn)_2$

研究中有趣的一点是，整个复合焊料在180℃时融化，低于焊料中Sn的熔点232℃。其原因是熔融态的In层与固态Sn层混合，并在界面形成了较薄的、低熔点的In-Sn层。由In-Sn二元合金相图可知，In-Sn共晶合金的熔点较低，为118℃。室温下，β相的Sn其溶解度在 (15～28)% (质量分数) 之间，γ相的In其溶解度接近于28% (质量分数)，这说明In和Sn相互扩散比较容易。因此，共晶合金In-Sn层的生长消耗了In层和Sn层，直到将这两种焊料层都消耗殆尽。

注意到已制得基芯片上整个初始In-Sn组成为共晶成分，即48% (质量分数) Sn。尽管In与Au也有反应，但只消耗很小一部分的In，这使得In-Sn组成稍微偏离了共晶点，即50% (质量分数) Sn。因此，由于近共晶液态层的存在，蒸镀后的表面粗糙度得到改善，如前文所述，焊点界面也较好。在焊接过程中，在界面处Cu和Sn或In不断地相互扩散，形成不同种类的IMC。中间层在180℃时为完全熔融态，这就解释了为什么在In和Au层之间存在Sn层的情况下，In原子依然能够与Au反应生成$AuIn_2$。

XRD分析结果表明，完成键合之后没有残余的In和Sn，焊料与Cu完全反应生成IMC。这说明低温条件下获得的键合焊点由于高熔点IMC的存在，能够承受较高的服役温度。根据合金相图可知，Cu_6Sn_5、$Cu_{11}In_9$以及$AuIn_2$的熔点分别为430℃、320℃、580℃。由此可以推测，焊点的耐热温度应高于300℃。

8.4 低温C2C键合 [(SiO_2/Ti/Cu/Au/Sn/In/Sn/Au) 到 (SiO_2/Ti/Cu/Sn/In/Sn/Au)]

8.4.1 测试模型

图8.12所示的焊点系统与前面所述的稍有不同[10,11]，其中采用Cu/Au作为UBM材料，键合后将形成混合的In-Cu与In-Au IMC相，由图8.13可以查出

IMC 的组成及其熔点。不同焊料单元的结构以及尺寸如图 8.12 所示，键合压强为 2.5MPa，键合温度为 180℃，键合时间为 20min。键合后焊点无孔洞产生，其截面形貌如图 8.14 所示，差示扫描量热法（DSC）测得的曲线表明焊点的重熔温度高于 400℃。

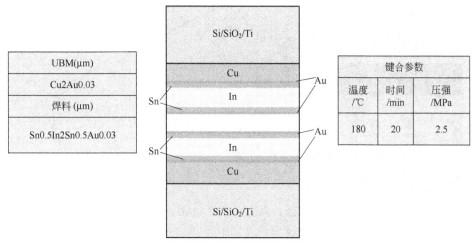

图 8.12 帽芯片与基芯片以及键合环的多层结构简图

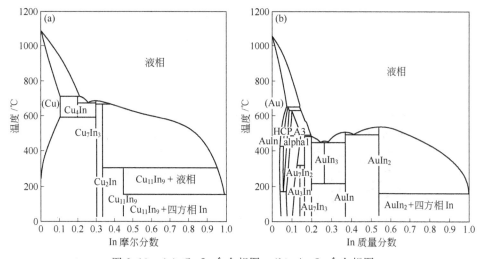

图 8.13 （a）Cu-In 合金相图；（b）Au-In 合金相图

8.4.2 测试结果评估

如图 8.15 所示，试样的氦气泄漏测试和剪切强度测试结果比较理想。此外，低温互连也通过了可靠性评估：如高压蒸煮试验（PCT）、高温存储试验（HTS）以及热循环试验（TCT）等。

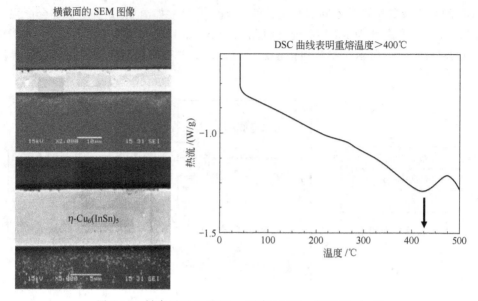

图 8.14 键合压强 2.5MPa、温度 180℃、时间 20min 得到的焊点截面图（左图）及其 DSC 曲线

	密封后	PCT	HTS	TCT
氦泄漏率 /(atm·mL/s)	$<5\times10^{-8}$	$<5\times10^{-8}$	$<5\times10^{-8}$	$<5\times10^{-8}$
剪切强度 /MPa	27	34	28	18

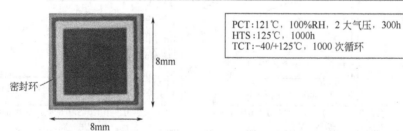

PCT：121℃，100%RH，2 大气压，300h
HTS：125℃，1000h
TCT：-40/+125℃，1000 次循环

图 8.15 氦气泄漏测试以及剪切强度测试后 C2C 键合的可靠性评估

8.5 低温 C2W 键合 [(SiO$_2$/Ti/Au/Sn/In/Au) 到 (SiO$_2$/Ti/Au)]

第一节中已经提到过，3D IC 芯片堆叠一般利用 C2C/C2W 键合方法把芯片键

合到基芯片或其他芯片上。由于IC芯片良率以及芯片尺寸的不同,一般很少采用W2W键合方法。大多数现有的键合技术其温度都超过300℃[1~3]。尽管Si C2C、C2W以及W2W的键合温度可以低至室温[13],然而这类键合技术要求键合表面足够地平整和洁净,使其无法满足大规模生产需要,本书也不予讨论。

当Cu填充的TSV芯片与其他芯片键合时,芯片之间的热失配会引起翘曲并进而导致微结构发生破坏,因此,需要键合温度低于200℃的低熔点焊料,例如InAg、InCu、InSn、InNi、InAu、InSnAu、InCuNi以及InSnCu等,以减少这种破坏。此外,低熔点焊料焊点中可以形成强度较高的IMC,IMC的重熔温度也较高,从而能够保证3D IC芯片堆叠的稳定性。

值得强调的是,对于熔点为156℃的In基焊料,选取合理组分的焊料以及UBM使其只形成高熔点的IMC是十分重要的。由于In原子与其他元素更趋于形成三元共晶相,而三元共晶相的熔点低于In的二元共晶相熔点,所以有必要对焊料层组合进行合理的设计,避免其形成三元共晶相。

8.5.1 焊料设计

对于3D集成的低温键合,已经研究过采用In与Ni、Cu或Au的复合焊料[4~7]。一般情况下,将In层和其他金属层之间嵌入诸如Ti的阻挡层,防止In层与其他金属层之间过多相互扩散[8]。

本节中选取In基焊层作为低温焊料,与薄Au层键合之后形成AuIn基IMC。由于Au扩散至In层的速度很快,即使在室温条件下也可以形成AuIn金属间化合物,因此需要在两者之间嵌入一层薄Sn层,用以减小键合前Au和In的相互扩散。亦即,在键合前Sn与Au反应暂时形成一个扩散阻挡层,阻止Au和In的相互扩散;同时,该阻挡层不会中断In与Au在键合过程中以及键合后完全反应生成IMC。

8.5.2 测试模型

3D IC芯片堆叠的测试模型如图8.16~图8.18所示,尺寸参数见表8.2所示。测试芯片尺寸为8mm×8mm,上面植有近1700个微凸点,每个微凸点由直径为100μm的Ti/Au(0.1/1μm)UBM、直径为80μm的Sn/In(0.5/2μm)焊料以及厚度为0.05μm的Au层组成。另一侧的芯片上面只有尺寸为120μm×120μm的Ti/Au(0.1/1μm)UBM。

测试模型在覆盖有SiO$_2$层的200mm硅晶圆上制成。首先在晶圆正面的SiO$_2$层上溅射0.1μm厚的Ti层以及1μm厚的Au层,然后形成UBM。晶圆背面的加工工艺为:①背磨至200μm;②在晶圆上涂覆SiO$_2$作为阻挡层;③溅射腔内在SiO$_2$层上沉积0.1μm厚的Ti层和1μm厚的Au层;④利用湿法刻蚀制作直径100μm的Ti/Au UBM衬垫;⑤利用20μm的干膜进行压膜;⑥显影制作微凸点开口;⑦蒸镀室内依次在UBM的Au层上沉积0.5μm厚的Sn层以及2μm厚的In

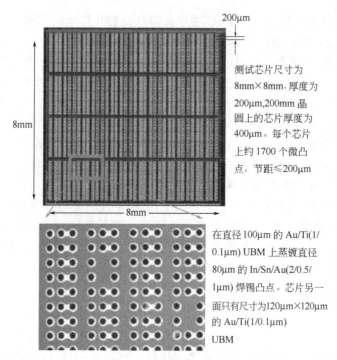

图 8.16 测试模型上表面焊锡凸点及焊盘

测试芯片尺寸为 8mm×8mm,厚度为 200μm,200mm 晶圆上的芯片厚度为 400μm。每个芯片上约 1700 个微凸点,节距≤200μm

在直径 100μm 的 Au/Ti(1/0.1μm) UBM 上蒸镀直径 80μm 的 In/Sn/Au(2/0.5/1μm) 焊锡凸点。芯片另一面只有尺寸为 120μm×120μm 的 Au/Ti(1/0.1μm) UBM

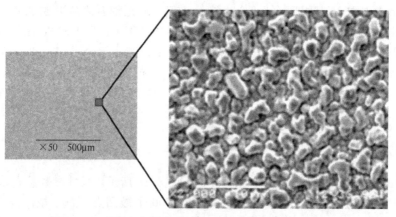

图 8.17 蒸镀法制得的复合涂层表面形貌的 SEM 图像

层;⑧去除干膜。图 8.17 所示为蒸镀法制得的复合涂层表面形貌的 SEM 图像,可以看出其表面比较粗糙,具有颗粒状结构,晶粒尺寸为 2~3μm。对这种粗糙的表面进行键合时,需利用熔融焊料层以及加压的方法填充键合面间的间隙。

表 8.2 测试模型的几何尺寸

几何参数	尺寸
晶圆尺寸	200mm
芯片尺寸	8mm

续表

几何参数	尺寸
堆叠芯片厚度	200μm
基芯片厚度	400μm
焊锡凸点直径	80μm
焊锡凸点节距	≤200μm
凸点数目	约1700
UBM(Ti/Au)	0.1/1μm
凸点中In的高度	2.5μm
凸点整体高度	3.6μm

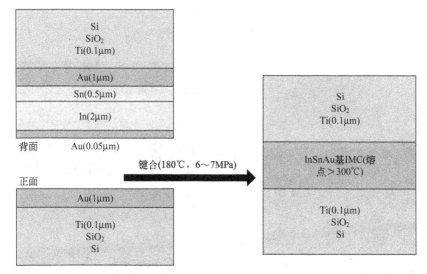

图 8.18 包含UBM、微凸点的芯片与只有UBM的芯片的结构简图（左图），键合后的焊点结构图（右图）

8.5.3 用于3D IC 芯片堆叠的 InSnAu 低温键合

图 8.19 所示为3D IC 芯片堆叠的低温键合组装工艺。第一层为 200mm 的晶圆，其厚度为 400μm，正面有 120μm×120μm 的 UBM。第二层芯片的背面与第一层晶圆的正面通过无铅 InSnAu 焊料凸点形成互连，键合温度为 180℃、压强为 6～7MPa、时间为 45s。键合结束后，无铅焊料转变成 InSnAu IMC，其重熔温度比 In 的熔点高数百摄氏度。然后采用同样的工艺条件，将第三个芯片的背面通过焊锡凸点与第二层芯片的正面进行互连。最后，将堆叠芯片在 120℃ 下退火 12h。采用现有的低温键合工艺制作的三层堆叠芯片如图 8.20 所示，从其剖面图可以看出，键合效果很好且未发现可见孔洞。通过实验设计给出了最佳的键合条件，详见参考文

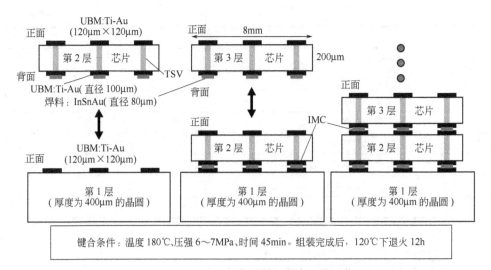

图 8.19 3D 堆叠封装的低温键合组装工艺

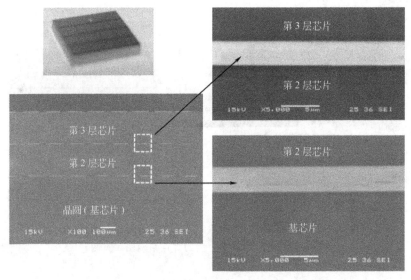

图 8.20 三层堆叠芯片的组装模块及其 SEM 截面图，每个互连层上包含 1700 个微焊点

献 [10]。

8.5.4 InSnAu IMC 层的 SEM、TEM、XDR、DSC 分析

由于 IMC 层相对较薄，所以采用 TEM 技术对其进行局部化学分析。硅基底附近区域、AuTi 与焊料中间层之间的界面反应区域的 TEM 图像如图 8.21 所示，各层之间的界面无孔洞、厚度均匀，且全部转化为 IMC。各测试点元素组成标注在图 8.21 上，主要反应区存在 InAu 和 InSnAu 两种 IMC，与 XRD 分析结果一致，

证明在键合界面处的确存在两种 IMC。图 8.22 所示为焊点的 DSC 曲线,可以看出焊点的重熔温度高于 365℃。此外,图 8.21 所示的 SEM 图像表明焊点键合均匀且未发现缺陷。

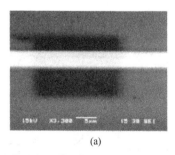

(a)

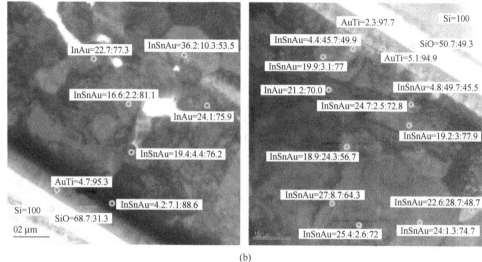

(b)

图 8.21 界面区域的 SEM 图像 (a) 以及 TEM 图像 (b)

8.5.5 InSnAu IMC 层的弹性模量和硬度

采用纳米压痕技术测定 AuSnIn 基 IMC 层的弹性模量及硬度。相对于光学显微镜而言,由于试样横截面上的 IMC 层厚度很小,纳米压痕仪的光学显微镜难以观察到,所以在完成剪切测试后,在芯片的断面上进行纳米压痕测量。测得 InSnAu 的弹性模量约为 81GPa,比 Au(78GPa)的弹性模量稍大[40],但远小于 Cu_6Sn_5(112.6GPa)和 Cu_3Sn(132.7GPa)的弹性模量[41],测得 InSnAu 的硬度接近于 1.5GPa。

8.5.6 三次回流后的 InSnAu IMC 层

至少有两种低温键合组装形式:一种针对芯片级互连,另一种针对板级互连。对于板级互连,由于它是最后一道组装工序,只要焊点满足可靠性要求,就可以利

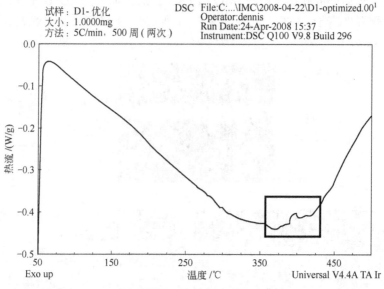

图 8.22　DSC 曲线显示焊点的重熔温度接近 370℃

用低熔点焊料将封装体贴装到基板上。然而对于芯片级封装，由于封装体还需采用无铅焊料组装到基板上，因此在组装到印刷电路板（PCB）上之前必须保证封装体的质量。下面列举了一些质量检测方法。

① 温度循环试验（-40~125℃，1000 个循环），封装需进行预处理，包括湿度暴露以及至少三次无铅 SMT 回流。

② 偏加速试验，130℃/85% 相对湿度，电压 1.8V，时间为 100h。

③ 高压蒸煮（蒸汽）测试，温度 121℃，相对湿度 100%，压强 2atm，时间 168h。

一般情况下，芯片级互连的质量和可靠性要求都高于板级互连，也就是说用于板级互连的低熔点焊料并不能用于芯片级互连。图 8.23 所示为检测封装质量的回流曲线，还给出了不同放大倍数下 IMC 的剖面图。从图中可以看出，经过三次回流之后，3D IC 芯片堆叠的 InSnAu IMC 并无显著变化。

8.5.7　InSnAu IMC 层的剪切强度

由于 InSnAu IMC 互连强度较高，所以剪切测试（施加水平推力）中多数是 $200\mu m$ 厚的芯片发生破坏。因此，为了测得 InSnAu IMC 互连的剪切强度，采用 $400\mu m$ 厚的芯片进行测试。图 8.24 所示为 InSnAu IMC 互连经过剪切测试后的典型失效模式，可以看出，断裂发生在 IMC 处以及 TiAu UBM 和 InSnAu IMC 的界面处。

图 8.25 所示分别为完成键合后和经过 3 次回流的 InSnAu IMC 互连的剪切强度测试结果。从图中可以看出：① 经过 3 次回流之后，位于基芯片与第二层芯片之

第 8 章 芯片到芯片、芯片到晶圆、晶圆到晶圆键合

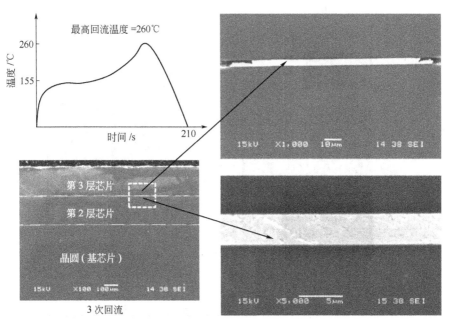

图 8.23 最高回流温度 260℃，经过 3 次回流后 3D IC 堆叠的截面图

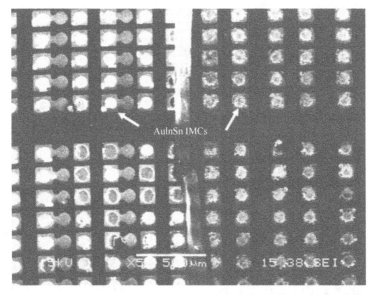

图 8.24 断裂发生在 IMC 层以及 TiAu UBM 和 InSnAu IMC 的界面处

间的第一层 IMC 焊点，以及位于第二层芯片与第三层芯片之间的第二层 IMC 焊点的剪切强度都减小了近 15%；②尽管 IMC 的剪切强度有所减小，但仍远高于所要求的剪切强度（20MPa）；③以上两种情况中，第一层 IMC 焊点的剪切强度都比第二层 IMC 焊点的剪切强度高 8%，这可能是由剪切测试的设置引起的。首先，对第三层芯片施加水平推力，夹持住下面两层芯片，从而得到第二层 IMC 的剪切强

261

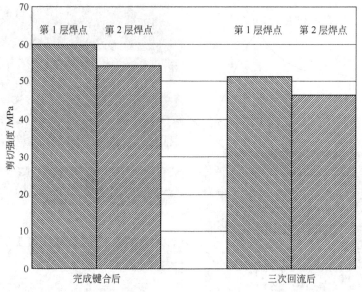

图 8.25 三层堆叠芯片的键合剪切强度：完成键合后和经历三次回流后

度；然后对第二层芯片施加水平推力，夹持住基芯片，从而得到第一层 IMC 的剪切强度。由于夹持夹具并非完全刚性，因此会释放部分推力，而且夹持夹具越大，释放的推力越多。

8.5.8 InSnAu IMC 层的电阻

采用开尔文四点测量法测量 InSnAu IMC 互连的电阻，如图 8.26 所示。制作了相距 $800\mu m$ 的两个微凸点，其直径为 $200\mu m$。测量结果如图 8.27 所示，可以看

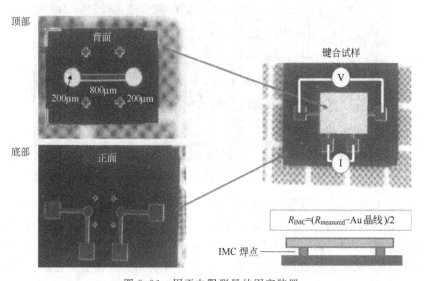

图 8.26 用于电阻测量的固定装置

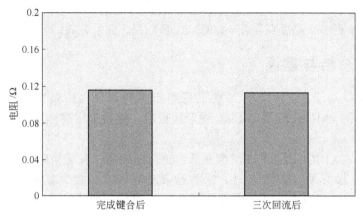

图 8.27 IMC 焊点的电阻：完成键合后和经过三次回流后

出 InSnAu IMC 焊点的电阻值约为 0.12Ω，并且经历三次回流后并未发生变化。

8.5.9 InSnAu IMC 层的热稳定性

采用差示扫描量热法（DSC）已经测得 InSnAu IMC 的重熔温度高于 365℃，找出 3D IC 芯片堆叠 IMC 互连的塌陷温度会是一个很有趣的问题。然而，由于 DSC 设备铂坩埚尺寸的限制，无法放置整个 3D IC 芯片堆叠进行测量，因此采用间接测量方法，即采用热机械分析仪（TMA）测量整个 3D IC 芯片堆叠的热膨胀与温度之间的关系，结果如图 8.28 所示。在 InSnAu IMC 互连发生重熔之前，曲线随着温度不断上升，近似于直线。然而，当 InSnAu IMC 互连发生重熔时，热膨胀曲线变得不稳定且开始下降，说明 3D IC 芯片堆叠完全失效。根据图 8.28 所示的 3 个试样测得的 TMA 曲线可知，3D IC 芯片堆叠中的 IMC 在 380～400℃ 之间

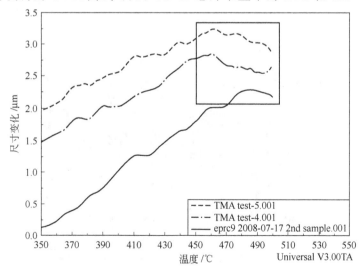

图 8.28 TMA 分析结果：400℃ 左右发生重熔

开始重熔，这个温度高于365℃，原因在于测量的是整个3D IC芯片堆叠的温度，其吸收的热量更多。温度在450～480℃之间时IMC发生失效。

8.5.10 总结与建议

① 设计了用于3D IC芯片堆叠的低熔点焊料系统（InSnAu）。键合温度为180℃、压强为6MPa、时间为45s。键合结束后，将整个组装体在120℃下退火处理12h。

② SEM、XDR以及TEM图像显示，键合比较均匀且无缺陷，所有键合面转化为InSnAu和InAu IMC。此外，DSC结果表明IMC的重熔温度高于365℃。

③ 三次回流后（最高回流温度为260℃），3D IC芯片堆叠中的InSnAu IMC无明显变化，IMC的电阻无变化。

④ 三次回流后（最高回流温度为260℃），InSnAu IMC的剪切强度减少了15%。尽管IMC的剪切强度有所减小，但仍高出所要求强度（20MPa）的两倍。

⑤ 利用纳米压痕方法测得InSnAu IMC的弹性模量为81GPa，硬度为1.5GPa。

⑥ 剪切测试中，InSnAu IMC互连的失效模式为沿IMC以及TiAu UBM和InSnAu IMC界面处发生断裂。

⑦ 对于3D IC芯片堆叠而言，由于芯片的厚度薄至$20\mu m$，很难或者不可能采用常规方法表征其特性，亟需新的测量方法和设备。

8.6 低温W2W键合 [TiCuTiAu 到 TiCuTiAuSnInSnInAu]

W2W键合技术（如图8.29所示）是半导体制造中最重要的技术之一。如前文所述：①W2W键合是3D Si集成中常见的组装技术；②对于3D IC集成来说，

图8.29 W2W键合机

W2W 键合除了一些特殊应用以外并不常见，这些特殊应用包括从 MEMS 高良率帽晶圆到 MEMS/ASIC（专用 IC）晶圆的键合、从高良率透明塑料透镜到发光二极管（LED）/驱动晶圆的键合等；③对于微通道应用而言，采用 W2W 键合技术将两个高良率 TSV 微通道晶圆键合起来。由于 3D Si 集成中 W2W 键合不使用微凸点，并且距离量产应用尚远，因此本节主要讨论基于焊料的 W2W 键合技术。焊料系统为 SnIn 和 CuTiAu 金属化层，键合温度≤180℃，且只用于 MEMS。

8.6.1 测试模型

图 8.30[21,22]所示为待键合的上下两个晶圆的示意图，以及晶圆上沉积的金属化层结构图：①密封环呈 119×119 阵列排布，尺寸为 11mm×11mm，高度为 300μm；②密封环由 SnIn 焊料组成，焊料的合金相图如图 8.31 所示；③UBM 采用的是 TiCuTiAu 结构；④空腔尺寸为 6mm×6mm，深度为 250μm。

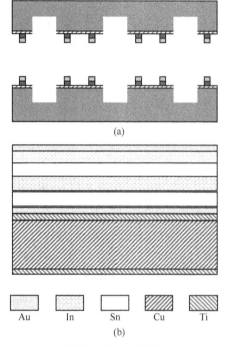

8.6.2 测试模型制作

首先采用热氧化与低压化学气相沉积工艺，在 200mm 硅晶圆上沉积 300Å 厚的 SiO_2 和 1500Å 厚的 Si_3N_4，作为空腔刻蚀的硬掩膜。采用干膜作为光阻材料制作光刻图形。接着对上下两晶圆采用湿法

图 8.30 待键合的上下晶圆示意图（a）与沉积在低温键合晶圆上的金属化层（b）

KOH 刻蚀工艺制作键合环内部的空腔，键合环通过在 Cu 基高温组分上涂覆 SnIn 焊料制得。然后在 $Si/SiO_2/Si_3N_4$ 基底上溅射厚度分别为 0.02μm、2μm、0.05μm、0.03μm 的 TiCuTiAu 金属化层。这里，第一层较薄的 Ti 层起到黏结层的作用，第二层 Ti 层作为缓冲层，Au 薄层也是必要的，可以在沉积焊料之前起到湿润以及防止金属氧化的作用。

为了在相对较短的时间内制得 InSn 合金，沉积了四层厚度分别为 1μm、1μm、0.7μm、0.8μm 的 Sn/In/Sn/In 焊料。根据该设计，InSn 中 Sn 原子数占 50%，接近于 InSn 合金的共晶组分。最后在焊料上面沉积 0.03μm 厚的 Au 层，防止后续工艺以及储存过程中焊料发生氧化。完成各材料沉积以及去除干膜之后，在键合之前采用 O_2 电浆除残渣法去除焊料表面的氧化层以及有机污染物。

8.6.3 低温 W2W 键合

采用 EVG520 晶圆键合机，在可控氮气氛中对焊料区施加 5.5MPa 的压强进

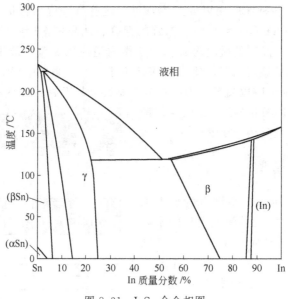

图 8.31 InSn 合金相图

行键合,如图 8.29 所示。两个晶圆在键合过程中所受最高温度分别为 180℃ 和 150℃,选取这两个温度的原因是:①Sn、In 以及 InSn 共晶合金的熔点分别为 232℃、156℃、118℃;②为了使焊料能够较好地进行回流,施加温度应该比焊料熔点高出 30~40℃。因此,180℃ 的键合温度远高于 In 和共晶合金 InSn 的熔点。对比发现,150℃ 的键合温度也比 InSn 共晶合金的熔点高出近 30℃。因此,这两个键合温度有助于确定采用 InSn 焊料层进行 W2W 密封键合的特性。在最高温度下保温 20min,确保焊料和高温组分之间能够充分扩散。W2W 键合的温度曲线如图 8.32 所示。图 8.33 所示为 180℃ 和 150℃ 下得到的键合模块典型的 C-SAM 图像。完成键合后,将晶圆切割成 13mm×13mm 的单个裸芯片,切割速率为

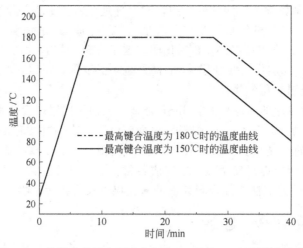

图 8.32 晶圆最高键合温度分别为 180℃ 和 150℃ 的温度曲线

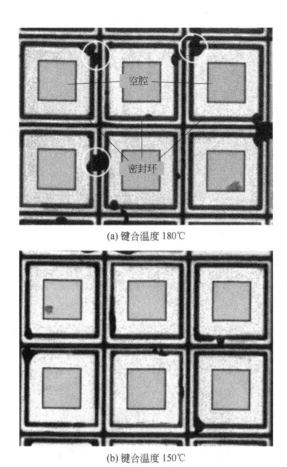

图 8.33　键合后器件的 C-SAM 图

2mm/s。

8.6.4　C-SAM 检测

晶圆切割完成后，采用 C-SAM 对裸芯片进行无损检测，观察 W2W 键合界面。图 8.33 所示为键合模块的典型 C-SAM 图像。若未观察到缺陷，则标注为"good die"，即合格芯片。键合良率定义为合格裸芯片个数占整个键合晶圆上裸芯片总数的百分比。基于该定义可知，180℃和150℃键合温度下得到的200mm晶圆的键合良率为100%，如图8.33所示。然而，由于C-SAM的分辨率为微米级，所以无法观测到极小的缺陷。键合温度的影响也可以通过C-SAM进行观察。相比于150℃键合温度情况，键合温度为180℃时，有更多的焊料从密封环处被挤出。这是由于键合温度越高，液态焊料的流动性越好，相同压力下有更多的熔融焊料从密封环处流出，尤其在边角区域更为明显，如图8.33(a)中圈出的部分所示。边角处的密封环被挤出的焊料最长达2mm，对于小型封装而言，溢出的焊料可能导致内部器件损坏。以下是一些简单有效的解决方法：①在保证产品良率的条件下，减小键合压强或焊料厚度;②在密封环中制作沟槽防止焊料溢出。

8.6.5 微结构的 SEM/EDX/FIB/TEM 分析

不同键合温度下各制取 5 个试样,利用 SEM 观察其剖面,研究焊点的微结构。利用 SiC 砂纸对试样进行减薄,采用 $1\mu m$ 的抛光钻和 $0.05\mu m$ 的 SiO_2 磨抛液对试样表面进行抛光。利用 X 射线能谱仪(EDX)分析键合焊点的组成成分,采用 FIB 制作 TEM 以及 EDX 所需的焊点薄试样。图 8.34 所示为 180℃下制得的密封环焊点的 SEM 图像,焊点厚度约为 $5.5\mu m$,其中 IMC 层厚度为 $1.5\mu m$,如图 8.34(a)、(b) 所示。键合后整个焊点的厚度变薄,但未发现可检测的孔洞或裂纹,溢出的焊料约 $10\mu m$,如图 8.34(c) 所示。焊点的组成元素为 55%~60% 的 Sn、30%~35% 的 In 以及 5%~10% 的 Au。

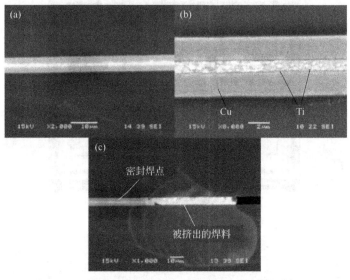

图 8.34 键合温度为 180℃时,密封焊点的界面微结构:
(a) 低倍放大;(b) 高倍放大;(c) 焊料从密封焊点边缘挤出

采用 TEM/EDX 测定密封焊点的精确组成成分。图 8.35 所示为焊点的 TEM 微结构图,EDX 分析结果见表 8.3。可以明显看出,在焊点中嵌有两条平行的 Ti 线。根据 EDX 的分析结果,c 点与 k 点处 Ti 层中含有少量的 In、Cu 和 Sn,说明这些元素可以透过 Ti 层进行扩散,从而证明了采用 Ti 层作为低温 W2W 键合缓冲层的合理性。距离 Cu 较近的 a 点、b 点和 l 点其组成成分也出现了上述现象。对于 a 点和 b 点来说,Cu 中出现了少量的 In 和 Sn,这正好与富 Cu 相对应。对于 l 点,Cu 含量与 SnIn 含量的比例为 1:1。根据 180℃下 CuInSn 系统的相平衡热力学计算结果,如图 8.36 所示[9],发现该系统中存在 η 相。实际上,在 CuIn 和 CuSn 两个二元系统里都存在 η 相,即 η-Cu_2In 和 η-Cu_6In_5,两者均为 NiAu 型结构[12]。Sommadossi 等人的研究表明,Sn 的溶解度较高时混合相为 η-$Cu_6(Sn-In)_5$[12]。

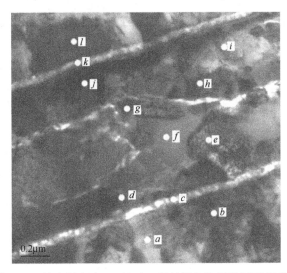

图 8.35 键合温度为 180℃时，密封焊点的 TEM/EDX 分析

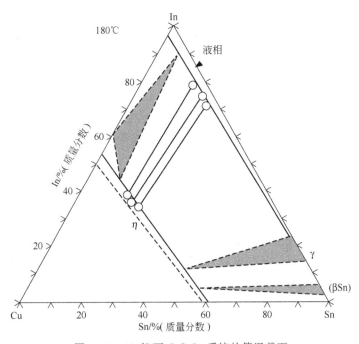

图 8.36 180℃下 CuInSn 系统的等温截面

表 8.3 所选位置组分的 TEM/EDX 分析结果

位置	组分/%				
	Cu	Ti	Sn	In	Au
a	88.1		7.1	4.7	
b	77.0		11.8	11.2	

续表

位置	组分/%				
	Cu	Ti	Sn	In	Au
c	5.3	79.6	9.9	5.1	
d	12.9	5.7	12.5	47.7	21.2
e	6.8		3	54.8	35.4
f			33.4	18.8	47.8
g	5.7		2.8	60.2	31.3
h	2.4		6.9	59.4	31.5
i	6.1		7.9	54.5	31.4
j	6.0	4.0	21.9	44.4	23.7
k	4.8	90.4	2.6	2.2	
l	50.3		30.7	19.1	

对于 e 点、j 点、h 点以及 i 点，其组成对应于 Au(InSn)$_2$ 相。在液/固相扩散键合过程中，少量的 Cu 可能溶解于混合相中。d 点处 CuSnInAu 来源于两个相，分别为 AuInSn 相和 CuSnIn 相。f 点其组成为 Au(SnIn) 相，该相形成的原因可能是由于一部分焊料被挤出使得 In 原子数量减少。

图 8.37 所示为 150℃下制得的密封环焊点的 SEM 图像，焊点厚度约为 6.5μm，其中 IMC 厚度约为 2μm。对比图 8.34 与图 8.37 发现，150℃下制得的焊点比 180℃下制得的焊点要厚，两者厚度的差异主要是键合温度不同导致的。在较高的键合温度下，焊锡材料融化得更快，其流动性也更好。在相同的压力下，键合温度较高时液态焊料更容易被挤出。因此，当键合温度为 150℃时，参与扩散键合的残留焊料更多，也就是说更多的焊料透过 Ti 层扩散至 Cu 基底[如图 8.37(b) 所示]，致使形成的焊点以及 IMC 较厚。当键合温度为 150℃时，由于键合温度较低，在密封焊点中心部位产生的裂纹可能不利于封装的气密性和可靠性。键合质量较差的原因有以下几点：①在焊料表面沉积薄 Au 层后，可能会形成一层很薄的 AuIn 层；②在键合过程中，所有键合对都必须穿透 AuIn 层从而实现键合；③In 和 Sn 的熔点分别为 156℃和 232℃；④理想情况下，最好在低至 118℃时，InSn 层能够形成共晶合金；⑤然而事实上并非如此，当温度达到 150℃时，InSn 需要经历更长时间的相互扩散以形成低温合金；⑥即使最终获得了共晶合金，其温度变化也仅仅约为 30℃；⑦这种条件下制得的焊料与高键合温度下制得的焊料相比，液态焊料的流动性以及润湿性都较差。因此，只有较少的液态合金透过 AuIn 层互相扩散并结合，结果在 AuIn 层附近形成孔洞，连续的孔洞导致裂纹的产生。

EDX 分析表明，Sn 和 In 原子穿透 Ti 层进入 Cu 以及 Cu 附近的区域，并形成 Cu_3(SnIn) 相。表 8.4 中的 EDX 分析结果表明，中心区域形成了 Cu_6(SnIn)$_5$ 相以及 AuIn 相。

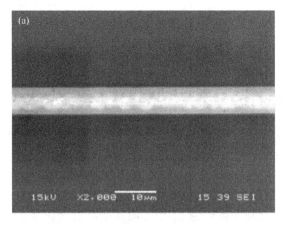

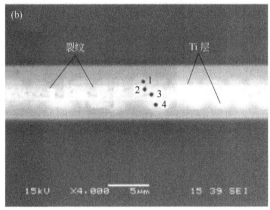

图 8.37 键合温度为 150℃时,密封焊点的界面微结构:
(a) 低倍放大;(b) 高倍放大

表 8.4 图 8.37(b) 中界面的组分

位置	元素/%				
	Cu	Ti	Sn	In	Au
1	78.01	5.05	3.48	8.94	4.52
2	72.51	5.49	13.60	8.40	0
3	60.17	3.31	22.88	13.64	0
4	51.21	2.02	20.86	18.21	7.70

8.6.6 氦泄漏率测试与结果

通过基于 MIL-STD-883 的氦泄漏率测试,对已成型裸芯片上密封环的气密性进行检测。由于键合裸芯片内部的空腔体积约为 $0.02cm^3$,故轰击室中氦气的压强应设置为 75psi,曝露时间为 2h。轰击结束后,将试样放入氦气泄漏检测仪中测量泄漏率，试样总数为 21 个。结果表明，180℃下制得的器件其泄漏率低于 2×10^{-8} atm·mL/s，低于 MILSTD-883E 的拒绝限制值（5×10^{-8} atm·mL/s）。然而对于 150℃下制得的器件，只有 86% 的试样其泄漏率低于 5×10^{-8} atm·mL/s，这可能是由密封环中间区域的孔洞/裂纹导致的。

8.6.7 可靠性测试与结果

由于 150℃下制得的封装体气密性差，所以在研究可靠性问题时只考虑键合温度 180℃下制得的封装体。可靠性测试项目以及结果如表 8.5 所示。试样分别经历高压蒸煮试验（121℃，2atm，300h）、高湿度存储试验（85℃/相对湿度 85%，1000h）、高温存储试验（125℃，1000h）以及温度循环测试（-40~125℃，1000 个循环），每种测试试样个数均为 21 个。测试前后对试样进行 C-SAM 检测、氦气泄漏率测试和剪切测试，若试样的氦气泄漏率超过 5×10^{-8} atm·mL/s 或密封环中出现大裂纹则视为失效。经过高压蒸煮试验、高湿度存储试验、高温存储试验、温度循环测试之后，氦气泄漏率小于 5×10^{-8} atm·mL/s 试样的比率分别为 90.5%、95.2%、100%、90.5%。封装的剪切强度要求为 6MPa，从表 8.5 可以看出，所有试样的剪切测试结果均高于最低要求。可靠性测试后对焊点界面微结构的分析如图 8.38 所示，可以看出：①微结构未发生明显变化。②SEM/EDX 分析发现密封环的组分也未发生明显变化。由于可靠性测试温度与 IMC 的熔点相比并不是很高，故该结果是合理的。③Ti 层也可以阻止 Cu 与 IMC 之间的扩散。然而，根据剪切测试结果发现，经过可靠性测试之后不同 IMC 间结合力有所降低。此外，在长时间的可靠性测试中，溢出的焊料与高温成分（例如 Cu 与 IMC）之间发生了扩散。如图 8.39 所示为高温存储测试后密封环边缘处的微结构图。根据 EDX 分析发现，所有溢出的焊料都转变成了高温化合物，消耗了几十微米的 Cu 基底生成 $Cu_6(SnIn)_5$。对于其他三项可靠性测试，由于测试温度不高并未发现较活跃的扩散现象。例如，温度循环试验中密封环两侧的 Cu 基板只有 10~30μm 被 20μm 的焊料消耗，由于扩散并没有形成新的化合物和裂纹，故在可靠性测试中不会引起密封环失效。

表 8.5 可靠性评估测试得到的试样特性

可靠性测试项目	结果	
	氦气泄漏率小于 5×10^{-8} atm·mL/s 试样的比率	剪切强度/MPa
高压蒸煮试验	90.5%	27.61
高湿度存储试验	95.2%	24.57
高温存储试验	100%	21.27
温度循环试验	90.5%	25.62
完成键合之后	100%	46.32

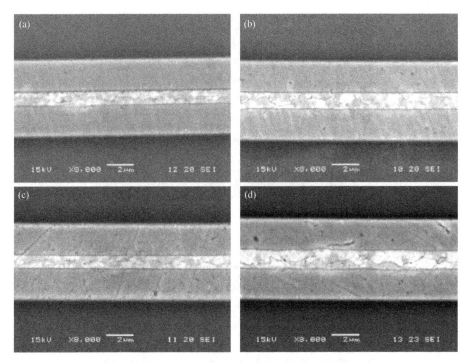

图 8.38 经过可靠性测试得到的合格晶圆上密封焊点的界面微结构：(a) 高压蒸煮试验；(b) 高湿度存储试验；(c) 高温存储试验；(d) 温度循环试验

同样对气密性较差的一些试样进行测试，以了解密封环的失效机制。这些试样的失效机制是一致的，如图 8.40 所示。试样中出现贯穿整个密封环的裂纹，沿着 InAu 化合物与 CuSnIn 化合物的键合界面扩展。当焊点内有不同的 IMC 生成时，温度循环与高压蒸煮试验时会产生应力，这就是为什么经过这两项测试之后试样气密性较差的原因。这说明密封焊料中，不同相之间的结合力是决定可靠性的关键因素。因此，至少有两种提高现有密封焊点可靠性的方法：①沉积薄 Au 层减少 AuIn 化合物的生成；②减小 Ti 缓冲层的厚度，提高焊料与 Cu 基底间的扩散速率。这样溢出的焊料更少，并且形成更厚的不连续 AuIn 相 IMC 焊点。

8.6.8 总结与建议

① 基于 C-SAM 观测结果，150℃ 和 180℃ 下制得的所有封装密封环中均未发现孔洞。当键合温度为 180℃ 时，所有试样的氦气泄漏率都较低，为 2×10^{-8} atm·mL/s；当键合温度为 150℃ 时，只有 86% 的试样其气密性合格。

② 在两种键合温度下（150℃ 和 180℃），In 原子和 Sn 原子可以透过 Ti 层与 Cu 结合形成焊点。密封焊点由高熔点 IMC 组成，使得封装在进行下一级互连时能够承受更高的温度。

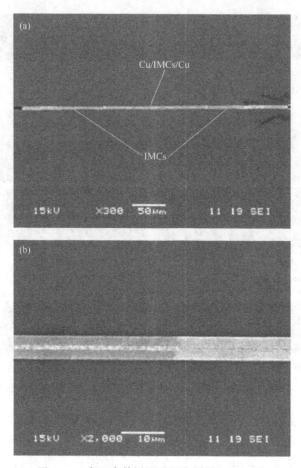

图 8.39 高温存储测试后溢出焊料与 IMC 的
扩散情况：(a) 低倍放大；(b) 高倍放大

③ 键合温度为 180℃时，被挤出的焊料更多。

④ 当键合温度较低时，例如 150℃时，由于该温度下液态合金流动性较差，所以得到的密封焊点较厚。然而，密封环中心区域形成的裂纹会引起气密性问题。

⑤ 两种键合温度（150℃和 180℃）下制得的密封环其剪切强度都比较高。

⑥ 经过四项可靠性测试之后，即高压蒸煮试验、高湿度存储试验、高温存储试验、温度循环测试，发现 180℃下键合得到的封装体具有良好的可靠性和气密性。

⑦ 经过可靠性测试之后，失效封装中的裂纹沿不同的 IMC 层扩展。因此，焊料中不同相之间的结合力是决定密封焊点可靠性的关键因素。以下是两种提高现有密封焊点可靠性的方法：沉积薄 Au 层减少 AuIn 化合物的生成；减小 Ti 缓冲层的厚度，提高焊料与 Cu 基底间的扩散速率。

⑧ 针对低温键合和气密性 MEMS 封装目前所用的材料系统（Cu/Ti/Au）及

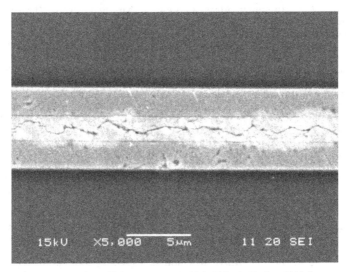

图 8.40　高压蒸煮试验后，失效密封焊点的界面微结构

其键合工艺参数（键合温度为 180℃、压强为 5.5MPa、时间为 20min）给出了一些建议。

8.7　参考文献

[1] Klumpp, A., "Vertical System Integration by Using Inter-Chip Vias and Solid Liquid Interdiffusion Bonding," *Jpn. J. Appl. Phys.*, Vol. 43, 2004, pp. L829–L830.
[2] Chen, K., "Microstructure Examination of Copper Wafer Bonding," *J. Electron. Mater.*, Vol. 30, 2001, pp. 331–335.
[3] Morrow, P. R., "Three-dimensional Wafer Stacking via Cu-Cu Bonding Integrated with 65-nm Strained-Si/low-k CMOS Technology," *IEEE Electron. Device Lett.*, Vol. 27, 2006, pp. 335–337.
[4] Fukushima, T., Y. Yamada, H. Kikuchi, T. Tanaka, and M. Koyanagi, "Self-Assembly Process for Chip-to-Wafer Three-Dimensional Integration," *IEEE/ECTC Proceedings*, Reno, NV, 2007, pp. 836–841.
[5] Sakuma, K., P. Andry, C. Tsang, K. Sueoka, Y. Oyama, C. Patcl, B. Dang, S. Wright, B. Webb, E. Sprogis, R. Polastre, R. Horton, and J. Knickerbocker, "Characterization of Stacked Die using Die-to-Wafer Integration for High Yield and Throughput," *IEEE/ECTC Proceedings*, Lake Beuna Vista, FL, May 2008, pp. 18–23.
[6] Wakiyama, S., H. Ozaki, Y. Nabe, T. Kume, T. Ezaki, and T. Ogawa, "Novel Low-Temperature CoC Interconnection Technology for Multichip LSI (MCL)," *IEEE/ECTC Proceedings*, Reno, NV, 2007, pp. 610–615.
[7] Liu, Y. M., and T. H. Chuang, "Interfacial Reactions Between Liquid Indium and Au-Deposited Substrates," *J. Electron. Mater.*, Vol. 29, No. 4, 2000, pp. 405–410.
[8] Zhang, W., A. Matin, E. Beyne, and W. Ruythooren, "Optimizing Au and In Micro-Bumping for 3D Chip Stacking," *IEEE/ECTC Proceedings*, Lake Beuna Vista, FL, May 2008, pp. 1984–1989.
[9] Liu, X., H. Liu, I. Ohnuma, R. Kainuma, K. Ishida, S. Itabashi, K. Kameda, and K. Yamaguchi, "Experimental determination and thermodynamic calculation of the phase equilibria in the Cu-In-Sn system," *J. Electron. Mater.*, Vol. 30, no. 9, 2001, pp. 1093–1103.

[10] Choi, W., C. Premachandran, C. Ong, L. Xie, E. Liao, A. Khairyanto, B. Ratmin, K. Chen, P. Thaw, and J. H. Lau, "Development of Novel Intermetallic Joints Using Thin Film Indium Based Solder by Low Temperature Bonding Technology for 3D IC Stacking," *IEEE/ECTC Proceedings*, San Diego, CA, May 2009, pp. 333–338.

[11] Choi, W., D. Yu, C. Lee, L. Yan, A. Yu, S. Yoon, J. H. Lau, M. Cho, Y. Jo, and H. Lee, "Development of Low Temperature Bonding Using In-based Solders," *IEEE/ECTC Proceedings*, Orlando, FL, May 2008, pp. 1294–1299.

[12] Sommadossi, S., and A. F. Guillermet, "Interface Reaction Systematic in the Cu/In-48Sn/Cu System bonded by diffusion soldering," *Intermetallics*, Vol. 15, 2007, pp. 912–917.

[13] Shigetou, A., T. Itoh, K. Sawada, and T. Suga, "Bumpless Interconnect of 6-μm Pitch Cu Electrodes at Room Temperature," *IEEE/ECTC Proceedings*, Lake Buena Vista, FL, May 27–30, 2008, pp. 1405–1409.

[14] Made, R., C. L. Gan, L. Yan, A. Yu, S. U. Yoon, J. H. Lau, and C. Lee, "Study of Low Temperature Thermocompression Bonding in Ag-In Solder for Packaging Applications," *J. Electron. Mater.*, Vol. 38, 2009, pp. 365–371.

[15] Yan, L.-L., C.-K. Lee, D.-Q. Yu, A.-B. Yu, W.-K. Choi, J. H. Lau, and S.-U. Yoon, "A Hermetic Seal Using Composite Thin Solder In/Sn as Intermediate Layer and Its Interdiffusion Reaction with Cu," *J. Electron. Mater.*, Vol. 38, 2009, pp. 200–207.

[16] Yan, L.-L., V. Lee, D. Yu, W. K. Choi, A. Yu, A.-U. Yoon, and J. H. Lau, "A Hermetic Chip to Chip Bonding at Low Temperature with Cu/In/Sn/Cu Joint," *IEEE/ECTC Proceedings*, Orlando, FL, May 2008, pp. 1844–1848.

[17] Yu, A., C. Lee, L. Yan, R. Made, C. Gan, Q. Zhang, S. Yoon, and J. H. Lau, "Development of Wafer Level Packaged Scanning Micromirrors," *Proc. Photon. West*, Vol. 6887, 2008, pp. 1–9.

[18] Lee, C., A. Yu, L. Yan, H. Wang, J. Han, Q. Zhang, and J. H. Lau, "Characterization of Intermediate In/Ag Layers of Low Temperature Fluxless Solder Based Wafer Bonding for MEMS Packaging," *J. Sensors Actuators* (in press).

[19] Yu, D.-Q., C. Lee, L. L. Yan, W. K. Choi, A. Yu, and J. H. Lau, "The Role of Ni Buffer Layer on High Yield Low Temperature Hermetic Wafer Bonding Using In/Sn/Cu Metallization," *Appl. Phys. Lett.* (in press).

[20] Yu, D. Q., L. L. Yan, C. Lee, W. K. Choi, S. U. Yoon, and J. H. Lau, "Study on High Yield Wafer to Wafer Bonding Using In/Sn and Cu Metallization," *Proceedings of the Eurosensors Conference*, Dresden, Germany, 2008, pp. 1242–1245.

[21] Yu, D., C. Lee, and J. H. Lau, "The Role of Ni Buffer Layer Between InSn Solder and Cu Metallization for Hermetic Wafer Bonding," *Proceedings of the International Conference on Electronics Materials and Packaging*, Taipei, Taiwan, October 22–24, 2008, pp. 335–338.

[22] Yu, D., L. Yan, C. Lee, W. Choi, M. Thew, C. Foo, and J. H. Lau, "Wafer Level Hermetic Bonding Using Sn/In and Cu/Ti/Au Metallization," *IEEE Proceeding of Electronics Packaging and Technology Conference*, Singapore, December 2008, pp. 1–6.

[23] Chen, K., C. Premachandran, K. Choi, C. Ong, X. Ling, A. Khairyanto, B. Ratmin, P. Myo, and J. H. Lau, "C2W Bonding Method for MEMS Applications," *IEEE Proceedings of Electronics Packaging Technology Conference*, Singapore, December 2008, pp. 1283–1287.

[24] Premachandran, C. S., J. H. Lau, X. Ling, A. Khairyanto, K. Chen, and Myo Ei Pa Pa, "A Novel, Wafer-Level Stacking Method for Low-Chip Yield and Non-Uniform, Chip-Size Wafers for MEMS and 3D SIP Applications," *IEEE/ECTC Proceedings*, Orlando, FL, May 27–30, 2008, pp. 314–318.

[25] Simic, V., and Z. Marinkovic, "Room Temperature Interactions in Ag-Metals Thin Film Couples," *Thin Solid Films*, Vol. 61, 1979, pp. 149–160.

[26] Lin, J.-C., "Solid-Liquid Interdiffusion Bonding Between In-Coated Silver Thick Films," *Thin Solid Films*, Vol. 61, 1979, pp. 212–221.

[27] Roy, R., and S. K. Sen, "The Kinetics of Formation of Intermetallics in Ag/In Thin Film Couples," *Thin Solid Films*, Vol. 61, 1979, pp. 303–318.

[28] Chuang, R. W., and C. C. Lee, "Silver-Indium Joints Produced at Low Temperature for High-Temperature Devices," *IEEE Trans. Components Packag. Technol. A*, Vol. 25, 2002, pp. 453–458.

[29] Chuang, R. W., and C. C. Lee, "High-Temperature Non-Eutectic Indium-Tin Joints Fabricated by a Fluxless Process," *Thin Solid Films*, Vol. 414, 2002, pp.

175–179.
- [30] Lee, C. C., and R. W. Chuang, "Fluxless Non-Eutectic Joints Fabricated Using Gold-Tin Multilayer Composite," *IEEE Trans. Components Packag. Technol. A*, Vol. 26, 2003, pp. 416–422.
- [31] Humpston, G., and D. M. Jacobson, *Principles of Soldering*, ASM International, Materials Park, MD, 2004
- [32] Chuang, T., H. Lin, and C. Tsao, "Intermetallic Compounds Formed During Diffusion Soldering of Au/Cu/Al$_2$O$_3$ and Cu/Ti/Si with Sn/In Interlayer," *J. Electron. Mater.*, Vol. 35, 2006, pp. 1566–1570.
- [33] Lee, C., and S. Choe, "Fluxless In-Sn Bonding Process at 140°C," *Mater. Sci. Eng.*, Vol. A333, 2002, pp. 45–50.
- [34] Lee, C. "Wafer Bonding by Low-Temperature Soldering," *Sensors & Actuators*, Vol. 85, 2000, pp. 330–334.
- [35] Vianco, P. T., "Intermetallic Compound Layer Formation Between Copper and Hot-Dipped 100 In, 50In-50Sn, 100Sn, and 63Sn-37Pb Coatings," *J. Electron. Mater.*, Vol. 23, 1994, pp. 583–594.
- [36] Frear, D. R., "Intermetallic Growth and Mechanical-Behavior of Low and High Melting Temperature Solder Alloys," *Metall. Mater. Trans. A*, Vol. 25, 1994, pp. 1509–1523.
- [37] Morris, J. W., "Microstructure and Mechanical Property of Sn-In and Sn-Bi Solders," *J. Miner. Metals Mater. Soc.*, Vol. 45, 1993, pp. 25–27.
- [38] Mei, Z., "Superplastic Creep of Low Melting-Point Solder Joints," *J. Electron. Mater.*, Vol. 21, 1992, pp. 401–407.
- [39] Chuang, T. H., "Phase Identification and Growth Kinetics of the Intermetallic Compounds Formed During In-49Sn/Cu Soldering Reactions," *J. Electron. Mater.*, Vol. 31, 2002, pp. 640–645.
- [40] http://www.webelements.com/WebElements: The periodic table on the website.
- [41] Kazumasa, T., "Micro Cu Bump Interconnection on 3D Chip Stacking Technology," *Jpn J. Appl. Phys.*, Vol. 43, No. 4B, 2004, pp. 2264–2270.

第 9 章 3D IC 集成的热管理

9.1 引言

3D 集成中存在很多关键问题[1~33]。例如，电子设计自动化（EDA）软件还未普及；缺乏测试方法及设备；需要已知合格芯片（KGD）；快速芯片与低速芯片掺杂；大芯片与小芯片掺杂；3D 集成需采用硅通孔技术（TSV）；加工过程中需要进行晶圆减薄和薄晶圆拿持技术；需要采用微凸点；热管理问题；对中时设备的精度问题；3D 封装的检测问题；缺乏 3D 集成的专业技术、基础设备及标准；TSV 技术的成本远高于引线键合技术；用于 TSV 量产的设备少且价格昂贵；缺乏 TSV 的设计规范和软件；缺乏 TSV 相关检测技术和软件；填充 Cu 有利于器件的热管理，但是增大了材料的热膨胀系数（CTE）；无孔洞填充 Cu 需要花费较长时间；TSV 晶圆良率要求苛刻（≥99.9%）；CTE 不匹配引起的 TSV 晶圆翘曲问题；制作高良率、高深宽比的 TSV 较为困难；缺乏 TSV 的专业技术、基础设备和标准，等等。

正如前面提到的，热管理是 3D IC 集成中的一个重要问题。原因在于：①小尺寸封装中的堆叠多功能芯片产生的热量很高；②3D IC 使得单位面积上的总功率增大；③若无有效的冷却处理，会造成 3D 堆叠芯片过热；④3D 堆叠芯片之间的间隙对于冷却通道来说过小，不能提供流体流动的通道；⑤薄芯片会导致芯片上产生过热点。

热管理是 3D IC 集成的第四个关键技术。然而应该指出的是，由于系统级封装（SiP）工作时产生的热量较大，或者构建热管理系统花费较大，所以热管理成为了 SiP 发展的障碍。因此，对于 3D IC 集成 SiP 的广泛应用而言，亟需低成本、高效率的热管理设计指引和解决方案。

本章基于热传导理论[34~36]，研究了含有 TSV 转接板的 3D 集成 SiP 和堆叠式存储芯片的热性能，所得结果以设计图表的形式给出，提出了一些实用的设计指引。采用第 3 章得到的不同 TSV 直径、节距、深宽比的 TSV 转接板/芯片的等效热传导率进行数值模拟[31]，最后设计了带埋入微通道和 TSV 的热管理系统，用于解决两个 100W 芯片的散热问题[19]。

9.2 TSV 转接板对 3D SiP 封装热性能的影响

9.2.1 封装的几何参数与材料的热性能参数

图 9.1 所示为含有 TSV 转接板 3D SiP 的一个简化模型,其中将 TSV 转接板等效为一个均匀块体(即无 TSV),等效热传导率根据本书第 3.4.1 节确定。该模型包括一个大芯片、等效 TSV 转接板、1-2-1 型堆积(BU)基板、印刷电路板(PCB)和焊点,具体尺寸以及热性能参数见表 9.1~表 9.3。

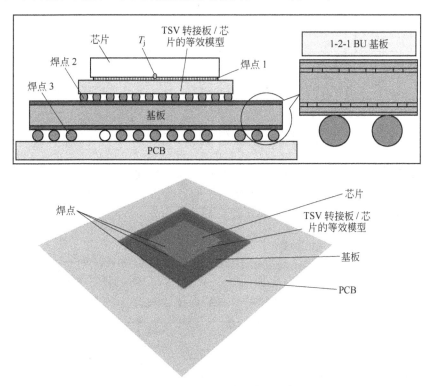

图 9.1 3D IC SiP 的结构简图及模型

表 9.1 TSV 转接板/芯片的几何尺寸与各材料的热导率

组件	转接板	TSV	TSV 填充料			
材料	Si	Cu	下填料微粒	下填料微粒	Al	Cu
热导率/(W/m·℃)	150	390	4.0	25	237	390
尺寸/mm	1.4×1.4×0.3	变量	局部通孔填充的电镀铜厚度(5~25μm)			
TSV 节距/mm			0.15~0.6			

表 9.2　3D IC SiP 的几何尺寸与各材料热导率

组件	芯片	TSV	凸点	下填料	PCB
材料	Si(TSV)	Cu	SnAg	聚合物	FR4
热导率/(W/m·℃)	经验公式	390	57	0.5	//0.8⊥0.3
尺寸/mm	5×5	φ0.05	φ0.2,高度0.15	5×5×0.15	76×114×1.6
功率/W	0.2	—	—	—	—

表 9.3　数值模拟材料参数

组件	芯片	焊料 1	转接板	焊料 2	基板	焊料 3	PCB
材料	Si	SnAg	Si+TSV (Cu)	SnAg	复合	SnAg	FR4
热导率/(W/m·℃)	150	57	k_{eq}	57	//100⊥0.5	57	//0.8⊥0.3
尺寸/mm	21×21×0.75	节距:0.15 高度:0.08 平均直径:0.08	可变	节距:0.5 高度:0.1 平均直径:0.1	1-2-1 (45×45×1)	节距:1.0 高度:0.6 平均直径:0.6	101×114×1.6

9.2.2　TSV 转接板对封装热阻的影响

通常将结点至环境热阻作为描述封装热性能的指标，热阻越高则封装的热性能越差。封装结点至环境热阻 R_{ja} 可以表示为 $R_{ja}=(T_j-T_a)/P$，其中 T_j 和 T_a 分别为结点温度与环境温度，P 为芯片的总耗散功率。

9.2.3　芯片功率的影响

芯片功率对封装热阻的影响如图 9.2 所示。可以看出：①随着芯片功率的升高，热阻减小，原因是高功率芯片使得封装温度也较高；②含有 TSV 转接板的封

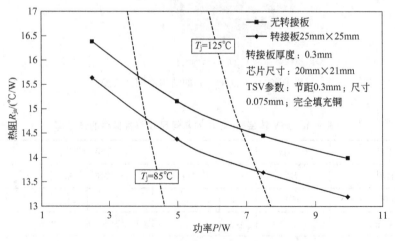

图 9.2　耗散功率与热阻的关系

装其热阻低于无 TSV 转接板的封装，这是由于含 TSV 的转接板更易于散热。图 9.2 中的两条虚线分别代表结温为 85℃和 125℃时的热阻。

9.2.4 TSV 转接板尺寸的影响

图 9.3 所示为 TSV 转接板尺寸对封装热性能的影响。从图中可以看出，当 TSV 转接板尺寸从 21mm×21mm 增大至 45mm×45mm 时，热阻 R_{ja} 降低了 14%，这说明可以通过增加 TSV 转接板的尺寸来提高封装的热性能。

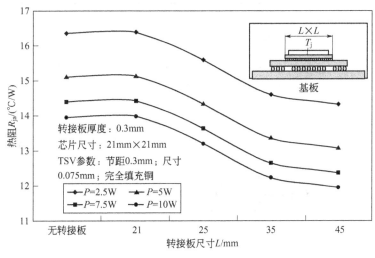

图 9.3 TSV 转接板尺寸与热阻的关系

9.2.5 TSV 转接板厚度的影响

TSV 转接板厚度对封装热性能的影响如图 9.4 所示。可以看出，TSV 转接板

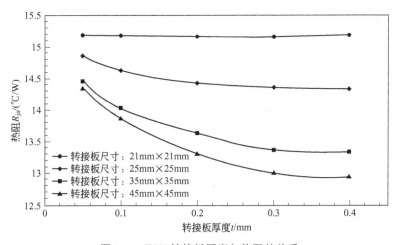

图 9.4 TSV 转接板厚度与热阻的关系

越厚则热阻越低,因为较厚的转接板传热效果较好。但是对于小尺寸的转接板而言,这种效应可以忽略不计。

9.2.6 芯片尺寸的影响

对于不同的 TSV 转接板与芯片尺寸比,芯片尺寸对封装热性能的影响如图 9.5 所示。从图中可以看出,在芯片与 TSV 转接板尺寸比相同的情况下,芯片的尺寸越小,封装的热阻反而越大。

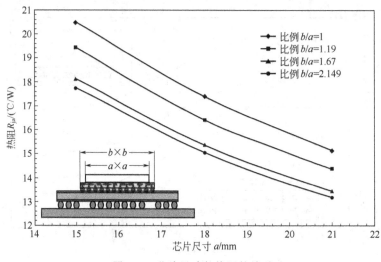

图 9.5 芯片尺寸与热阻的关系

9.3 3D 存储芯片堆叠封装的热性能

9.3.1 均匀热源 3D 堆叠 TSV 芯片的热性能

图 9.6 所示为芯片堆叠数与封装最大结温的关系。数值模拟时,芯片尺寸均为 5mm×5mm×0.05mm,每个芯片上各布有 15×15 个 Cu 填充 TSV,节距均为 0.2mm,每个芯片的功率均为 0.2W,并假设所有芯片上功率均匀分布,环境温度为 25℃。从图中可以看出,封装的最大结温随着芯片堆叠数目呈线性增加。结果表明,当最大允许结温为 85℃时,最多可以堆叠 7 个芯片。

图 9.7 所示为 TSV 芯片堆叠封装中每层的最大结温。可以看出,每层芯片的最大结温几乎相同。由于假设每层芯片的功率耗散均匀分布,所以得到的不同层芯片的温度分布也是均匀的。图 9.8 所示为芯片上最高温度的等高线图。

9.3.2 非均匀热源 3D 堆叠 TSV 芯片的热性能

图 9.6~图 9.8 为假设芯片上耗散功率均匀分布时的分析结果。然而,多数应

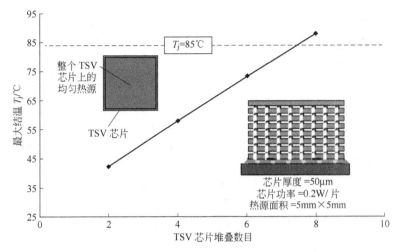

图 9.6　TSV 芯片堆叠数目与最大结温的关系（均匀热源）

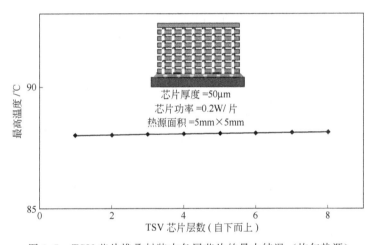

图 9.7　TSV 芯片堆叠封装中各层芯片的最大结温（均匀热源）

用中每层芯片的耗散功率并非均匀分布，相应地，含有 TSV 的 3D IC 芯片堆叠封装的热性能会有所不同。此外，由于 Si 材料具有较高的热传导率，故普通 Si 芯片在平行于其表面方向上传热很好。但是，为了获得小外形的 3D IC 芯片堆叠封装，每层芯片的厚度都必须减薄至 $50\mu m$ 甚至更薄，因而，平行扩散效应将受到超薄芯片的抑制，致使芯片上过热点会十分集中。非均匀热源与过热点共同作用是 3D IC 集成热管理面临的一个重要挑战。

9.3.3　各带一个热源的两个 TSV 芯片

含有两层 Cu 填充 TSV 堆叠芯片的 3D 集成封装的热性能如图 9.9 和图 9.10 所示。两个芯片尺寸为 $5mm \times 5mm$，每层芯片中间位置设置面积为 $0.2mm \times 0.2mm$、功率为 0.2W 的热源。从图中可以看出，无论是只有一层芯片还是两层芯

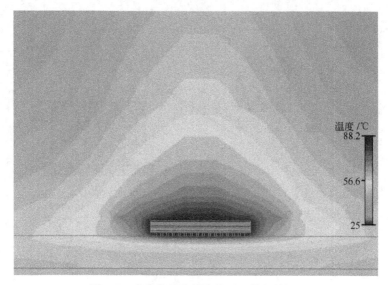

图 9.8　芯片的温度等高线图（均匀热源）

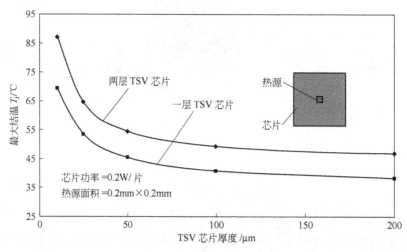

图 9.9　TSV 芯片厚度与最大结温的关系（热源位于芯片中心）

片堆叠：①对于非均匀热源，芯片厚度对 3D IC 集成热性能的影响比较明显；②当 3D IC 集成中芯片厚度减小到 $50\mu m$ 以下时，其影响更加明显；③随着芯片厚度的增加，最大结温与热阻均减小。

9.3.4　各带两个热源的两个 TSV 芯片

芯片的尺寸为 $5mm\times 5mm\times 0.05mm$，每层芯片上都有两个功率为 0.1W、面积为 $0.2mm\times 0.2mm$ 的热源。图 9.11 和图 9.12 所示为芯片上两个热源间距对 3D IC 集成热性能的影响。从图中可以看出：当 $b/a \leqslant 0.7$ 时，两个热源的间距越大，封装的热性能越好，即最大结温越低、热阻越小；当两个热源间距 b 大于 $0.7a$ 时，

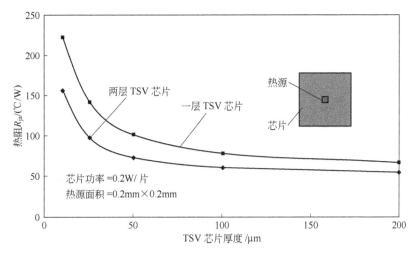

图 9.10 TSV 芯片厚度与热阻的关系（热源位于芯片中心）

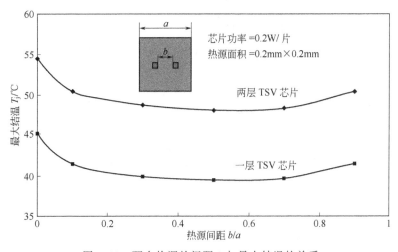

图 9.11 两个热源的间距 b 与最大结温的关系

即热源太靠近芯片边缘时，封装的热性能较差，这是因为芯片边缘附近的热扩散受到了抑制。

9.3.5 交错热源作用下的两个 TSV 芯片

前面所述为两层芯片上的两个热源重叠布置在 z 方向上，现在将它们在 z 方向交错排布，研究对 3D 堆叠芯片热性能的影响，结果如图 9.13 和图 9.14 所示。可以看出：①与重叠热源的分析结果相似，当 $b/a \leqslant 0.7$ 时，两对交错热源间距越大，最大结温越低、热阻越小；②当两对交错热源间距 b 大于 $0.7a$，即热源与芯片边缘十分接近时，封装的热性能较差；③相比于热源重叠布置的情况，交错排布时 3D 堆叠芯片封装的最大结温和热阻都较低，这是因为交错布置的热源避免了叠加

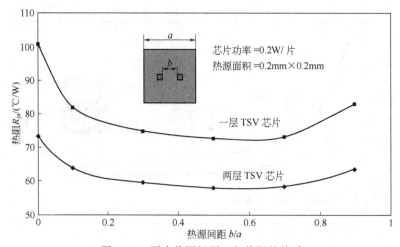

图 9.12　两个热源间距 b 与热阻的关系

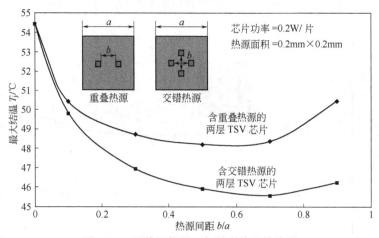

图 9.13　两热源间距 b 与最大结温的关系

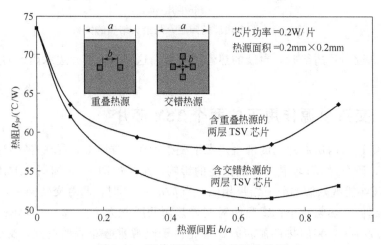

图 9.14　两热源间距 b 与热阻的关系

效应，因而具有更好的热性能。该结果有助于 3D 芯片堆叠的设计和布局，即可以通过热源位置重布或转换芯片方位改善封装的热性能。

9.4 TSV 芯片厚度对热点温度的影响

不同芯片厚度（10~200μm）所对应的 TSV 芯片温度分布如图 9.15 所示。芯片尺寸为 5mm×5mm，每层芯片中间都有面积为 0.2mm×0.2mm、功率为 0.2W 的热源。可以看出，当热源功率为 0.2W、芯片厚度为 100~200μm 时，芯片表面的热量耗散情况较好。芯片厚度为 200μm 时，其温度近于均匀分布，约为 35℃。然而当芯片厚度减少至 10μm 时，芯片热点温度达到 69℃，并且热点区域十分明显。因此，对于 DRAM 中常见的功率为 0.4W、厚度为 10μm、面积为 5mm×5mm 的芯片，其热点温度可以达到 138℃，已经远远超过大多数 Si 芯片的最大允许结温（一般为 85℃）。

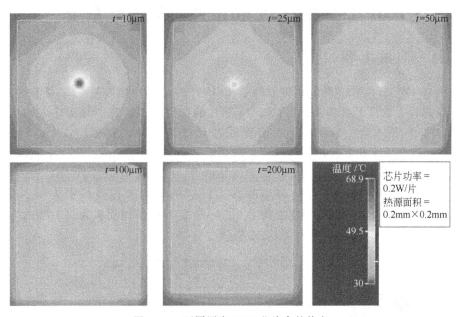

图 9.15 不同厚度 TSV 芯片中的热点

9.5 总结与建议

采用数值模拟研究了 TSV 转接板对 3D SiP 和芯片堆叠封装热性能的影响，一些重要的结果总结如下[31]。

① TSV 的散热效应可以提高 3D SiP 的热性能。

② 当TSV转接板尺寸从21mm×21mm增大至45mm×45mm时，SiP的热阻减少了14%。

③ 当TSV转接板厚度从50μm增加到400μm时，SiP的热阻减少了11%。

④ 堆叠芯片数目越多，最大结温越高。因此，堆叠芯片的数目受到允许结温的限制。

⑤ 芯片厚度对3D SiP的热性能具有重要影响。芯片越薄，最大结温和热阻越高。

⑥ 芯片越薄热点越集中。当芯片厚度减小至50μm或更小时，热点对芯片的厚度变化十分敏感。

⑦ 对于重叠布置的两个热源，其间距越大，热性能越好，但热源不能太靠近芯片边缘。

⑧ 与重叠布置的两个热源相比，两对交错排布时热源的热性能更好。

9.6 3D SiP封装的TSV和微通道热管理系统

9.6.1 测试模型

图9.16所示为带有热管理微通道和TSV的Si载芯片结构图[19]。该载芯片由两个硅芯片组成且无其他器件，两个Si芯片的区别在于底部芯片无输入输出口，如图9.17~图9.19所示。TSV沿着载芯片外围排列，共有144个，间距为0.5μm。键合后，载芯片通过经壁面金属化的TSV实现电互连。将一个或两个输入口与一个或两个输出口连接起来形成流体通道，流体通道与每个TSV周围都有密封环，从而将流体与电互连隔绝开来。TSV的深度和直径分别为400μm和150μm，微流体管道的深度和宽度均为350μm。密封环所用焊料为Au20Sn，凸点下金属层（UBM）所用材料为TiCuNiAu。

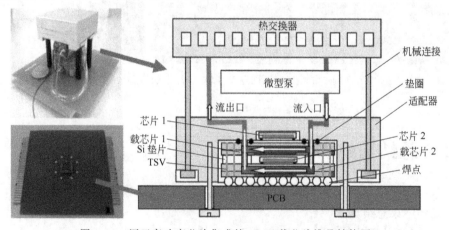

图9.16 用于高功率芯片集成的3D Si载芯片堆叠结构图

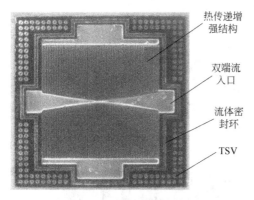

图 9.17 微通道的输入/输出口、密封环以及 TSV 的设计

9.6.2 测试模型制作

图 9.20 所示为在 200mm 晶圆上制作测试模型的工艺流程。首先采用 Novellus PECVD 系统进行等离子体增强化学气相沉积,制作 3μm 厚的 SiO_2 层,然后刻蚀制作通孔和冷却通道图形作为硬掩膜,接着采用深反应离子刻蚀(DRIE)工艺对 TSV 和冷却通道进行刻蚀。由于 TSV 与冷却通道结构的深度不同,故将 DRIE 工艺分两步进行:首先,采用光刻胶作为刻蚀掩膜刻蚀深度为 100μm 的通孔;然后去除光刻胶,同时对 TSV 和通道进行刻蚀,刻蚀深度为 350μm。第一步中沉积的 SiO_2 层可作为刻蚀掩膜。由于 DRIE 工艺只能刻蚀盲孔,所以完成 DRIE 工艺之后要将 Si 晶圆背面减薄和抛光至 450μm,从而制得通孔。然后在晶圆两面沉积钝化层,采用 Plasma-Therm SLR720 系统沉积厚度为 1μm 的 SiO_2 钝化层。

完成钝化沉积后,在晶圆两面的 TSV 和冷却通道周围溅射并制作 UBM 图形。采用 Balzer 系统完成溅射工艺,UBM 所用材料为 Ti/Cu/Ni/Au,厚度分别为 0.1μm、2μm、0.5μm、0.1μm。为了确保 TSV 侧壁完全覆盖有用于电互连的金属薄膜,需在晶圆两面溅射一层较厚的 Cu 层,制作 UBM 层时也采用了这种方法。对于这种特殊应用,溅射工艺比电镀 Cu 工艺更加简单且耗时少,这种 TSV 的电互连制作性价比较高。

通常情况下,较厚的 UBM 层难以进行湿法刻蚀。以下是一个可行的工艺流程:①采用三碘化物刻蚀最上面的 Au 层;②采用 AC-100 刻蚀 Ni 层;③采用 A95 刻蚀 Cu 层;④使用去离子水(DI)稀释的 49%HF 溶液(即 49%HF:H_2O = 1:15)刻蚀 Ti 层。制作完 UBM 图形之后,采用激光钻孔在上层芯片上制作两个或四个矩形孔,即流体的输入和输出口。

制作 Au20Sn 焊料层的方法有很多种,包括蒸镀法、电镀法、黏合法以及焊料预成型法等。这些方法中,蒸镀法制作的焊料层尺寸以及位置控制更加精确。因此制作完 UBM 图形后,采用蒸镀法制得交替间隔的 Sn/Au 层,共八层 Sn 和八层 Au。同时利用剥离工艺制作焊料图形,每层 Sn 和 Au 的厚度分别为 0.2μm 和

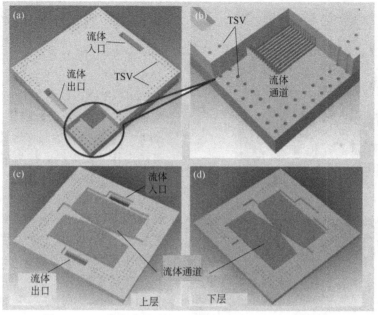

图 9.18 Si 载芯片的结构图：(a) 俯视图；(b) 剖视图；
(c) 上层 Si 芯片；(d) 下层 Si 芯片

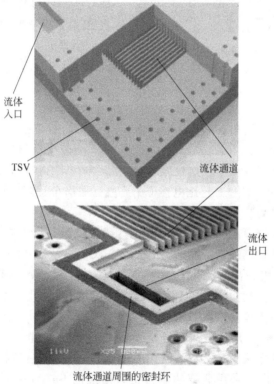

图 9.19 微通道的放大图像

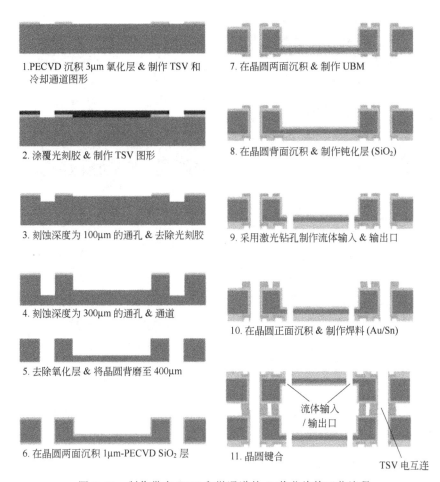

图 9.20　制作带有 TSV 和微通道的 Si 载芯片的工艺流程

$0.4\mu m$。采用 Semicore Equipment 公司生产的 Temescal 电子束蒸镀机（型号为 VES-2550）完成蒸镀工艺，蒸镀速率为 5Å/s。剥离工艺采用厚度为 $20\mu m$ 的干膜作为模具。

制得的晶圆如图 9.21 所示。AuSn 焊料层厚度为 $3.5\mu m$，UBM 厚度为 $2.7\mu m$，其中包括 $0.1\mu m$ 厚的 Ti 层、$2\mu m$ 厚的 Cu 层、$0.5\mu m$ 厚的 Ni 层以及 $0.1\mu m$ 厚的 Au 层。TSV 周围的环状密封环宽度为 $100\mu m$，Si 晶圆之间的密封环宽度为 $300\mu m$。TSV 的间距为 $500\mu m$，其内部未填满 Cu，而是通过双面金属薄膜溅射制作孔壁金属化层。

9.6.3　晶圆到晶圆键合

焊料层形成后，采用晶圆级键合方法对两个晶圆进行键合。键合温度为 150℃，最高温度下保温时间为 15min，键合压力为 8kN（静态压强为 4.7MPa）。完成键合之后，将晶圆切割成 $15.1\text{mm}\times 15.1\text{mm}\times 0.8\text{mm}$ 的单个载芯片。晶圆切

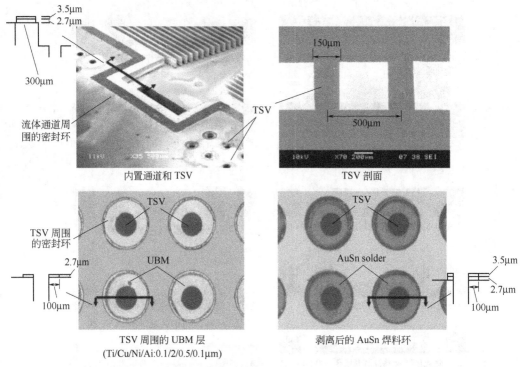

图 9.21　带有 TSV 和微通道 Si 载芯片的几何尺寸和材料

割采用的是迪斯科公司生产的 DAD651 型切割机，主轴转速为 30000r/min，进给速率为 5mm/s。

9.6.4　热性能与电性能

制得的 UBM 层如图 9.21 左下图所示，其边缘较为平滑。图 9.22 所示为压降

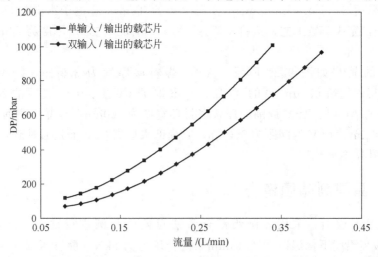

图 9.22　压降测试结果（双输入/输出口晶圆的热性能优于单输入/输出口情况）

测试结果。可以看出，相比于只有一个输入/输出口的载芯片，有两个输入/输出口的载芯片其热性能更好。测试了不同直径 TSV 的导电性，结果见表 9.4。当直径大于 $150\mu m$ 时，TSV 的电阻小于 0.5Ω，因此，采用直径为 $150\mu m$ 的 TSV 以获取更高的输入/输出速率。图 9.23 所示为经过金属化处理的直径为 $150\mu m$ 的 TSV 剖面图。从图中可以清楚地看到，TSV 侧壁覆盖有连续的金属层。

表 9.4 不同直径 TSV 的电阻

TSV 直径/μm	电阻/$k\Omega$	平均电阻/$k\Omega$
100	1.50~12.00	6.75
125	1.50~6.50	4.00
150	0.46~0.52	0.49
200	0.38~0.44	0.41
300	0.38~0.44	0.41

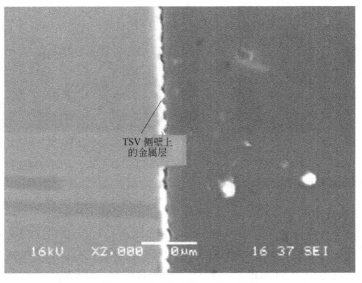

图 9.23 TSV 侧壁的薄金属层

9.6.5 品质与可靠性

采用商业剪切试验机（Dage-SERIES-4000-T）对切割好的封装试样进行剪切强度测试。测得 20 个不同试样的剪切强度变化范围为 18.7~35.1MPa，平均值为 27.2MPa，标准差为 2.2MPa。剪切测试过程中发现断裂普遍发生在 UBM 层上，这表明焊点的键合强度高于 UBM 层的黏结强度。图 9.24 所示为剪切测试后的断面，可以看出金属层从表面剥落，说明键合强度较好。图 9.25 所示为键合界面的剖面图，可以看出截面比较均匀。

在 AuSn 键合过程中，压强是获得良好键合效果的重要参数。键合温度为

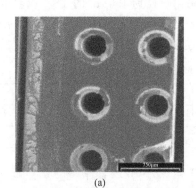

(a)

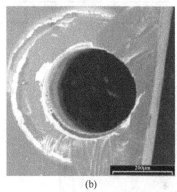

(b)

图 9.24 (a) 剪切测试后焊点表面的 SEM 图像；(b) 放大图像

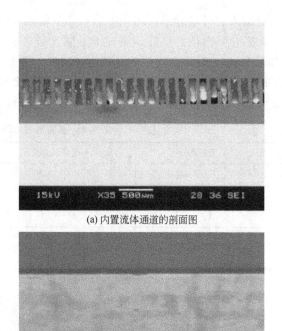

(a) 内置流体通道的剖面图

(b) 键合后 AnSn 焊料密封环的剖面图

图 9.25 键合界面的剖面图

350℃，最高温度下保温 15min，键合压力为 3kN（静态压强为 1.7MPa）时，得到的 AuSn 键合环质量较差，其截面如图 9.26 所示。可以看到，在界面中间出现了明显的裂纹，这是键合压强不够引起的。无论从宏观还是微观角度来说，较高的压强可以增大两键合界面的接触面积。一般来说，Si 晶圆表面并非十分平整，尤其经过背磨、抛光以及精加工之后。因此，施加高压强可以使两个键合面紧密地贴合在一起。同时，蒸镀后 AuSn 表面粗糙度的均方根值为 24nm，如图 9.27 所示。粗糙的表面会减小两键合面的接触面积，因而需要施加高压强克服表面粗糙问题，并增大两键合表面的接触面积。

图 9.28 所示为 Si 晶圆的键合对，分别为单个载芯片的上下表面，其两边的金属图形用于提供电互连以及再布线。切割后芯片的良率超过 98%，说明键合强度足够高，可以完成切割工艺。

对试样进行 1000 次 -40~125℃ 的热循环测试，高低温各保持 15min，升/降温速率为 15℃/min。测试完成后，20 个试样的平均剪切强度为 26MPa，基本与测试前的剪切强度一样。

第 9 章　3D IC 集成的热管理

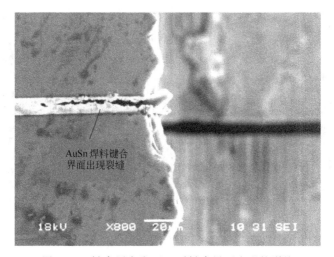

图 9.26　键合压力为 3kN 时键合界面出现的裂纹

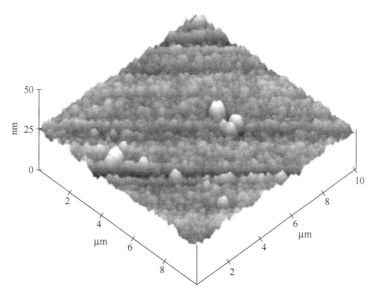

图 9.27　UBM 上蒸镀所得 AuSn 焊料的表面粗糙度，
粗糙度的均方根值为 25nm

键合载芯片的密封性是整个系统的一个关键指标。针对这个问题，设计了一个测试模型，即在载芯片中通入高压循环水流，流速逐渐增加至 350mL/min，结果未发现泄漏，载芯片中最大压降值为 $6 \times 10^4 Pa$。

9.6.6　总结与建议

① 提出了一套制作含有内置流体通道以及电互连 TSV 的 Si 载芯片的集成工艺。载芯片可以互相堆叠，中间还可以嵌入 Si 转接板，从而得到用于高功率芯片

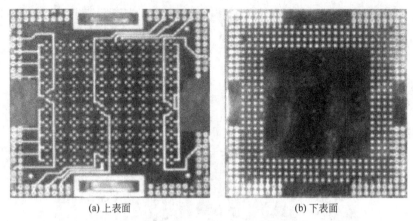

(a) 上表面 (b) 下表面

图 9.28　切割得到的单个 Si 载芯片

散热的堆叠式冷却模块。

② 直径大于 150μm、深度小于 400μm 的 TSV 经过壁面金属化后，其电阻为 0.49Ω。键合后载芯片上下表面间的电阻为 1.16Ω，可以通过增大金属层厚度或 TSV 直径来减小电阻。

③ 单个载芯片中通入高流速（350mL/min）水流未观察到泄漏现象，其最大压降值为 6×10^4 Pa。

④ 相比于单输入/输出口的载芯片，双输入/输出口 Si 载芯片的热性能更好。

⑤ 较高的键合压强可以提高 AuSn 焊料的键合质量。键合压强为 4.7MPa 时，芯片的剪切强度高于 27MPa，已经满足 Si 载芯片切割/拿持的要求，且能够提供无泄漏的流体通道。经过 1000 次 −40～125℃ 的热循环测试后，试样的平均剪切强度为 26MPa，未发生明显的退化。

9.7　参考文献

[1] Andry, P. S., C. K. Tsang, B. C. Webb, E. J. Sprogis, S. L. Wright, B. Bang, and D. G. Manzer, "Fabrication and Characterization of Robust Through-Silicon Vias for Silicon-Carrier Applications," *IBM Journal of Research and Development*, Vol. 52, No. 6, 2008, pp. 571–581.

[2] Knickerbocker, J. U., P. S. Andry, B. Dang, R. R. Horton, C. S. Patel, R. J. Polastre, K. Sakuma, E. S. Sprogis, C. K. Tsang, B. C. Webb, and S. L. Wright, "3D Silicon Integration," *IEEE/ECTC Proceedings*, Orlando, FL, May 2008, pp. 538–543.

[3] Kumagai, K., Y. Yoneda, H. Izumino, H. Shimojo, M. Sunohara, and T. Kurihara, "A Silicon Interposer BGA Package with Cu-Filled TSVs and Multilayer Cu-Plating Interconnection," *IEEE/ECTC Proceedings*, Orlando, FL, May 2008, pp. 571–576.

[4] Sunohara, M., T. Tokunaga, T. Kurihara, and M. Higashi, "Silicon Interposer with TSVs (Through-Silicon Vias) and Fine Multilayer Wiring," *IEEE/ECTC Proceedings*, Orlando, FL, May 2008, pp. 847–852.

[5] Lee, H. S., Y.-S. Choi, E. Song, K. Choi, T. Cho, and S. Kang, "Power Delivery Network Design for 3D SIP Integrated over Silicon Interposer Platform," *IEEE/ECTC Proceedings*, Reno, NV, May 2007, pp. 1193–1198.

[6] Matsuo, M., N. Hayasaka, and K. Okumura, "Silicon Interposer Technology for High-Density Package," *IEEE/ECTC Proceedings*, Las Vegas, NV, May 2000, pp. 1455–1459.

[7] Selvanayagam, C., J. H. Lau, X. Zhang, S. Seah, K. Vaidyanathan, and T. Chai, "Nonlinear Thermal Stress/Strain Analysis of Copper Filled TSVs (Through-Silicon Vias) and Their Flip-Chip Microbumps," *IEEE/ECTC Proceedings*, Orlando, FL, May 2008, pp. 1073–1081.

[8] Wong, E., J. Minz, and S. K. Lim, "Effective Thermal Via and Decoupling Capacitor Insertion for 3D System-on-Package," *IEEE/ECTC Proceedings*, San Siego, CA, May 2006, pp. 1795–1801.

[9] Lau, J. H., C. Lee, C. Premachandran, and A. Yu, *Advanced MEMS Packaging*, McGraw-Hill, New York, 2010.

[10] Lau, J. H., *Reliability of RoHS-Compliant 2D and 3D IC Interconnects*, McGraw-Hill, New York, 2011.

[11] Lau, J. H., "Critical Issues of TSV and 3D IC Integration," *IMAPS Transactions, Journal of Microelectronics and Electronic Packaging*, First Quarter Issue, 2010, pp. 35–43.

[12] Lau, J. H., "Design and Process of 3D MEMS System-in-Package (SiP)," *IMAPS Transactions, Journal of Microelectronics and Electronic Packaging*, First Quarter Issue, 2010, pp. 10–15.

[13] Lau, J. H., R. Lee, M. Yuen, and P. Chan, "3D LED and IC Wafer-Level Packaging," *Journal of Microelectronics International*, Vol. 27, No. 2, 2010, pp. 98–105.

[14] Lau, J. H., "State-of-the-Art and Trends in 3D Integration," *Chip Scale Review*, March–April 2010, pp. 22–28.

[15] Lau, J. H., Y. S. Chan, and R. S. W. Lee, "3D IC Integration with TSV Interposers for High-Performance Applications," *Chip Scale Review*, September–October 2010, pp. 26–29.

[16] Lau, J. H., "Overview and Outlook of Through-Silicon Via (TSV) and 3D Integrations," *Journal of Microelectronics International* Vol. 28, No. 2, 2011, pp. 8–22.

[17] Yu, A., J. H. Lau, S. Ho, A. Kumar, H. Yin, J. Ching, V. Kripesh, D. Pinjala, S. Chen, C. Chan, C. Chao, C. Chiu, M. Huang, and C. Chen, "Three-Dimensional Interconnects with High Aspect Ratio TSVs and Fine-Pitch Solder Microbumps," *IEE Proceedings*, San Diego, CA, May 2009, pp. 350–354. Also, accepte lication in *IEEE Transactions in Advanced Packaging*.

[18] Yu, A., A. Kumar, S. Ho, H. ınn, J. H. Lau, J. Ching, V. Kripesh, D. Pinjala, S. Chen, C. Chan, C. Chao, C. Chiu, M. Huang, and C. Chen, "Development of Fine-Pitch Solder Microbumps for 3D Chip Stacking," *IEEE/EPTC Proceedings*, Singapore, December 2008, pp. 387–392. Also, accepted for publication in *IEEE Transactions in Advanced Packaging*.

[19] Yu, A., N. Khan, G. Archit, D. Pinjalal, K. Toh, V. Kripesh, S. Yoon, and J. H. Lau, "Fabrication of Silicon Carriers with TSV Electrical Interconnects and Embedded Thermal Solutions for High Power 3D Packages," *IEEE Transactions on Components and Packaging Technology*, Vol. 32, No. 3, September 2009, pp. 566–571.

[20] Tang, G., O. Navas, D. Pinjala, J. H. Lau, A. Yu, and V. Kripesh, "Integrated Liquid Cooling Systems for 3D Stacked TSV Modules," *IEEE Transactions on Components and Packaging Technology*, Vol. 33, No. 1, 2010, pp. 184–195.

[21] Zhang, X., T. Chai, J. H. Lau, C. Selvanayagam, K. Biswas, S. Liu, D. Pinjala, G. Tang, Y. Ong, S. Vempati, E. Wai, H. Li, B. Liao, N. Ranganathan, V. Kripesh, J. Sun, J. Doricko, and C. Vath, "Development of Through Silicon Via (TSV) Interposer Technology for Large Die (21- × 21-mm) Fine-Pitch Cu/Low-k FCBGA Package," *IEEE/ECTC Proceedings*, May 2009, pp. 305–312. Also, *IEEE Transactions in Advanced Packaging* (in press).

[22] Kumar, A., X. Zhang, Q. Zhang, M. Jong, G. Huang, V. Kripesh, C. Lee, J. H. Lau, D. Kwong, V. Sundaram, R. Tummula, and M. Georg, "Evaluation of Stresses in Thin Device Wafers Using Piezoresistive Stress Sensors," *IEEE Proceedings of EPTC*, December, pp. 1270–1276. Also, *IEEE Transactions in Components and Packaging Technology*, Vol. 1, No. 6, June 2011, pp. 841-850.

[23] Vempati1, S. R., S. Nandar, C. Khong, Y. Lim, K. Vaidyanathan, J. H. Lau, B. P. Liew, K. Y. Au, S. Tanary, A. Fenne, R. Erich, and J. Milla, "Development of 3D Silicon Die Stacked Package Using Flip-Chip Technology with Microbump

Interconnects," *IEEE/ECTC Proceedings*, San Diego, CA, May 2009, pp. 980–987.

[24] Lim, S., V. Rao, H. Yin, W. Ching, V. Kripesh, C. Lee, J. H. Lau, J. Milla, and A. Fenner, "Process Development and Reliability of Microbumps," *IEEE/ECTC Proceedings*, December 2008, pp. 367–372. Also, *IEEE Transactions in Components and Packaging Technology*, Vol. 33, No. 4, December 2010, pp. 747–753.

[25] Selvanayagam, C., J. H. Lau, X. Zhang, S. Seah, K. Vaidyanathan, and T. Chai, "Nonlinear Thermal Stress/Strain Analysis of Copper-*Filled TSVs (Through-Silicon Vias) and Their Flip-Chip Microbump*s," *IEEE Transactions in Advanced Packaging*, Vol. 32, No. 4, Nov. 2009, pp. 720–728.

[26] Made, R., C. L. Gan, L. Yan, A. Yu, S. U. Yoon, J. H. Lau, and C. Lee, "Study of Low Temperature Thermocompression Bonding in Ag-In Solder for Packaging Applications," *Journal of Electronic Materials*, Vol. 38, 2009, pp. 365–371.

[27] Yan, L.-L., C.-K. Lee, D.-Q. Yu, A.-B. Yu, W.-K. Choi, J. H. Lau, and S.-U. Yoon, "A Hermetic Seal Using Composite Thin Solder In/Sn as Intermediate Layer and Its Interdiffusion Reaction with Cu," *Journal of Electronic Materials*, Vol. 38, 2009, pp. 200–207.

[28] Lee, C., A. Yu, L. Yan, H. Wang, J. Han, Q. Zhang, and J. H. Lau, "Characterization of Intermediate In/Ag Layers of Low-Temperature Fluxless Solder–Based Wafer Bonding for MEMS Packaging," *Journal of Sensors & Actuators* (in press).

[29] Yu, D.-Q., C. Lee, L. L. Yan, W. K. Choi, A. Yu, and J. H. Lau, "The Role of Ni Buffer Layer on High-Yield Low-Temperature Hermetic Wafer Bonding Using In/Sn/Cu Metallization," *Applied Physics Letters*, Vol. 94, No. 3, January 2009, pp. 34105–34105-3.

[30] Zhang, X., A. Kumar, Q. X. Zhang, Y. Y. Ong, S. W. Ho, C. H. Khong, V. Kripesh, J. H. Lau, D.-L. Kwong, V. Sundaram, Rao R. Tummula, and Georg Meyer-Berg, "Application of Piezoresistive Stress Sensors in Ultrathin Device Handling and Characterization," *Journal of Sensors & Actuators: A. Physical*, Vol. 156, November 2009, pp. 2–7.

[31] Lau, J. H., and G. Tang, "Thermal Management of 3D IC Integration with TSV (Through Silicon Via)", *IEEE/ECTC Proceedings*, San Diego, May 2009, pp. 635–640.

[32] Yamaji, Y., "Thermal Characterization of Baredie Stacked Modules with Cu Through-Vias," *IEEE/ECTC Proceedings*, 2001, pp. 730–732.

[33] Wong, E., "Effective Thermal Via and Decoupling Capacitor Insertion for 3D System-on-Package," *IEEE/ECTC Proceedings*, 2006, pp. 1795–1801.

[34] Bar-Cohen, A., and A. Kraus, *Advances in Thermal Modeling of Electronic Components and Systems*, Vol. 2, ASME, New York, 1988.

[35] Kraus, A., and A. Bar-Cohen, *Thermal Analysis and Control of Electronic Equipments*, Hemisphere, New York, 1983.

[36] Simons, R. E., "Simple Formulas for Estimating Thermal Spreading Resistance," *Electronics Cooling*, Vol. 10, 2004, pp. 6–8.

第 10 章 3D IC 封装

10.1 引言

第 1 章曾提到,3D 集成包括 3D IC 集成、3D Si 集成和 3D IC 封装三部分。本书前 9 章主要讨论了 3D IC 集成和 3D Si 集成相关技术,因为它们是未来的重要技术。然而,如图 10.1 所示,3D IC 集成和 3D Si 集成技术虽然取得了一些新的进展,但离量产尚有距离。相反,3D IC 封装技术已经比较成熟。

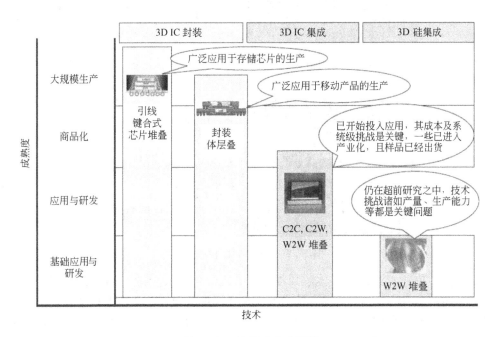

图 10.1 3D 集成技术概览

本章主要讨论 3D IC 封装技术,包括:①Cu-low-k 堆叠芯片的引线键合技术;②芯片到芯片的面对面堆叠;③扇出嵌入式晶圆级封装到芯片的面对面组装。本章最后,给出了引线键合时引线的受力与变形分析结果。

本章首先简要讨论 TSV 技术的成本问题。

10.2 TSV 技术与引线键合技术的成本比较

如前所述,TSV 是一项颠覆性技术,正试图取代十分成熟的、高良率低成本的引线键合技术。第 2 章曾阐明,TSV 制程包括 6 个关键工艺步,如图 10.2 所示。以 300mm 晶圆为例,每一工艺步使用的自动化设备及其配件所需成本通常都超过 100 万美元,而且这些设备都必须放置在 10 级或更低的净化间。并且,由于技术不断更新,这些设备使用寿命短,贬值快。此外,建设和维护每天 24h 运行的净化间的成本也非常高。而引线键合机只需安装在 1000 级(有些甚至只需 10000 级)的净化间内,因此,与引线键合技术相比,TSV 技术十分昂贵,需要认真考虑其成本问题。

全自动引线键合是一种很成熟的技术,其良率达 99.99%,且成本低、可靠性好

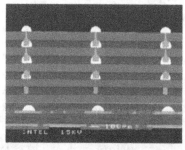

加工 TSV 的六个关键步骤:
1. 通孔刻蚀 (DRIE)
2. 绝缘层 (PECVD)
3. 阻挡层/种子层 (PVD)
4. 通孔填镀(电镀铜)
5. CMP
6. TSV Cu 外露
所耗成本高低:
PVD＞PECVD＞CMP＞电镀＞刻蚀

图 10.2　TSV 取代引线键合

目前,超过 85% 的半导体芯片都采用引线键合技术。例如,图 10.3 所示为苹果 iPhone4 使用的基于引线键合技术的三星八层堆叠闪存芯片,包括基板在内整个封装的厚度大约为 0.93mm,其中芯片堆叠的高度约为 670μm,各个芯片的厚度从 55~70μm 不等,厚度最大的芯片位于最底层。

然而,如同 IBM 公司 40 年前开发的焊锡凸点倒装芯片的情况一样[1,2],虽然刚开始时得不到市场认可,但由于其具有独特的优势,TSV 技术终将在低功耗、宽带宽、高性能以及高密度等领域得到应用。与引线键合技术相比,TSV 技术可以实现芯片与芯片之间最短互连、最小的焊盘尺寸和节距,进而能实现更好的电性能、更低的功耗、更宽的数据带宽、更高的密度、更小的尺寸以及更轻的质量。

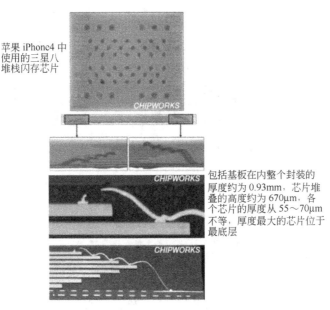

图 10.3　苹果 iPhone4 中使用的三星八层堆叠闪存芯片

10.3　Cu-low-k 芯片堆叠的引线键合

10.3.1　测试模型

图 10.4 给出了两种测试模型，它们最底层的芯片都是 65nm Cu-low-k 芯片，并粘接在有机基板上。测试模型 1（TV1）中的 Cu-low-k 母芯片上方堆叠了两个子芯片，两个子芯片之间采用内置引线薄膜（WEF）粘接，如图 10.4（a）所示。测试模型 2（TV2）中的 Cu-low-k 母芯片上方也堆叠了两个子芯片，但两个子芯片之间采用芯片贴膜（DAF）粘接，如图 10.4（b）所示。对于这两种测试模型，下层子芯片与 Cu-low-k 芯片之间也采用 DAF 粘接，Cu-low-k 芯片与基板之间采用芯片粘接膏粘接。

测试模型中的 65nm Cu-low-k 芯片具有 6 个金属层、408 个 I/O，尺寸为 7mm×7mm，厚度为 0.3mm；子芯片的尺寸为 3mm×5mm×0.075mm；有机基板由两个金属层以及一个塑料球栅阵列（PBGA）基板组成。最终整体封装的尺寸为 17mm×17mm×1.1mm。基板和塑封材料使用的都是绿色环保材料，符合环境保护相关法规。

10.3.2　Cu-low-k 焊盘上的应力

Cu-low-k 材料的力学性能较差，在封装制程中容易出现 low-k 剥落、键合焊

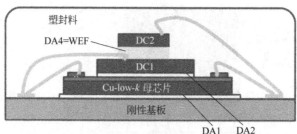

(a) 测试模型 1(TV1) 使用内置引线薄膜 (WEF)

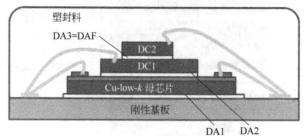

(b) 测试模型 2(TV2) 使用芯片贴膜
(DAF) 将两个子芯片以合适的角度粘结

图 10.4　65nm Cu-low-k 堆叠芯片封装示意图

盘坑洞以及钝化层分层等问题[3~5]。切割时，low-k 晶圆常见的失效形式包括金属/中间绝缘层（ILD）分层、金属/ILD 剥落以及金属层变色[6~13]。新加坡微电子研究所开发了聚合物封装切割技术（PEDL）[6,14]。与传统的切割方法相比，该技术可以解决 Cu-low-k 封装时的问题，同时通过减小 Cu-low-k 芯片的角隅应力，进而提高封装的可靠性。

应力集中位于硅芯片的最外侧拐角处，这是由于不同材料间热膨胀系数（CTE）不匹配以及芯片拐角处应力奇异性的缘故。芯片拐角处的切应力是导致 low-k 层与氟硅酸玻璃层（FSG）分层失效的主要原因之一[14]。测试结果表明，在硅芯片的边缘开一道 45°角、35μm 长的凹槽之后，最大应力可以减少 34% 或者更多[14]。

封装结构如图 10.4(b) 所示，材料属性见表 10.1。PEDL 技术采用斜边切口与 BCB 涂覆减小 low-k 层的应力，如图 10.5 所示。图 10.6 所示为 175~25℃ 降温过程中整体封装结构在四种不同切口与 BCB 涂覆组合下的有限元分析结果[15]。图 10.6(a) 表示的是靠近粘接层 2 的 low-k 层中的切应力分布，而图 10.6(b) 表示的是 low-k 拐角处的切应力分布。通过观察可以发现，四种情况下最大应力几乎位于同一个位置。

表 10.1　用于有限元建模与分析的材料属性

材料	CTE/(ppm/℃)	E/GPa	泊松比
BCB	52	3	0.35
low-k(BC)	10	8	0.30
FTEOS	2	105.44	0.30

续表

材料	CTE/(ppm/℃)	E/GPa	泊松比
TEOS	0.57	66	0.18
Si	2.7	131	0.28
Cu	17	110	0.34
塑封料	8.9/40.4(T_g=147.8℃)	20.7(25℃) 0.296(250℃)	0.30
芯片粘接膏	50/100(T_g=150℃)	5.9(25℃) 0.56(260℃)	0.30
DAF	80/170(T_g=128℃)	1.66(25℃) 0.037(200℃)	0.30
基板	15	29	0.30
阻焊层	52	4	0.40

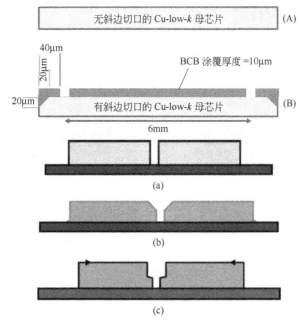

图 10.5 （图 A）无斜边切口、无 BCB 的堆叠芯片示意图；
（图 B）带有斜边凹槽和 BCB 的堆叠芯片示意图。不同切口形式：
(a) 纵向切口；(b) 斜边切口；(c) 阶梯切口

图 10.7 所示为采用或者未采用斜边切口和 BCB 涂覆时，low-k 层、粘接层以及环氧树脂塑封料中的切应力情况。由于 low-k 层很容易发生分层，因此仔细观察了 low-k 层拐角处以及靠近粘接层 2 处这两个位置的切应力。结果发现，斜边切口能够有效减小 low-k 层拐角处的切应力。当采用有斜边切口、无 BCB 涂覆这个方案取代初始方案（即无斜边切口、无 BCB 涂覆）时，low-k 层拐角处的切应力由 100.90MPa 减小到 57.68MPa。当既有斜边切口又有 BCB 涂覆时，low-k 层拐角处

的切应力由 100.90MPa 减小到 23.75MPa。另一方面还发现，BCB 涂覆可以有效减小粘接层附近 low-k 层内的切应力。当既有斜边切口又有 BCB 涂覆时，靠近粘接层的 low-k 层内的切应力由 32.30MPa 减小到 11.90MPa。

总之，如图 10.7 所示，斜边切口以及 BCB 涂覆都能够减小 low-k 层的应力。因此，将 PEDL 技术应用到封装中不仅可以减小芯片拐角处的应力，而且能够减小子芯片 1 下方 low-k 层的应力。

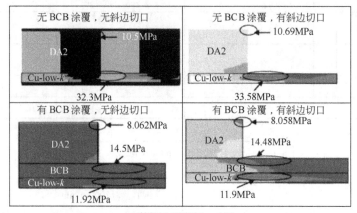

(a) 粘接层 2 附近 low-k 层

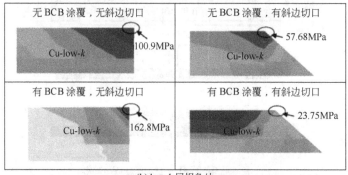

(b) low-k 层拐角处

图 10.6　切应力分布

10.3.3　组装与工艺

正如第 4 章和第 5 章所提到的，所有 3D 集成都需要薄晶圆/芯片。在本研究中，Cu-low-k 晶圆减薄至 300μm 以下，Al 晶圆（子芯片）减薄至 75μm。所有晶圆都是先经过粗磨，然后再进行精磨。为了消除减薄之后晶圆背面的微裂纹，需要增加另一道工序，即干法抛光。所有 Cu-low-k 晶圆的 TTV 为 2.21μm，而 Al 晶圆的 TTV 为 2.28μm。

（1）切割方法评估

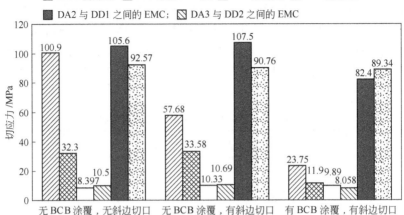

图 10.7 不同切口与 BCB 涂覆情况下的切应力分布

芯片切割分离是每一个 IC 芯片都必须经历的工艺过程。随着封装向 3D SiP 以及超薄、low-k 晶圆方向发展,切割分离过程变得更为重要。本研究比较了纵向切口法[见图 10.5(a)]、斜边切口法[见图 10.5(b)]以及阶梯切口法[见图 10.5(c)]对芯片强度和碎裂结果的影响。三种切割法的切割参数如表 10.2 所示。

表 10.2 三种切割方法的切割参数

切割方法	参数	刀片型号	切割速度/(r/min)	进给速度/(mm/s)	刀片高度/mm
纵向切口		27HCAA	30000	15	0.060
斜边切口	刀片 1	2050D-T1	30000	30	0.115
	刀片 2	HDCC-B	30000	30	0.060
阶梯切口	刀片 1	27HCCC	30000	20	0.100
	刀片 2	27HAAA	30000	20	0.060

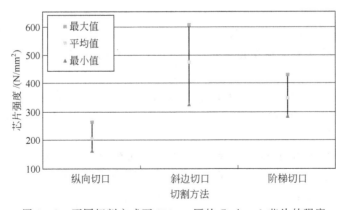

图 10.8 不同切割方式下 300μm 厚的 Cu-low-k 芯片的强度

弯曲试验被证明是确定芯片强度的可靠测量方法[16]。本研究中采用三点弯曲试验测定芯片强度，结果见图10.8和表10.3，每组试验的样本数均为33。结果发现：①斜边切口得到的芯片强度要高于其他两种方法；②斜边切口得到的芯片强度高出纵向切口130%；③斜边切口比其他两种方法产生的碎裂更少；④斜边切口产生的正面碎裂低于纵向切口10%；⑤斜边切口产生的正面碎裂低于阶梯切口15%。

表10.3 三种切割方法下芯片的碎裂情况

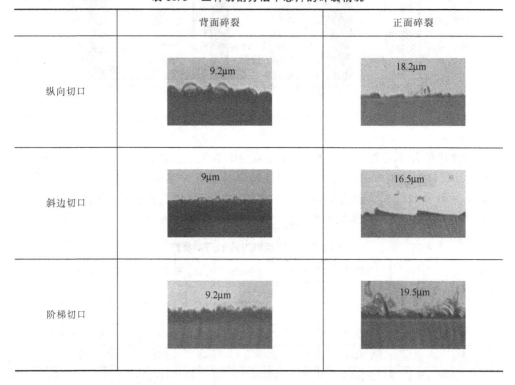

（2）芯片粘接工艺

薄芯片非常容易弯曲，所以在拿持和放置过程中需要完全支撑，以避免芯片弯曲和确保粘接层厚度一致，这两点对于封装可靠性来说十分重要。利用薄膜材料可以更容易地控制薄芯片粘接层，因此TV2采用DAF粘接上层两个子芯片，而TV1则采用WEF粘接上层两个子芯片（见图10.4）。无论是TV1还是TV2，子芯片1与Cu-low-k母芯片的粘接都是采用DAF。Cu-low-k母芯片与基板的粘接采用的是芯片粘接膏，通常其表面都不平整，这就要求粘接膏材料具有较好的流动性以覆盖整个表面，做到既无孔洞也不发生分层。

表10.4（a）给出了DAF的粘接工艺参数和粘接结果，发现第4组工艺参数可以避免产生孔洞，可将其作为最终的粘接工艺参数。DAF工艺的关键问题是薄膜孔洞，典型的DAF工艺温度为120～150℃[17]，本研究中的温度为150℃。一旦DAF完全固化，采用压铸模工艺很难有效地将DAF中的气泡（孔洞）排出。

表 10.4 DAF、WEF 工艺参数

	序号	工艺参数	后固化后的 CSAM (一步后固化法,160℃,60min)
(a) DAF 的粘接 工艺参数	1	键合力:1.0kgf 键合温度:150℃ 保温时间:1s	孔洞
	2	键合力:1.0kgf 键合温度:150℃ 保温时间:2s	孔洞
	3	键合力:2.0kgf 键合温度:150℃ 保温时间:2s	孔洞
	4	键合力:2.5kgf 键合温度:150℃ 保温时间:2s	OK
(b) WEF 的后固 化工艺参数	后固化参数	不同后固化及芯片 剪切测试后的图像	观察结果
	一步后固化法: 160℃,60min		剪切测试后的光亮区
	两步后固化法: 120℃,60min 140℃,60min		剪切测试后两薄膜的对比

WEF是一种可以将同样大小的引线键合芯片互相粘接实现堆叠的芯片粘接材料,适用于在芯片中心和外围布置有键合焊盘的情况。最终得到的WEF粘接参数如下:①粘接力为0.3kgf;②粘接温度为150℃;③150℃下保温1s。WEF工艺的一步后固化法与两步后固化法的比较如表10.4(b)所示。由剪切试样的图像可以看出,两步后固化法得到的外形轮廓要好于一步后固化法的外形轮廓,因此选取两步后固化法用于WEF。

图10.9所示为采用最终的WEF工艺参数得到的TV1中两个子芯片间引线的X射线图(利用CT重构技术)。WEF可以润湿引线并将其完全封装在内部,没有观察到引线变形。本例中,WEF共有两层,上层与顶部芯片的背面相贴,其流动性受到限制;底层与底部芯片的引线键合部分相对,其黏度较小,以避免引线变形并将底部芯片的间隙完全填充。

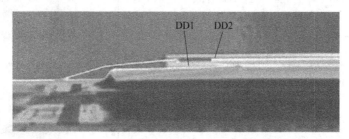

图10.9 TV1中两个子芯片间引线的X射线图

(3) 引线键合工艺

在3D系统级封装(SiP)中,引线键合是十分关键的互连方式之一。先进的引线键合机采用软件控制,并具有以下功能:对于不同尺寸芯片的堆叠,具有对复杂路径细长引线键合与精确定位能力;具有对悬臂薄芯片的引线键合能力;具有对低拱高引线的正反向键合能力;具有对low-k介电焊盘以及有源电路焊盘的键合能力。

图10.10(a)表示的是TV1中第一个子芯片上的超低拱高(50μm)引线键合结果[15],键合时采用了一种全新的正向运动键合法和特别设计的键合路径[图10.10(b)]。引线为0.8mil的NL4 Au线,引线键合机选用ASM Eagle 60,引线间距为53μm,钝化层开口为43$\mu m \times$100μm,一些关键的引线键合参数见表10.5。对于芯片堆叠封装,正向键合法具有许多优点。除了能够实现更低的线拱高度以及较少的颈部损伤以外,与反向键合

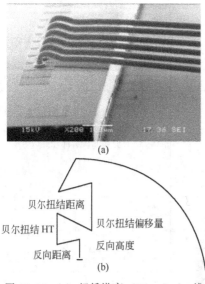

图10.10 (a) 超低拱高(50μm) Au线键合;(b) 超低拱高键合路径:贝尔环

相比，正向键合还可以减少焊盘损伤、提高产量并有利于模塑。另外，由于一次键合的变形小，正向键合能够对更小的焊盘节距实现键合。

表 10.5 关键的引线键合参数

键合参数	设置	
	一次键合	二次键合
键合时间/ms	10	8
键合功率/Dac	36	30
键合力/gf	8	27
EFO 电流/mA	3800	
EFO 时间/ms	223	
键合温度/℃	170	

引线键合完成后，试样还需进行引线拉力测试与剪切推球测试。结果发现：①所有推球测试的读数超过最低要求（8gf）时，都以预想的模式失效；②在剪切推球测试过程中，所有试样都未观察到剥落现象；③引线拉力测试的读数超过最低要求时，都以预想的模式即辊颈折断方式失效（见图 10.11）。引线键合测试模型（TV1，TV2）的实物如图 10.12 所示。

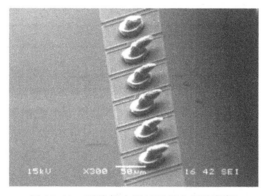

图 10.11 引线拉力测试后发生辊颈折断

（4）模塑工艺

随着 3D 芯片堆叠中引线密度及长度的不断增加，与传统的单个芯片相比，多芯片的模塑工艺更为困难。不同层次的引线键合回路受模塑化合物流动拖曳力的影响会导致引线偏移量不同，这就增加了引线短路的可能性。因此，需要认真选取合适的塑封材料（MC）以提高 3D 芯片堆叠封装的可靠性。

为选取 MC 材料，对四种 MC 材料（见表 10.6）进行附着力测试（拉拔实验）和湿气敏感性等级测试（MSL）（MSL2 及 MSL3）。试样准备如下：①将哑 low-k 芯片粘接到有机基板上；②分别采用四种 MC 材料进行塑封；③切割成 17mm×17mm 的封装体用于 MSL 测试，切割成 8mm×8mm 的封装体用于拉拔测试。先

 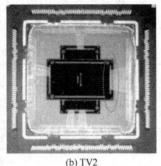

(a) TV1　　　　　　　　　(b) TV2

图 10.12　引线键合测试模型

对试样进行最高温度为 260℃ 的 3 次回流,然后进行 MSL1 测试,之后再进行拉拔实验(附着力测试)。由表 10.6 第四列可以看出,四种 MC 所对应的拉力值并无显著差异[15]。

表 10.6　塑封材料的选取

MC 编号	MSL3(失效试样/总体试样)	MSL2(失效试样/总体试样)	拉力值/gf	翘曲值/μm
MC1	0/7	1/6	2.33	377
MC2	0/7	0/7	2.00	504
MC3	0/7	5/7	2.33	692
MC4	0/7	3/7	1.33	620

对于 MSL 测试,需对试样分别完成 MSL3 及 MSL2 测试,且三次回流的最高温度均为 260℃。然后利用 C-SAM 以及透射扫描分析分别对四种 MC 的封装体进行检测,采用 JEDEC 推荐的以发生 10% 的分层作为失效准则。如表 10.6 中第二列所示,四种 MC 经过 MSL3 测试之后都未产生孔洞或分层。对于 MSL2 测试,发现 MC2 经过测试后 7 个试样均未发生失效(见表 10.6 第 3 列),而 MC1 的 6 个试样中有 1 个发生失效,MC3 的 7 个试样中有 5 个发生失效,MC4 的 7 个试样中有 3 个发生失效。因此,选择 MC2 用于后续封装。

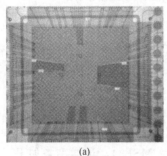 

(a)　　　　　　　　　(b)

图 10.13　(a) TV1 最大引线偏移为 9.51%;
(b) TV2 最大引线偏移为 9.04%

完成了 MC 材料的选取之后,需要对 TV1 与 TV2 进行模塑。如图 10.13 所示,对 TV1 和 TV2 的引线偏移进行测量,最大偏移量均小于 10%,在允许范围之内。

(5)可靠性测试与结果

根据 JEDEC 制定的元器件标准,最后需对 TV1 及 TV2 进行可靠性测试(见表 10.7),利用 C-SAM 与透射扫描方法检测是否发生分层。在完成 1000 个热循环(TC)测试以及 1000h 的高温储存(HTS)测试之后,所有试样均未观察到分层(见图 10.14)。

表 10.7 65nm Cu-low-k 芯片堆叠封装的热循环及高温储存可靠性测试结果

可靠性测试	组别	样本大小	电气测试 失效	C-SAM 分层
热循环(−40℃/150℃)	TV1	22	0	0
	TV2	21	0	0
高温储存(150℃)	TV1	22	0	0
	TV2	21	2	0

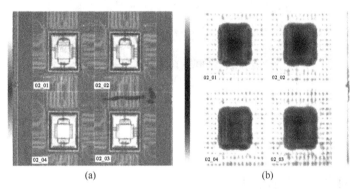

图 10.14 (a) C-SAM 结果表明 1000 次热循环(−40～150℃)后未发生分层;
(b) 透射扫描结果表明 1000 次热循环(−40～150℃)之后未发生分层

此外,还需通过电气测试观察热循环与高温储存测试过程中沿着菊花链路径的电阻特性。经过 1000 个热循环(−40～150℃)以及 1000h 的高温储存(150℃)测试之后,只有两个 TV2 试样在 HTS 测试中发生失效,其余试样均未发生电气失效(见表 10.7)。这两个失效的试样都是菊花链发生开路。将失效试样解封装后进行失效分析,拉拔及剪切推球测试都表明引线键合强度较低。最坏的情况为球形键合接头与焊盘分离,为典型的柯肯达尔孔洞失效(见图 10.15)。失效的引线键合位于顶部芯片(即子芯片 2)的右侧。需要注意的是,TV2 试样中子芯片 2 与子芯片 1 之间有 1mm 的悬空部分 [见图 10.12(b)],而 TV1 则没有。可能正是由于这 1mm 的悬空导致引线键合强度较低。但是 TV2 试样的剪切强度和引线拉拔力

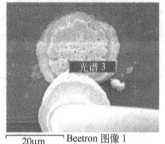
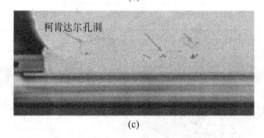

图 10.15 (a) MC 分层后 TV2 模型中引线键合失效的位置；
(b) 失效引线键合的俯视图；(c) TV2 模型中出现的柯肯达尔孔洞

都高于最低要求。

10.3.4 总结与建议

一些重要的结论及建议总结如下[15]。

① PEDL 技术被用于减小 Cu-low-k 层的应力，进而减少分层。Cu-low-k 层的应力减小了 4.2 倍。

② 对切割方法的评估表明，斜边切口切割方法优于纵向切割方法；采用斜边切口法，芯片正面碎裂减小了 15%，芯片强度提高了 2.3 倍。

③ 建立了超低拱高（50μm）引线键合工艺。

④ 建立了 WEF 工艺，允许特定的芯片贴膜在 Au 线之间渗透，以便于移除哑芯片（垫片）。

⑤ 开发了 65nm Cu-low-k 芯片堆叠 PBGA 封装（即 TV1），且顺利通过了 1000 次热循环（-40~150℃）和 JEDEC 元件级 1000h HTS（150℃）测试。

10.4 芯片到芯片的面对面堆叠

正如第 1 章所提到的，面对面芯片堆叠是 PoP 3D IC 封装技术之一[18~56]。本节主要讨论采用 AuSn 无铅焊料进行面对面芯片堆叠的互连可靠性，同时讨论晶圆微凸点和 PoP 封装的芯片到晶圆（C2W）键合组装工艺。

10.4.1 用于 3D IC 封装的 AuSn 互连

AuSn 焊料特别适用于光电及医疗器件封装，大多数的医疗和光电器件都需要无助焊剂工艺互连，因而使用 AuSn 焊料较为合适，如图 10.16 所示。其中，80%Au20%Sn 焊料得到了广泛应用，这是因为它具有高强度、高抗腐蚀性、高抗疲劳性。因此，Au20Sn 系焊料被选为无助焊剂倒装芯片的主要互连材料。除此以外，也有选用其他 AuSn 焊料和 SnAg 焊料的情况。

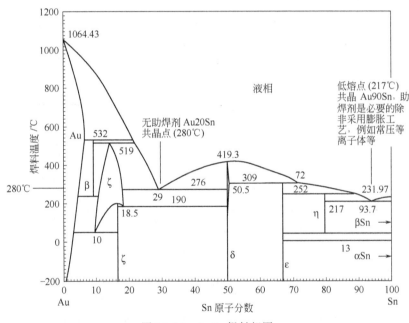

图 10.16 AuSn 焊料相图

10.4.2 测试模型

（1）测试模型

测试模型为一个 Si 堆叠模块，如图 10.17 所示[49]。该模块由一个顶部芯片（子芯片）与一个底部芯片（母芯片）组成，两个芯片都由晶圆制得。子芯片上制有 Cu 柱涂覆 AuSn 焊料的凸点，母芯片上制有用于连接子芯片凸点的化学镀 NiAu

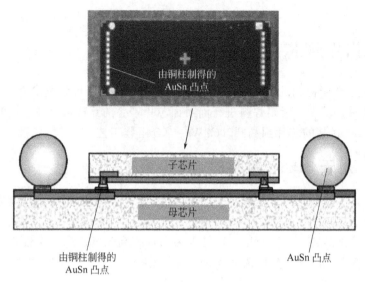

图 10.17 堆叠模块

UBM 以及用于下一级互连（如基板）的 AuSn 焊锡凸点。两个芯片的焊盘在堆叠后互连，形成菊花链。子芯片上有两排 20 个焊盘，如图 10.17 所示。Si 堆叠模块的材料及尺寸见表 10.8。贴装到刚性或柔性基板上的堆叠模块如图 10.18 所示。

表 10.8 堆叠模块的材料参数与几何参数

封装形式	堆叠模块	
测试芯片	子芯片	母芯片
芯片尺寸/mm	3.405×1.34×0.06	4.793×1.34×0.13
焊盘开孔	FC 焊盘 30μm	FC 焊盘 30μm
焊盘节距	100μm	100μm
凸点类型(高度)	铜柱涂覆 Au20Sn(25μm)	Au20Sn(125μm)
UBM	Al	化学镀 NiAu

图 10.18 所示的堆叠模块可以理解为广义的 PoP 封装，是一种低成本但可以满足某些应用对高性能、高密度、低功耗以及宽带宽需求的封装。其中，散热器/热沉是可选的。芯片到芯片的互连是最短距离的面对面方式，这样信号、功率、接地等都可以通过焊锡凸点传送至下一级互连。对于第 1 章中所提到的宽 I/O 存储芯片（见图 1.9），将存储芯片作为子芯片、逻辑芯片作为母芯片的 PoP 封装是最经济的封装形式。然而需要强调的是，片上系统（SoC）/逻辑芯片及存储芯片中并无 TSV。

(2) 测试模型的制作

AuSn 凸点可以采用多种方法制造，如电子束蒸镀法、电化学沉积以及焊膏等。这些方法当中，蒸镀法制得的凸点可以更好地控制其尺寸及位置。因此，在

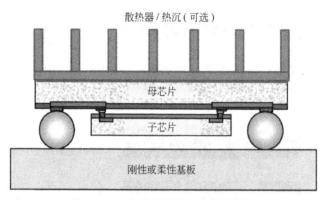

图 10.18 贴装在基板上的堆叠模块

UBM 图案制作之后,通过沉积一系列交替的 Au 层与 Sn 层可以形成 AuSn 层状结构。利用电子束蒸镀法进行 AuSn 沉积有如下一些优点[49]:

① 可以减少沉积后氧化物的生成;

② 可以精确地控制凸点的厚度及位置;

③ Au 沉积速率高;

④ 可以在 200mm 晶圆上得到厚度均匀的 AuSn 层。

使用直径 200mm、带有 5000Å SiO_2 介电薄膜的 p 型硅晶圆(100)制作测试模型。图 10.19 给出了在子芯片上制造 AuSn 焊锡凸点的过程工艺。第一步,先在硅晶圆上沉积 AlCu 层,然后形成菊花链状的金属焊盘并刻蚀 AlCu 层;第二步,先沉积 5000Å SiO_2 和 5000Å Si_3N_4 钝化层,接着在钝化层上利用干法刻蚀露出金属焊盘;第三步,通过溅射沉积得到 Ti/Cu 种子层,作为微 Cu 柱电镀工艺的导电层;第四步,利用贴膜机以及接触式光罩对准曝光机的紫外线(UV)将 20μm 厚的干膜光刻胶压制成薄片,然后通过显影光刻胶界定 UBM 衬垫;第五步,利用 RENA 200mm 晶圆电镀设备填充衬垫至 18～20μm,该电镀设备以 Sperolyte $CuSO_4$ 为主电解液;第六步,利用电子束蒸镀机交替沉积 2200Å 厚的 Au 层以及 2000Å 厚的 Sn 层,一共沉积 16 层,最后在微 Cu 柱上形成厚度为 3.5μm 的 AuSn 凸点;第七步,整个晶圆都有 AuSn 层覆盖,可以利用剥离工艺去除干膜层上的 AuSn,但由于与 Cu 之间的黏附作用,最后只在 Cu 柱上留下 AuSn。

完成干膜的剥离之后,利用选择性湿法刻蚀化学剂对 Ti/Cu 种子层进行逐一刻蚀,从而完成 AuSn 焊锡凸点的制作,如图 10.19 所示。在回流焊过程中,这些 AuSn 层熔化并形成均匀的 Au20Sn 焊点。图 10.20 所示为 AuSn 焊锡凸点,采用蒸镀工艺将 3.5μm 厚的 AuSn 沉积到 20μm 厚的铜柱上。AuSn 焊锡凸点以 100μm 的节距均匀排布在子芯片两侧,如图所示。

图 10.21 所示为母芯片上的 AuSn 焊锡凸点(本节不讨论)与化学镀 NiAu UBM。在将 AlCu 金属衬垫制成菊花链和金属衬垫钝化层开口之前,母芯片的制作工艺与子芯片是完全相同的。

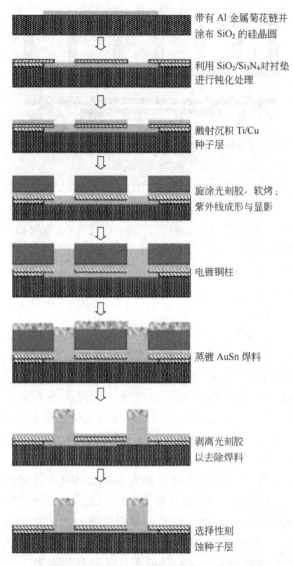

图10.19　子芯片上AuSn焊锡凸点的制造工艺

10.4.3　C2W组装

利用AuSn凸点将子芯片与母芯片组装起来的整个工艺采用的是C2W键合方法，即首先将子芯片晶圆切割成单个的芯片，然后选取完好的子芯片键合到由母芯片晶圆切割得到的完好母芯片上，如图10.22所示。

键合工艺由卡尔苏斯（Karl Suss）公司的FC150型倒装芯片键合机完成，如图10.23所示。采用氮气保护及非回流凸点进行组装，在键合之前需对子芯片与母芯片进行氩离子溅射清洗。AuSn倒装芯片键合的主要挑战是既要使AuSn在NiAu

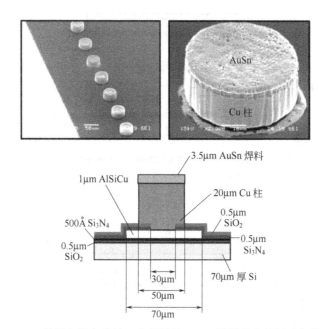

图 10.20 利用电子束蒸镀工艺制得的 AuSn 焊锡凸点的尺寸与均匀度

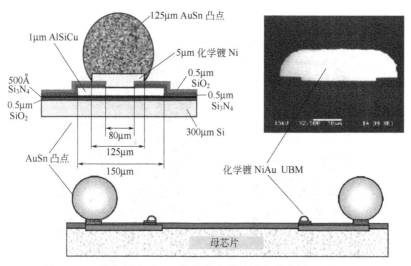

图 10.21 母芯片上 AuSn 焊锡凸点的尺寸与化学镀 NiAu UBM

UBM 焊盘上有较好的湿润,又要确保键合之后 AuSn 焊料受到 Cu 柱与 UBM 的挤压最小。除了凸点高度的一致性以外,还对组装工艺的对中精度、键合力、键合温度、与接触电阻相关的键合时间、AuSn 湿润情况以及 AuSn 凸点形状等参数进行了优化。

(1) 凸点高度的一致性

在进行倒装芯片键合之前,表征 AuSn 凸点高度十分重要,因为凸点高度不一致会造成薄弱连接甚至开路。图 10.24 所示为芯片级的凸点高度测量结果,可以接

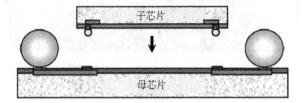

蒸镀法形成的 AuSn 凸点

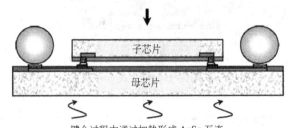

利用倒装芯片键合机将子芯片与母芯片进行对中

键合过程中通过加热形成 AuSn 互连

图 10.22 堆叠模块组装工艺

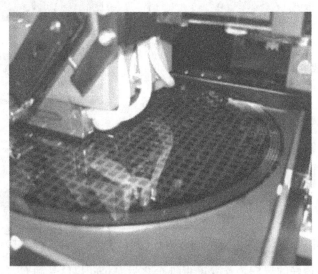

图 10.23 FC150 C2W 键合机

受的最大高度差为 $2.5\mu m$，W2W 级凸点可以接受的最大高度差为 $1.5\mu m$。

（2）对中精度

在倒装芯片传统的钎焊工艺中，进行回流焊之前需将焊锡凸点浸入到助焊剂中，然后再粘接到基板的焊盘上。在回流焊过程中，芯片上的焊锡凸点能够自动对中基板焊盘，并形成良好的互连。然而采用热压键合时，在组装过程中无法实现自

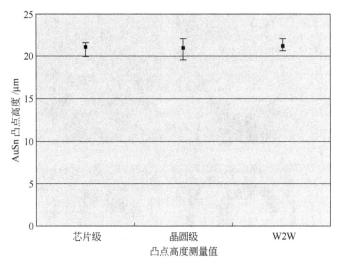

图 10.24　AuSn 凸点的高度测量结果

对中。因此键合对中精度变得重要，尤其对于带有焊锡凸点的大尺寸芯片而言。为了形成接触电阻稳定、湿润性良好的 AuSn 互连，需采用具有高精度对中功能的键合设备。

对中精度与互连接触电阻的关系如图 10.25 所示。在倒装芯片键合过程中，采用的键合力为 0.8kgf。相比于 AuSn 凸点与化学镀 NiAu 焊盘完全对中情况，AuSn 凸点偏离 NiAu 焊盘时，互连接触电阻有显著差别，如图 10.25 所示。对中不良导致 AuSn 与焊盘湿润较差（如图 10.26 所示），接触电阻较高。图 10.27 所示为对中准确的互连剖面图。

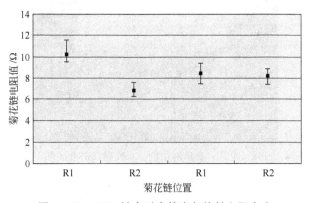

图 10.25　C2W 键合对中精度与接触电阻大小

10.4.4　C2W 实验设计

(1) 三因子 DoE

对无底部填充料的测试模型进行三因子实验设计（DoE）。如表 10.9 所示，设

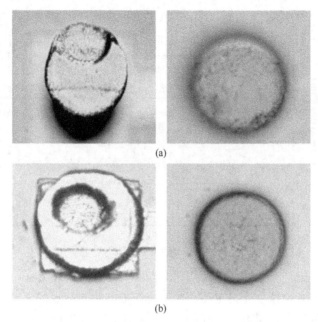

图 10.26 AuSn 湿润失效模式，(a) 对中不良；(b) 对中准确。左列图为子芯片侧，右列图为母芯片侧

图 10.27 准确对中的 AuSn 焊点的截面图

计参数为 3 个水平的键合力、2 个水平的键合温度和 1 个水平的键合时间。通过实验设计可以确定合适的键合力与键合温度，以得到具有良好湿润性和形状的 AuSn 焊点。3 种键合力大小在 0.4~1kgf 之间，见表 10.10。通过一个菊花链电路测量焊点电阻，如图 10.28 所示。

(2) DoE 结果

键合力对互连电阻以及 AuSn 焊点形状有显著的影响。图 10.29 表明，当键合力为 0.8kgf 或 1kgf 时，能够得到稳定的接触电阻。然而 1kgf 的键合力并不可取，

表 10.9 评估 C2W 键合的三因子 DoE

批次	键合力/kgf	键合温度/℃	键合时间/s
A	0.4	290	15
B		315	15
C	0.8	290	15
D		315	15
E	1.0	290	15
F		315	15

表 10.10 不同键合力与键合温度组合下的 C2W 键合质量

因子	值			响应
键合力/kgf	0.4	0.8	1.0	接触电阻
键合温度/℃	290		315	AuSn 焊点形状
键合时间/s	15			AuSn 湿润情况

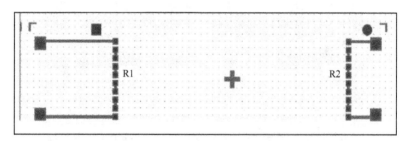

图 10.28 母芯片上测量电阻的菊花链

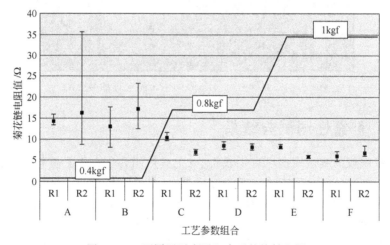

图 10.29 不同因子水平组合下的接触电阻

因为会导致芯片钝化层在 Al 焊盘与 UBM 界面处开裂，如图 10.30 所示。图 10.31 所示为采用 1kgf 键合力时的截面图，明显看到，过高的键合力会挤压出 Cu 柱与键合焊盘之间的 AuSn 共晶焊料，造成焊盘上的 AuSn 湿润性较差。另外，与键合温度和键合时间相比，键合力对接触电阻影响较大。

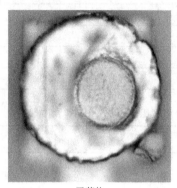
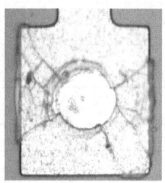

子芯片　　　　　　　　　母芯片

图 10.30　高键合力（1.0kgf）导致钝化层开裂

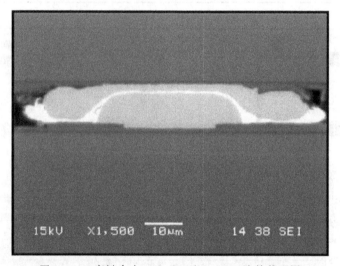

图 10.31　高键合力（1kgf）下 AuSn 互连的截面图

10.4.5　可靠性测试与结果

对组装完成的测试模型（采用表 10.9 中参数组合 D，无下填料）进行温度循环测试，以测试 AuSn 互连的热疲劳特性[49]。测试前测量一次接触电阻，然后进行 100 次循环测试，测试中每循环 25 次测量 1 次接触电阻，以菊花链开路或电阻无穷大为判断失效的标准。如图 10.32 所示，该组试样经过 100 次温度循环测试后并未发生开路。键合力为 0.8kgf 时，观察到接触电阻的最小变化低于 10%，这表明直接将均匀的 AuSn 共晶焊料钎焊到化学镀 NiAu UBM 焊盘上的互连结构具有

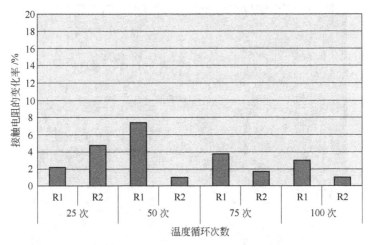

图 10.32 温度循环（-40~150℃）过程中接触电阻的变化率

足够的强度，并且能够承受温度循环载荷。测试结果表明表 10.9 中的 D 组合（键合力为 0.8kgf，键合温度为 315℃，键合时间为 15s）能够得到稳定的接触电阻和良好的热循环可靠性。

10.4.6 用于 3D IC 封装的 SnAg 互连

图 10.33 所示为将与上节相同的子芯片与母芯片采用 Sn3Ag 焊料涂覆 Cu 柱（见图 10.34）进行互连。母芯片上的焊球采用 Sn37Pb。重点关注子芯片与母芯片之间 SnAg 焊点的可靠性[50]。在组装之前，需将带有高度为 40μm SnAg 铜柱微凸

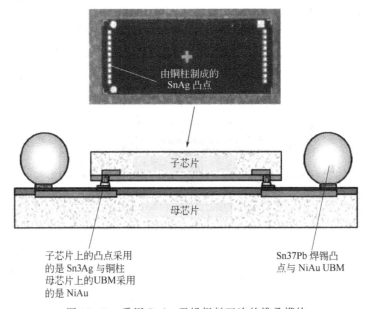

图 10.33 采用 SnAg 无铅焊料互连的堆叠模块

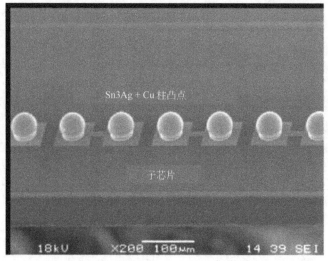

图 10.34 子芯片上的无铅 Sn3Ag 与 Cu 柱（40μm）凸点

点的子芯片晶圆减薄至 70μm，将带有高度为 200μm CSP SnPb UBM 微凸点的母芯片晶圆减薄至 300μm，如图 10.35 所示。图 10.36 为采用下填料的组装截面图。应当指出，所有焊点都采用正常的回流焊工艺，由于熔融焊料表面张力的作用，使得焊点形状比较美观和光滑，这与采用 C2W 键合的 AuSn 焊点不同。

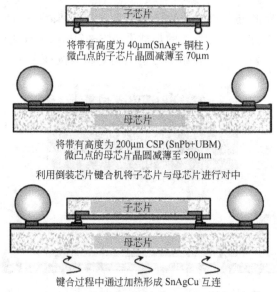

图 10.35 采用 SnAg 焊料的 PoP 模块组装工艺

完成组装之后，将试样进行如下测试：非偏置高加速应力测试（uHAST），条件为：130℃，相对湿度 85%RH，96h；MST L3 测试：30℃，60%RH，192h，以 260℃完成 3 次回流；MST L1 测试：85℃，85%RH，168h，以 260℃完成 3 次回流；HTS 测试：125℃，1000h；TC 测试：125～－55℃，驻留时间 15min，升降温速率

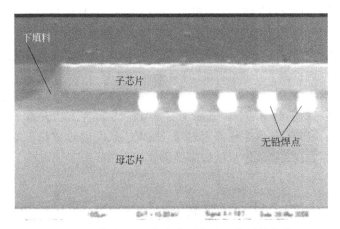

图 10.36　采用 SnAg 焊点将子芯片组装到母芯片上

15℃/min，循环 1000 次。图 10.37 为经过 1000 次温度循环之后失效的试样，裂纹在金属间化合物 Cu_6Sn_5 与焊料体的界面处萌生，然后扩展至整个焊点。

图 10.37　HST、MST L1、MST L3 和 1000 次热循环测试后失效的试样

10.4.7　总结与建议

一些重要的结论总结如下[49,50]。

① 为了保证测试芯片均匀受压并形成良好的 AuSn 互连，要求凸点共面性达到芯片级（±3.5μm）。

② 键合时采用 0.8kgf 的键合力并保证对中准确，可以得到稳定的接触电阻（±2Ω）和良好的 AuSn 湿润性。

③ 组装结果表明，过高的键合力（1.0kgf）并不可取，因为会导致芯片钝化层在 Al 焊盘与 UBM 界面处开裂。过高的键合力也会挤压出凸点与键合焊盘之间的 AuSn 共晶焊料，造成焊盘上的 AuSn 湿润性较差。

④ 温度循环测试结果表明，采用 0.8kgf 的键合力且无下填充料可以得到稳定的接触电阻（±9Ω）以及良好的可靠性（>100 次温度循环）。

⑤ 更为重要的是，这一部分的研究结果可以作为相关行业采用 AuSn 无助焊剂焊锡凸点技术进行倒装芯片组装时的参考。

⑥ 展示了利用高度为 40μm、节距为 100μm 的 SnAg 微凸点完成 C2W 倒装芯片的回流焊组装工艺（精确对中），此外还展示了节距为 100μm 微凸点的底部填充工艺（无孔洞）。所有堆叠封装都通过了 JEDEC 标准的封装级可靠性测试，如热循环、高温储存、1 级和 3 级湿度敏感性测试以及 uHAST 测试等。

10.5 用于低成本、高性能与高密度 SiP 封装的面对面互连

10.5.1 用于超细节距 Cu-low-k 芯片的 Cu 柱互连技术

10.3 节和 10.4 节分别介绍和讨论了利用引线键合机进行 Au 线键合、采用

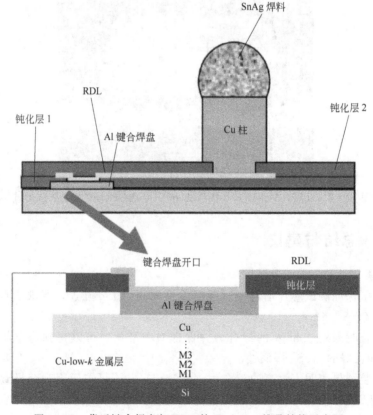

图 10.38 靠近键合焊盘与 RDL 的 Cu-low-k 堆叠结构示意图

C2W（热压法）键合实现 AuSn 互连以及利用回流焊实现 SnAg 互连的 3D IC 封装。对于具有超细节距 Cu-low-k 焊盘的摩尔芯片而言，也许 Cu 线互连技术更为适用。线互连技术（WIT）来自于 Love 等人的一项专利——"用于连接集成电路与基板的引线互连结构"（美国专利号 5334804，1994 年 8 月 2 日授权，也可见参考文献 [2] 的第 14 章），现在也称之为 Cu 柱。图 10.38 为具有 Cu-low-k 金属层的摩尔芯片焊盘的示意图。通常，需要通过 RDL 连接焊盘、Cu 柱及焊料。

10.5.2 可靠性评估

图 10.39（a）、(b) 所示为 JSR 公司生产的 JSR151N 光刻胶薄膜以及用于制造 Cu 柱的开口（尺寸为 $100\mu m \times 160\mu m$）。图 10.39（c）、(d) 分别为回流焊前后的电镀 Cu 柱与焊料。图 10.39（e）、(f) 为 TC 测试结果，可以看出失效模式为 $1\mu m$ 厚的 RDL 发生开裂，因此需要更厚的 RDL 以满足 TC 可靠性。欲了解更多信息，参见参考文献 [56]。

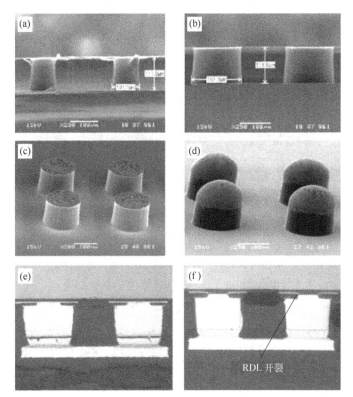

图 10.39 Cu 柱互连的 SEM 图像：(a) $110\mu m$ 厚的 JSR151N 光刻胶薄膜中直径为 $100\mu m$ 的开口；(b) $110\mu m$ 厚的 JSR151N 光刻胶薄膜中直径为 $160\mu m$ 的开口；(c) 电镀 Cu 柱；(d) 涂覆 SnAg 焊料的 Cu 柱；(e) 测试之前的试样；(f) 测试之后的失效形式

10.5.3 一些新的设计

图 10.40 所示为一些能满足低成本、高密度、高性能等要求的芯片到芯片面对面互连新设计。例如，Cu-low-k 芯片上系统（SoC）能够支撑采用焊锡凸点或 Cu 柱/焊料互连的宽 I/O 端口存储芯片；SoC 可以同时支撑存储芯片与图形芯片，还可以支撑存储芯片堆叠以及其他芯片；电源、接地、信号等都可以通过焊锡凸点或 Cu 柱/焊料互连传递至下一级；散热器/热沉是可选的。所有这些都是传统的、已经过验证的封装技术，并且这些芯片中都无 TSV。

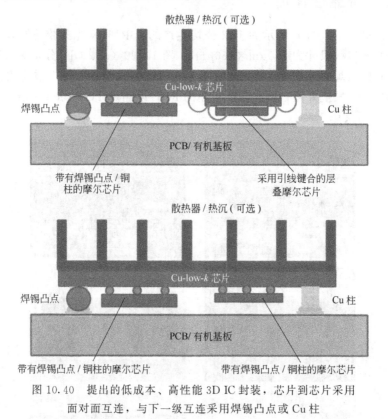

图 10.40 提出的低成本、高性能 3D IC 封装，芯片到芯片采用面对面互连，与下一级互连采用焊锡凸点或 Cu 柱

10.6 埋入式晶圆级封装（eWLP）到芯片的互连

10.6.1 2D eWLP 与再布线芯片封装（RCP）互连

在过去两年中，2D eWLP 已经十分普遍。该封装形式由英飞凌（Infineon）与飞思卡尔（Freescale）两家半导体公司开发，ST Microelectronics、ASE、Stac-

ChipPack 等公司也在进行量产，Freescale 公司将其称为再布线芯片封装（RCP）[57~59]。

10.6.2　3D eWLP 与再布线芯片封装（RCP）互连

2D eWLP/RCP 可以通过封装通孔（TPV）扩展为 3D IC 封装。不同于前九章所提到的 TSV，TPV 可以将封装内部与封装体进行 3D 互连，还可以对 eWLP/RCP 进行双面再布线和 TPV 工艺，这样可以改进传统的 PoP 结构并且能够扩展组装工艺[57]。

图 10.41 所示为 Freescale 公司以 3D RCP 单元为基础的 3D RCP SiP 封装的顶部与底部视图。在 RCP 封装中内置了 MEMS 器件和一对控制芯片。包含 MEMS 与 ASIC 器件的 RCP SiP 部分，其厚度大约为 $600\mu m$。其内的面对面芯片到芯片互连只采用了 C4 微凸点和单层 RCP 布线层。

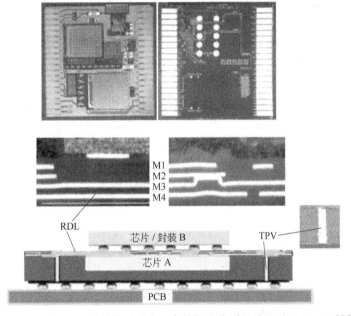

图 10.41　Freescale 公司的低成本、高性能芯片到芯片面对面 3D IC 封装，采用了类似于 TSV 的 TPV 垂直互连技术，但制程中无需使用任何半导体设备

10.6.3　总结与建议

① 介绍并讨论了一些采用传统技术完成 3D IC 封装的低成本芯片到芯片面对面互连的实例。

② 由于成本及可靠性等原因，封装行业应当开发更多的 3D IC 封装技术，例如芯片到芯片的面对面组装技术，不需要采用 TSV。

③ 对于高性能、高密度、低功耗以及宽频带等要求，TSV 技术仍是必要的。

10.7 引线键合可靠性

一个迅速、粗略判断引线键合质量和可靠性的方法就是拉拔实验（pull test）[60]。本节主要讨论微电子技术中的 Au 线和 Cu 线的拉力与位移。主要内容为：①推导了一组计算引线在拉拔实验过程中引线内力与变形的方程；②基于一些失效准则确定引线断裂时的最大拉力；③通过实验验证这些方程。

10.7.1 常用芯片级互连技术

有 3 种主要的芯片级互连技术[61,62]，即引线键合[63]、焊锡凸点[1,2]以及载带自动键合（TAB）[64]。世界上超过 85% 的芯片采用的都是引线键合技术。自 2006 年开始，采用引线键合技术的半导体芯片每年的产量已经超过了世界的总人口（70 亿）。

检测引线键合质量和可靠性最简单的方法就是拉拔实验，参考文献［62］已经给出了拉拔实验过程中引线中的内力，但只考虑了作用在未变形引线上的力平衡。本研究则考虑施加在变形引线上的力平衡，如图 10.42 与图 10.43 所示，并且考虑了引线的材料属性。因此，本研究得到的拉力关系应更为准确，并适用于不同材料的引线，如 Au 线和 Cu 线等。

目前大多数采用引线键合的芯片使用的都是 Au 线。然而，由于黄金价格的上涨以及 Cu 引线键合技术的发展[65~103]，许多公司都在寻求低成本的解决方案，并开始转向 Cu 引线键合技术。本研究给出了与 Au 线及 Cu 线的有关结果，这些结果都得到了实验验证。

10.7.2 力学模型

如图 10.42 所示，引线在 Q 点受到外力 F 的作用，使得 Q 点移动到了 Q' 点。现在的问题是如何确定 CQ' 与 BQ' 中的内力 f_d 和 f_t，以及 Q' 的位置。

在图 10.42 中，l_d 和 l_t 为引线的初始长度，θ_d 和 θ_t 为引线的初始角度，d 为引线键合点（支座）间的距离，H 为引线支座间的高度，h 为线的拱高，ϕ 为外力 F 与纵轴之间的夹角。

图 10.43 为引线受力图。外力 F 与内力 f_d、f_t 应满足平衡方程。这里一共有 6 个未知量：CQ' 中的内力 f_d，BQ' 中的内力 f_t，CQ' 的长度 L_d，BQ' 的长度 L_t，CQ' 的角度 φ_d，BQ' 的角度 φ_t，如图 10.42 与图 10.43 所示。需要 6 个方程才能求解这 6 个未知量（f_d，f_t，L_d，L_t，φ_d，φ_t）。

平衡方程

$$\frac{f_t}{F} = \frac{\cos\phi\cos\varphi_d + \sin\phi\sin\varphi_d}{\sin\varphi_t\cos\varphi_d + \cos\varphi_t\sin\varphi_d} \tag{10.1}$$

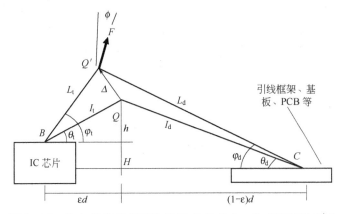

图 10.42 拉力 F 作用在引线 BQC 的 Q 点上，Q 点移动到 Q' 点

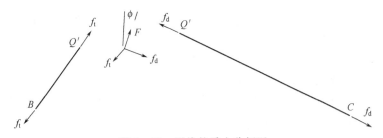

图 10.43 引线的受力分析图

$$\frac{f_d}{F} = \frac{\cos\phi\cos\varphi_t - \sin\phi\sin\varphi_t}{\sin\varphi_t\cos\varphi_d + \cos\varphi_t\sin\varphi_d} \tag{10.2}$$

物性方程

$$L_t - l_t = \frac{f_t l_t}{AE} \tag{10.3}$$

$$L_d - l_d = \frac{f_d l_d}{AE} \tag{10.4}$$

式中，A 为引线的横截面积；E 为弹性模量。

变形协调方程

$$\Delta^2 = L_t^2 + l_t^2 - 2L_t l_t \cos(\varphi_t - \theta_t) = L_d^2 + l_d^2 - 2L_d l_d \cos(\varphi_d - \theta_d) \tag{10.5}$$

几何方程

$$L_t = \frac{-g\cos(X+Y)}{\sin(90+\varphi_t)} \tag{10.6}$$

其中

$$g = L_t \cos(X-Y) + d\cos(X+Y)$$

$$X = 90 - \frac{\varphi_d}{2}$$

$$Y = \arctan\left(\frac{L_d - d}{L_d + d}\right) \cot\frac{\varphi_d}{2}$$

如果有可能的话，要得到上述 6 个方程的解析解是非常困难的。然而利用计算机软件，根据给定的 F、h、H、d、ε、ϕ、l_d、l_t、θ_d 以及 θ_t 就可以确定 f_d、f_t、L_d、L_t、φ_d 以及 φ_t 的值，Q 点的位移 Δ 也可以利用方程 (10.5) 计算得到。

若 $L_d = l_d$，$L_t = l_t$，$\varphi_d = \theta_d$，$\varphi_t = \theta_t$，即引线未发生变形，则方程 (10.3)～(10.6) 都不存在，方程 (10.1)、(10.2) 也退化为[62]：

$$\frac{f_t}{F} = \frac{\sqrt{h^2 + \varepsilon^2 d^2}}{(h + \varepsilon H)}\left[(1-\varepsilon)\cos\phi + \frac{(h+H)}{d}\sin\phi\right]$$

$$\frac{f_d}{F} = \frac{\sqrt{1 + \frac{(1-\varepsilon)^2 d^2}{(h+H)^2}}}{(h+\varepsilon H)}\left[(h+H)\left(\varepsilon\cos\phi - \frac{h}{d}\sin\phi\right)\right]$$

10.7.3 数值结果

应当重申的是，对于任何施加在 Q 点的外力 F 及其与纵轴的夹角 ϕ，都可以利用计算机软件求解非线性方程 (10.1)～(10.6)，从而得到内力 f_d、f_t 以及 Q 点的位移 Δ。然而对于工程实践而言，预测引线断裂时的最大拉力更为有用。为此，必须首先定义引线的失效准则。若选取引线的抗拉强度 σ_s 作为引线的失效准则，那么当引线断裂时其内力等于 $A\sigma_s$。例如，如果 BQ' 中的内力 f_t 先达到抗拉强度，那么方程 (10.3) 变为

$$L_t - l_t = \frac{\sigma_s l_t}{E} \tag{10.3'}$$

如果 CQ' 中的内力先达到抗拉强度，方程 (10.4) 变为

$$L_d - l_d = \frac{\sigma_s l_d}{E} \tag{10.4'}$$

现在考虑如图 10.44 所示的一种特殊情况（$H=0$，$\varepsilon=0.5$，$\varphi=0$，$d=5\text{mm}$），引线是抗拉强度为 25gf、弹性模量为 130GPa 的硬质 Cu 引线。通过求解方程

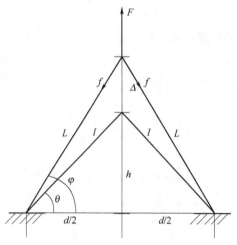

图 10.44 特例：$H=0$，$\varepsilon=0.5$，$\varphi=0$

(10.1)~(10.6)，得到引线断裂时（以抗拉强度为失效准则）的最大拉力 F 随拱高 h 的变化情况如图 10.45 所示。从图中可以看出拱高越大，最大拉力值越大。

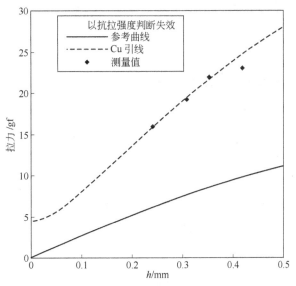

图 10.45　最大拉力与线拱高 h 的关系曲线以及根据参考文献 [62] 中的方程绘制的参考曲线

10.7.4　实验结果

引线拉拔实验采用的单轴拉伸机（electro force），可以施加与测量很小的力和位移，如图 10.46 所示。实验时加载速率为 0.025mm/s，夹头（支座）间的距离 d

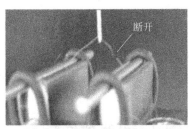

图 10.46　拉拔实验装置与结果

为 1.5mm。拉拔实验结果如图 10.45 与图 10.46 所示。可以看出：①大多数引线都在拉力作用点与夹头之间断开；②实验结果与通过方程预测的结果吻合较好。

图 10.45 还给出了参考文献 [62] 中的方程的求解结果，与实验结果并不相同。其原因是：①参考文献 [62] 未考虑材料强度；②参考文献 [62] 中的方程不适用于变形引线。

10.7.5　关于 Cu 引线的更多结果

图 10.47 所示为不同 Cu 引线（硬质 Cu 引线与退火 Cu 引线）的最大拉力与线拱高的关系曲线，可以看出：①对两种 Cu 引线而言，线拱高越大，最大拉力值越大；②抗拉强度越大，最大拉力值越大。

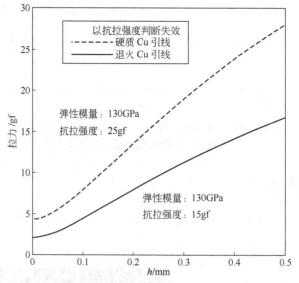

图 10.47　不同 Cu 引线的最大拉力与线拱高的关系曲线

图 10.48 所示为 Cu 引线的最大位移，可以看出：①线拱高越大，位移越小；②线拱高越小（$h<0.3$mm），最大位移的变化越显著；③硬质 Cu 引线的位移大于退火 Cu 引线的位移。

10.7.6　关于 Au 引线的结果

不同 Au 线的最大拉力与线拱高的关系曲线如图 10.49 所示，可以看出：①对所有 Au 线而言，线拱高越大，最大拉力值越大；②抗拉强度与弹性模量越大，最大拉力值越大。

图 10.50 所示为不同 Au 线（硬质 Au 线、去应力 Au 线以及退火 Au 线）的最大位移，可以看出：①线拱高越大，位移越小；②当 $h>0.3$mm 时，位移值趋近于常数；③抗拉强度与弹性模量越大，最大拉力值越大。

根据这些结果可以发现，引线材料（Cu 或 Au 线）的强度对最大拉力的影响

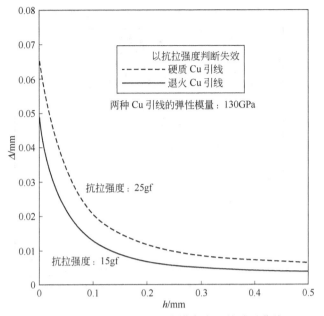

图 10.48　最大位移 Δ 与线拱高度 h 的关系曲线

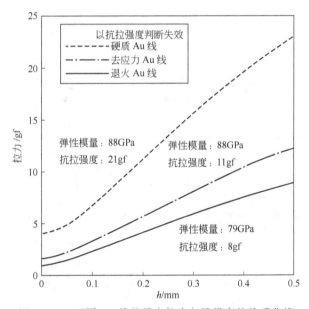

图 10.49　不同 Au 线的最大拉力与线拱高的关系曲线

最显著。对于较小的线拱高而言，位移的影响也十分重要。

10.7.7　Cu 引线与 Au 引线的应力应变关系

图 10.51 所示为测定 Cu 线与 Au 线载荷位移曲线的装置，图 10.52 所示为单轴拉伸实验前后的引线，可以看出所有失效的引线都是在引线中点附近断开。

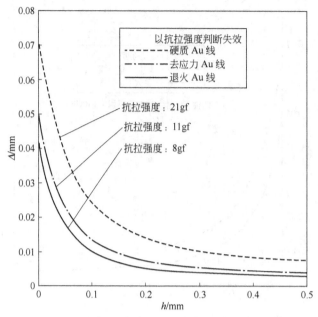

图 10.50　不同 Au 线的最大位移与线拱高的关系曲线

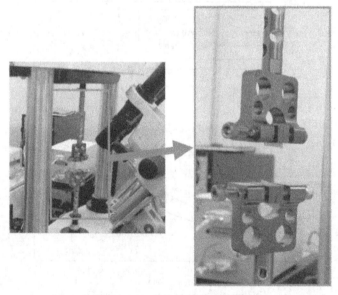

图 10.51　单轴拉伸实验装置

图 10.53 和图 10.54 分别为典型的 Cu 线与 Au 线的载荷位移曲线，可以看出：两者十分相似；两者大小不同，通常 Cu 线的抗拉强度更高。

10.7.8　总结与建议

本节主要研究 Au 线与 Cu 线的拉力及位移，一些重要的结论和建议总结如下。

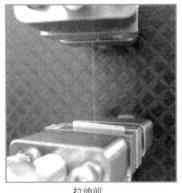 

拉伸前　　　　　　　　　　　　　拉伸后

图 10.52　单轴拉伸实验前后的引线

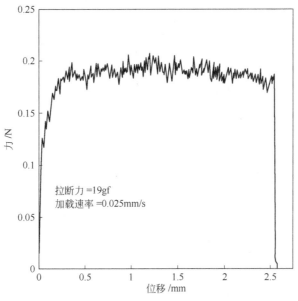

图 10.53　Cu 线的载荷位移曲线

① 建立了求解引线受拉时内力与位移的 6 个必要方程。

② 根据引线的失效准则（即抗拉强度），可以确定或者预测引线的最大拉力及位移。

③ 通过实验测得了引线的最大拉力，且与预测结果吻合较好。

④ 无论是 Cu 线还是 Au 线，线拱高越大，最大拉力值越大，而最大位移值越小。

⑤ 无论是 Cu 线还是 Au 线，抗拉强度越大，最大拉力值和位移越大。

⑥ 无论是 Cu 线还是 Au 线，材料强度对预测最大拉力和位移的影响最重要。

⑦ 无论是 Cu 线还是 Au 线，引线的变形对预测最大拉力和位移的影响也很重要。

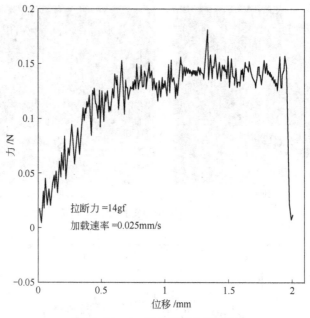

图 10.54　Au 线的载荷位移曲线

⑧ 一般来说，Cu 线的抗拉强度与弹性模量要大于 Au 线的抗拉强度与弹性模量。

⑨ 考虑引线的大变形、横截面积的改变（颈缩）以及非线性应力应变关系，可以得到更为精确的方程。

⑩ 需要对 Au 线甚至 Cu 线进行额外的测试。

10.8　参考文献

[1] Lau, J. H., *Low-Cost Flip Chip Technology*, McGraw-Hill, New York, 2000.
[2] Lau, J. H., *Flip Chip Technology*, McGraw-Hill, New York, 1995.
[3] van Driel, W. D., G. Wisse, A. Y. L. Chang, J. H. J. Jansen, X. Fan, G. Q. Zhang, and L. J. Ernst, "Influence of Material Combinations on Delamination Failures in a Cavity-Down TBGA Package," *IEEE Transactions of Components and Packaging Technology*, Vol. 27, No. 4, December 2004, pp. 651–658.
[4] Harman, G. G., and C. E. Johnson, "Wire Bonding to Advanced Copper, Low-*k* Integrated Circuits, the Metal/Dielectric Stacks, and Materials Considerations," *IEEE Transactions of Components and Packaging Technology*, Vol. 25, No. 4, December 2002, pp. 677–683.
[5] Zhang, J., and J. Huneke, "Stress Analysis of Spacer Paste Replacing Dummy Die in a Stacked CSP Package", *Proceedings of IEEE/ICEPT*, 2003, pp. 82–85.
[6] Yoon, S. W., D. Wirtasa, S. Lim, V. Ganesh, A. Viswanath, V. Kripesh, and M. K. Iyer, "PEDL Technology for Copper/Low-*k* Dielectrics Interconnect," *IEEE Proceedings of EPTC*, Decemeber 2005, pp. 711–715.
[7] Hartfield, C. D., E. T. Ogawa, Y.-J. Park, T.-C. Chiu, and H. Guo, "Interface Reliability Assessments for Copper/Low-*k* Products," *IEEE Transactions of Device Materials Reliability*, Vol. 4, No. 2, June 2004, pp. 129–141.
[8] Alers, G. B., K. Jow, R. Shaviv, G. Kooi, and G. W. Ray, "Interlevel Dielectric

Failures in Copper/Low-*k* Structures," *IEEE Transactions of Device Materials Reliability*, Vol. 4, No. 2, June 2004, pp. 148–152.

[9] Chiu, C. C., H. H. Chang, C. C. Lee, C. C. Hsia, and K. N. Chiang, "Reliability of Interfacial Adhesion in a multilevel copper/low-*k* interconnect structure," *Microelectronic Reliability*, Vol. 47, Nos. 9–11, 2007, pp. 1506–1511.

[10] Wang, Z., J. H. Wang, S. Lee, S. Y. Yao, R. Han, and Y. Q. Su, "300-mm Low-*k* Wafer Dicing Saw Development," *IEEE Transactions of Electronic Packaging & Manufacturing*, Vol. 30, No. 4, October 2007, pp. 313–319.

[11] Chungpaiboonpatana, S., and F. G. Shi, "Packaging of Copper/Low-*k* IC Devices: A Novel Direct Fine-Pitch Gold Wire-Bond Ball Interconnects onto Copper/Low-*k* Terminal Pads," *IEEE Transactions of Advanced Packaging*, Vol. 27, No. 3, August 2004, pp. 476–489.

[12] Wang, T. H., Y.-S. Lai, and M.-J. Wang, "Underfill Selection for Reducing Cu/Low-*k* Delamination Risk of Flip-Chip Assembly," *IEEE Proceedings of EPTC*, December 2006, pp. 233–236.

[13] Zhai, C. J., U. Ozkan, A. Dubey, S. Sidharth, R. Blish II, and R. Master, "Investigation of Cu/Low-*k* Film Delaminaiton in Flip Chip Packages," *IEEE/ECTC Proceedings*, June 2006, pp. 709–717.

[14] Ong, J. M. G., A. A. O. Tay, X. Zhang, V. Kripesh, Y. K. Lim, D. Yeo, K. C. Chan, J. B. Tan, L. C. Hsia, and D. K. Sohn, "Optimization of the Thermo-Mechanical Reliability of a 65-nm Cu/Low-*k* Large-Die Flip Chip Package," *IEEE Transactions of Components and Packaging Technology*, Vol. 32, No. 4, December 2009, pp. 838–848.

[15] Zhang, X., J. H. Lau, C. Premachandran, S. Chong, L. Wai, V. Lee, T. Chai, V. Kripesh, V. Sekhar, D. Pinjala, and F. Che, "Development of a Cu/Low-*k* Stack Die Fine Pitch Ball Grid Array (FBGA) Package for System in Package Applications," *IEEE Transactions on CPMT*, Vol. 1, No. 3, March 2011, pp. 299–309.

[16] Chen, S., C. Z. Tsai, E. Wu, I. G. Shih, and Y. N. Chen, "Study on the Effects of Wafer Thinning and Dicing on Chip Strength," *IEEE Transactions of Advanced Packaging*, Vol. 29, No. 1, February 2006, pp. 149–157.

[17] Toh, C., M. Gaurav, H. Tan, and P. Ong, "Die Attached Adhesives for 3D Same-Size Dies Stacked Packages," *IEEE/ECTC Proceedings*, May 2008, pp. 1538–1543.

[18] Carson, F., K. Ishibashi, and Y. Kim, "Three-Tier PoP Configuration Utilizing Flip Chip Fan-In PoP Bottom Package," *IEEE/ECTC Proceedings*, San Diego, CA, 2009, pp. 313–318.

[19] Xie, D., D. Shangguan, D. Geiger, D. Gill, V. Vellppan, and K. Chinniah, "Head in Pillow (HIP) and Yield Study on SIP and PoP Assembly," *IEEE/ECTC Proceedings*, San Diego, CA, 2009, pp. 752–758.

[20] Sze, T., M. Giere, B. Guenin, N. Nettleton, D. Popovic, J. Shi, R. Ho, R. Drost, D. Douglas, and S. Bezuk, "Proximity Communication Flip-Chip Package with Micron Chip-to-Chip Alignment Tolerances," *IEEE/ECTC Proceedings*, San Diego, CA, 2009, pp. 966–971.

[21] Kumar, A., V. Sekhar, S. Lim, C. Keng, G. Sharma, S. Vempati, V. Kripesh, J. H. Lau, D. Kwong, and X. Dingwei, "Wafer Level Embedding Technology for 3D Wafer Level Embedded Package," *IEEE/ECTC Proceedings*, San Diego, CA, 2009, pp. 1289–1296.

[22] Kim, D., Y. Kim, K. Seong, J. Song, B. Kim, C. Hwang, and C. Lee, "Evaluation for UV Laser Dicing Process and Its Reliability for Various Designs of Stack Chip Scale Package," *IEEE/ECTC Proceedings*, San Diego, CA, 2009, pp. 1531–1536.

[23] Sharma, G., S. Vempati, A. Kumar, S. Nandar, Y. Lim, K. Houe, S. Lim, N. Vasarla, R. Ranjan, V. Kripesh, and J. H. Lau, "Embedded Wafer Level Packages with Laterally Placed and Vertically Stacked Thin Dies," *IEEE/ECTC Proceedings*, San Diego, CA, 2009, pp. 1537–1543.

[24] Pendse, R., "Flip Chip Package-in-Package (fcPiP): A New 3D Packaging Solution for Mobile Platforms," *IEEE/ECTC Proceedings*, Reno, NV, May 2007, pp. 1425–1430.

[25] Carson, F., S. M. Lee, and L. S. Yoon, "The Development of the Fan-in Package-on-Package," *IEEE/ECTC Proceedings*, Lake Buena Vista, FL, May 2008, pp. 956–963.

[26] Kazuo, I., "PoP (Package-on-Package) Stacking Yield Loss Study," *IEEE/*

ECTC Proceedings, Reno, NV, May 2007, pp. 1403–1408.

[27] Carson, F., S. Lee, and N. Vijayaragavan, "Controlling Top Package Warpage," IEEE/ECTC Proceedings, Reno, NV, May 2007, pp. 737–742.

[28] Vijayaragavan, N., F. Carson, and A. Mistri, "Package on Package Warpage—Impact on Surface Mount Yields and Board Level Reliability," IEEE/ECTC Proceedings, Lake Buena Vista, FL, May 2008, pp. 389–396.

[29] IPC/JEDEC, "Moisture/Reflow Sensitivity Classification for Nonhermetic Solid-State Surface Mounted Devices," JSTD-020D, June 2007.

[30] Geiger, D., D. Shangguan, S. Tam, and D. Rooney, "Package Stacking in SMT for 3D PCB Assembly," IEEE Proceedings of IEMT, San Jose, CA, July 2003, p. S207P6.

[31] Carson, F., "Package-on-Package Variations on the Horizon," Semiconductor International, May 1, 2008.

[32] Wei, L., A. Yoshida, and M. Dreiza, "Control of the Warpage for Package-on-Package (PoP) Design," Proceedings of SMTAI, Chicago, IL, 2007, pp. 320–327.

[33] Dongji, X., D. Geiger, D. Shangguan, B. Hu, and J. Sjoberg, "Yield Study of Inline Package on Package (PoP) Assembly," IEEE Proceedings of Electronic Packaging Technology Conference, Singapore, Lake Buena Vista, FL, May 2008, pp. 1202–1208.

[34] Kazuo, I., "PoP (Package-on-Package) Stacking Yield Loss Study," IEEE/ECTC Proceedings, Reno, NV, 2007, pp. 1043–1048.

[35] Jonas, S., D. Geiger, and D. Shangguan, "Package on Package Process Development and Reliability Evaluation," IEEE/ECTC Proceedings, Lake Beuna Vista, FL, May 2008, pp. 2005–2010.

[36] Souriau, J., O. Lignier, M. Charrier, and G. Poupon, "Wafer Level Processing of 3D System in Package for RF and Data Applications," IEEE/ECTC Proceedings, Lake Beuna Vista, FL, 2005, pp. 356–361.

[37] Brunnbauer, M., E. Furgut, G. Beer, T. Meyer, H. Hedler, J. Belonio, E. Nomura, K. Kiuchi, K. Kobayashi, "An Embedded Device Technology Based on a Molded Reconfigured Wafer," IEEE/ECTC Proceedings, San Diego, CA, 2006, pp. 547–551.

[38] Keser, B., C. Amrine, T. Duong, Owen Fay, S. Hayes, G. Leal, W. Lytle, D. Mitchell, R. Wenzel "The Redistributed Chip Package: A Breakthrough for Advanced Packaging," IEEE/ECTC Proceedings, Reno, NV, 2007, pp. 286–290.

[39] Fillion, R., C. Woychik, T. Zhang, and D. Bitting, "Embedded Chip Build-Up Using Fine Line Interconnect," IEEE/ECTC Proceedings, Reno, NV, 2007, pp. 49–53.

[40] Kripesh, V., V. S. Rao, A. Kumar, G. Sharma, K. C. Houe, Z. Xiaowu, K. Y. Mong, N. Khan, and J. H. Lau, "Design and Development of a Multi-Die Embedded Micro Wafer Level Package," IEEE/ECTC Proceedings, Lake Beuna Vista, FL, May 2008, pp. 1544–1549.

[41] Hamano, T., T. Kawahara, and J. I. Kasai, "Super CSPTM: WLCSP Solution for Memory and System LSI," Proceedings of International Symposium on Advanced Packaging Materials, 1999, pp. 221–225.

[42] Braun, T., K. Becker, M. Koch, V. Bader, U. Oestermann, D. Manessis, R. Aschenbrenner, and H. Reichl, "Wafer Level Encapsulation—A Transfer Molding Approach to System in Package Generation," IEEE/ECTC Proceedings, San Diego, CA, 2002, pp. 235–244.

[43] Ostmann, A., A. Neumann, S. Weser, E. Jung, L. Bottcher, and H. Reichl, "Realization of a Stackable Package Using Chip in Polymer Technology," IEEE Proceedings of Polytronic, 2002, pp. 160–164.

[44] Ko, C., S. Chen, C. W. Chiang, T.-Y. Kuo, Y. C. Shih, and Y.-H. Chen, "Embedded Active Device Packaging Technology for Next-Generation Chip-in-Substrate Packages, CiSP," IEEE/ECTC Proceedings, San Diego, CA, 2006, pp. 322–329.

[45] Takyu, S., T. Kurosawa, N. Shimizu, and S. Harada, "Novel Wafer Dicing and Chip Thinning Technologies Realizing High Chip Strength," IEEE/ECTC Proceedings, San Diego, CA, 2006, pp. 1623–1627.

[46] Werner K., and F., Mariani, "Thinning and Singulation of Silicon: Root Causes of the Damage in Thin Chips," IEEE/ECTC Proceedings, San Diego, CA, 2006, pp. 1317–1322.

[47] Steen, M., *Laser Material Processing*, 3rd ed., Springer, New York, pp. 107–125.
[48] Li, J., H. Hwang, E. C. Ahn, Q. Chen, P. W. Kim, T. H. Lee, M. K Chung, and T. G. Chung, "Laser Dicing and Subsequent Die Strength Enhancement Technologies for Ultra-thin Wafer," *IEEE/ECTC Proceedings*, Reno, NV, 2007, pp. 761–766.
[49] Lim, S., V. Rao, H. Yin, W. Ching, V. Kripesh, C. Lee, J. H. Lau, J. Milla, and A. Fenner, "Process Development and Reliability of Microbumps," *IEEE/ECTC Proceedings*, December 2008, pp. 367–372. Also, *IEEE Transactions on Components and Packaging Technology*, Vol. 33, No. 4, 2010, pp. 747–753.
[50] Vempati, S., S. Nandar, C. Khong, Y. Lim, V. Kripesh, J. H. Lau, B. P. Liew, K. Y. Au, S. Tamary, A. Fenner, R. Erich, and J. Milla, "Development of 3D Silicon Die Stacked Package Using Flip-Chip Technology with Micro Bump Interconnects," *IEEE/ECTC Proceedings*, San Diego, CA, 2009, pp. 980–987.
[51] Zhang, F., M. Li, W. T. Chen, and K. S. Chan, "An Investigation into the Effects of Flux Residues on Properties of Underfill Materials for Flip-Chip Packages," *IEEE Transactions of Components and Packaging Technology*, Vol. 26, No. 1, June 2003, pp. 233–237.
[52] Kloeser, J., E. Zakel, F. Bechtold, and H. Reichl, "Reliability Investigation of Fluxless Flip-Chip Interconnections on Green Tape Ceramic Substrates," *IEEE Transactions of Components and Packaging Technology A*, Vol. 19, No. 1, March 1996, pp. 24–33.
[53] Hutter, M., H. Oppermann, G. Engelmann, L. Dietricj, and H. Reichl, "Precise Flip-Chip Assembly Using Electroplated $AuSn_{20}$ and $SnAg_{3.5}$ Solder," in *Proceedings of the 57th Electronic Components and Technology Conference*, 2006, pp. 1087–1094.
[54] Hutter, M., F. Hohnke, H. Oppermann, M. Klein, and G. Engelmann, "Assembly and Reliability of Flip-Chip Solder Joints Using Miniaturized Au/Sn Bumps," in *Proceedings of the 54th Electronic. Components and Technology Conference*, 2004, pp. 49–56.
[55] Baggerman, A. F. J., and M. J. Batenburg, "Reliable Au-Sn Flip-Chip Bonding on Flexible Prints," *IEEE Transactions of Components and Packaging Technology B*, Vol. 18, No. 2, May 1995, pp. 257–263.
[56] Rao, V., X. Zhang, S. Ho, R. Rajoo, C. Premachandran, V. Kripesh, S. Yoon, and J. H. Lau, "Design and Development of Fine Pitch Copper/Low-k Wafer Level Package", *IEEE Transactions on Advanced Packaging*, Vol. 33, No. 2, May 2010, pp. 377–388.
[57] Hayes, S., N. Chhabra, T. Duong, Z. Gong, D. Mitchfell, and J. Wright, "System-in-Package Opportunities with the Redistributed Chip Package (RCP)," *Proceedings of International Wafer Level Packaging Conference*, San Jose, CA, October 2011, pp. 1–7.
[58] Keser, B., C. Amrine, T. Duong, S. Hayes, G. Leal, W. Lytle, D. Mitchell, and R. Wenzel, "Advanced Packaging: The Redistributed Chip Package," *IEEE Transactions on Advanced Packaging*, Vol. 31, No. 1, February 2008, pp. 39–43.
[59] Trotta, S., M. Wintermantel, J. Dixon, U. Moeller, R. Jammers, T. Hauck, T., Samulak, B. Dehlink, K. Shun-Meen, H. Li, A. Ghazinour, Y. Yin, S. Pacheco, R. Euter, S. Majied, D. Moline, T. Aaron, V. P. Trivedi, D. J. Morgan, and J. John, "An RCP Packaged Transceiver Chipset for Automotive LRR and SRR Systems in SiGe BiCMOS Technology," *IEEE Transactions on Microwave Theory and Techniques*, Vol. PP, No. 99, 2012, pp. 1–17.
[60] Lau, J. H., *Reliability of RoHS-Compliant 2D and 3D IC Interconnects*, McGraw-Hill, New York, 2011.
[61] Lau, J. H., C. K. Lee, C. S. Premachandran, and A. Yu, *Advanced MEMS Packaging*, McGraw-Hill, New York, 2010.
[62] Harman, G., *Wire Bonding in Microelectronics*, McGraw-Hill, New York, 2010.
[63] Prasad, S., *Advanced Wirebond Interconnection Technology*, Springer, New York, 2004.
[64] Lau, J. H., *Hankbook of Tape Automated Bonding*, Van Nostrand Reinhold, New York, 1992.
[65] Onuki, J., M., Koizumi, and I. Araki, "Investigation of the Reliability of Copper Ball Bonds to Aluminum Electrodes," *IEEE Transactions on Components, Hybrid, and Manufacturing Technology*, Vol. 12, No. 4, 1987, pp. 550–555.
[66] Toyozawa, K., K. Fujita, S. Minamide, and T. Maeda, "Development of Copper

Wire Bonding Application Technology," *IEEE Transactions on Components, Hybrid, and Manufacturing Technology*, Vol. 13, No. 4, 1990, pp. 667–672.

[67] Khoury, S., D. Burkhard, D. Galloway, and T. Scharr, "A Comparison of Copper and Gold Wire Bonding on Integrated Circuit Devices," *IEEE Transactions on Components, Hybrid, and Manufacturing Technology*, Vol. 13, No. 4, 1990, pp. 673–681.

[68] Caers, J., A. Bischoff, J. Falk, and J. Roggen, "Conditions for Reliable Ball-Wedge Copper Wire Bonding," *Proceedings of 1993 Japan International Electronic Manufacturing Technology Symposium*, Kanazawa, Japan,1993, pp. 312–315.

[69] Nguyen, L., D. McDonald, A. Danker, and P. Ng, "Optimization of Copper Wire Bonding on Al-Cu Metallization," *IEEE Transactions on Components, Hybrid, and Manufacturing Technology A*, Vol. 18, No. 2, 1995, pp. 423–429.

[70] Tan, J., B. Toh, and H. Ho, "Modelling of Free Air Ball for Copper Wire Bonding," *Proceedings of the 6th IEEE Electronics Packaging and Technology Conference*, Singapore, 2004, pp. 711–717.

[71] Xu, H., C. Liu, V. Silberschmidt, and H. Wang, "Effects of Process Parameters on Bondability in Thermosonic Copper Ball Bonding," *IEEE/ECTC Proceedings*, Lake Buena Vista, FL, 2008. 1424-1430.

[72] Uno, T., S. Terashima, and T. Yamada, "Surface-Enhanced Copper Bonding Wire for LSI," *IEEE/ECTC Proceedings*, San Diego, CA, 2009, pp. 1486–1495.

[73] Deley, M., and L. Levine, "Copper Ball Bonding Advances for Leading Edge Packaging," *Proceedings of Semicon Singapore 2005*, Singapore, 2005, pp. 1–6.

[74] Shah, A., M. Mayer, Y. Zhou, S. Hong, and J. Moon, "In Situ Ultrasonic Force Signals During Low-temperature Thermosonic Copper Wire Bonding," *Microelectronics Engineering*, Vol. 85, No. 9, 2008, pp. 1851–1857.

[75] Hang, C. J., I. Lum, J. Lee, M. Mayer, C. Q. Wang, Y. Zhou, S. J. Hong, and S. M. Lee, "Bonding Wire Characterization Using Automatic Deformability Measurement," *Microelectronics Engineering*, Vol. 85, No. 8, 2008, pp. 1795-1803.

[76] Kim, H., J. Lee, K. Paik, K. Koh, J. Won, S. Choe, J. Lee, J. Moon, and Y. Park, "Effects of Cu/Al Intermetallic Compound (IMC) on Copper Wire and Aluminum Pad Bondability," *IEEE Transactions on Component Packaging Technology*, Vol. 26, No. 2, 2003, pp. 367–374.

[77] Murali, S. et al., "An Analysis of Intermetallics Formation of Gold and Copper Ball Bonding on Thermal Aging," *Material Research Bulletant*, Vol. 38, No. 4, 2003, pp. 637–646.

[78] Murali, S. et al., "Effect of Wire Size on the Formation of Intermetallics and Kirkendall Voids on Thermal Aging of Thermosonic Wire Bonds," *Materials Letters*, Vol. 58, No. 25, 2004, pp. 3096–3101.

[79] Wulff, F., C. Breach, D. Stephan, A. Saraswati, and K. Dittmer, "Characterization of Intermetallic Growth in Copper and Gold Ball Bonds on Aluminum Metallization," *IEEE/ECTC Proceedings*, Singapore, 2004, pp. 348–353.

[80] Ratchev, P., S. Stoukatch, and B. Swinnen, "Mechanical Reliability of Au and Cu Wire Bonds to Al, Ni/Au and Ni/Pd/Au Capped Cu Bond Pads," *Microelectronic Reliability*, Vol. 46, No. 8, 2006, pp. 1315–1325.

[81] Murali, S., N. Srikanth, and C. Vath, "An Evaluation of Gold and Copper Wire Bonds on Shear and Pull Testing," *ASME Transactions, Journal of Electronic Packaging*, Vol. 128, No. 3, 2006, pp. 192–201.

[82] Ibrahim, M. R., Y. Choi, L. Lim, J. Lu, L. Poh, and P. Ai, "The Challenges of Fine Pitch Copper Wire Bonding in BGA Packages," *IEEE Proceedings of 31st International Electronic Manufacturing Technology Symposium*, Kuala Lumpur, Malaysia, 2006, pp. 347–353.

[83] Hang, C. J., C. Wang, M. Mayer, Y. Tian, Y. Zhou, and H. Wang, "Growth Behavior of Cu/Al Intermetallic Compounds and Cracks in Copper Ball Bonds During Isothermal Aging," *Microelectronic Reliability*,Vol. 48, No. 3, 2008, pp. 416–424.

[84] Xu, H., C. Liu, V. Silberschmidt, S. Pramana, T. White, and Z. Chen, "A Re-examination of the Mechanism of Thermosonic Copper Ball Bonding on Aluminium Metallization Pads," *Scripta Materialia*, Vol. 61, No. 2, 2009, pp. 165–168.

[85] Zhang, S., C. Chen, R. Lee, A. Lau, P. Tsang, L. Mohamed, C. Chan, and M. Dirkzwager, "Characterization of Intermetallic Compound Formation and

Copper Diffusion of Copper Wire Bonding," *IEEE/ECTC Proceedings*, San Diego, CA, 2006, pp. 1821–1826.

[86] Yeoh, L. S., "Characterization of Intermetallic Growth for Gold Bonding and Copper Bonding on Aluminum Metallization in Power Transistors," *IEEE/ECTC Proceedings*, Singapore, 2007, pp. 731–736.

[87] Murali, S., N. Srikanth, and C. Vath, "Effect of Wire Diameter on the Thermosonic Bond Reliability," *Microelectronic Reliability*, Vol. 46, Nos. 2–4, 2006, pp. 467–475.

[88] Tan, C. W., A. Daud, and M. Yarmo, "Corrosion Study at Cu-Al Interface in Microelectronics Packaging," *Appied Surface Science*, Vol. 191, Nos. 1–4, 2002, pp. 67–73.

[89] England, L., and T. Jiang, "Reliability of Cu Wire Bonding to Al Metallization," *IEEE/ECTC Proceedings*, Reno, NV, 2007, pp. 1604–1613.

[90] Kaimori, S., T. Nonaka, and A. Mizoguchi, "The Development of Cu Bonding Wire with Oxidation-Resistant Metal Coating," *IEEE Transactions on Advanced Packagine*, Vol. 29, No. 2, 2006, pp. 227–231.

[91] Murali, S., N. Srikanth, and C. Vath, "Grains, Deformation Substructures, and Slip Bands Observed in Thermosonic Copper Ball Bonding," *Materials Characterization*, Vol. 50, No. 1, 2003, pp. 39–50.

[92] Murali, S., and N. Srikanth, "Acid Decapsulation of Epoxy Molded IC Packages with Copper Wire Bonds," *IEEE Transactions on Electronic Packaging Manufacturing*, Vol. 29, No. 3, 2006, pp. 179–183.

[93] Chylak, R., "Developments in Fine Pitch Copper Wire Bonding Production," *IEEE/ECTC Proceedings*, Singapore, 2009, pp. 1–6.

[94] Appelt, B., A. Tseng, and Y. Lai, "Fine Pitch Copper Wire Bonding—Why Now?" *IEEE/ECTC Proceedings*, Singapore, 2009, pp. 469–472.

[95] Yauw, O., H. Clauberg, K. Lee, L. Shen, and B. Chylak, "Wire Bonding Optimization with Fine Copper Wire for Volume Production," *IEEE/EPTC Proceedings*, Singapore, December 2010, pp. 467–472.

[96] Leng, E., C. Siong, L. Seong, P. Leong, K. Gunase, J. Song, K. Mock, C. Siew, and K. Siva, "Ultra Fine Pitch Cu Wire Bonding on C45 Ultra Low k Wafer Technology," *IEEE/EPTC Proceedings*, Singapore, December 2010, pp. 484–488.

[97] Teo, J., "17.5-μm Thin Cu Wire Bonding for Fragile Low-*k* Wafer Technology," *IEEE/EPTC Proceedings*, Singapore, December 2010, pp. 355–358.

[98] Kim, S., J. Park, S. Hong, and J. Moon, "The Interface Behavior of the Cu-Al Bond System in High Humidity Conditions," *IEEE/EPTC Proceedings*, Singapore, December 2010, pp. 545–549.

[99] Song, M., G. Gong, J. Yao, S. Xu, S. Lee, M. Han, and B. Yan, "Study of Optimum Bond Pad Metallization Thickness for Copper Wire Bond Process," *IEEE/EPTC Proceedings*, Singapore, December 2010, pp. 597–602.

[100] Boettcher, T., M. Rother, S. Liedtke, M. Ullrich, M. Bollmann, A. Pinkernelle, D. Gruber, H. Funke, M. Kaiser, K. Lee, M. Li, K. Leung, T. Li, M. Farrugia, and O. O'Halloran, "On the Intermetallic Corrosion of Cu-Al Wire Bonds," *IEEE/EPTC Proceedings*, Singapore, December 2010, pp. 585–590.

[101] Tang, L., H. Ho, Y. Zhang, Y. Lee, and C. Lee, "Investigation of Palladium Distribution on the Free Air Ball of Pd-Coated Cu Wire," *IEEE/EPTC Proceedings*, Singapore, December 2010, pp. 777–782.

[102] Kumar, B., M. Sivakumar, R. Malliah, M. Li, and S. Yew, "Process Characterization of Cu & Pd Coated Cu Wire Bonding on Overhang Die: Challenges and Solution," *IEEE/EPTC Proceedings*, Singapore, December 2010, pp. 859–867.

[103] Breach, C., N. Shen, T. Mun, T. Lee, and R. Holliday, "Effects of Moisture on Reliability of Gold and Copper Ball Bonds," *IEEE/EPTC Proceedings*, Singapore, December 2010, pp. 44–51.

第 11 章 3D 集成的发展趋势

11.1 引言

本章给出了至 2020 年，3D 集成技术的发展趋势。图 11.1 所示为 3D 集成发展的路线图[1,2]，可以看到，无源 TSV 转接板到 2020 年将是应用最多的 TSV 产品。

11.2 3D Si 集成发展趋势

第 1 章中曾提到，3D Si 集成的实现还面临技术、电子设计自动化（EDA）以及产业生态等方面的问题。但毫无疑问，3D Si 集成是超越摩尔定律的正确道路。产业界应该立即行动起来，力争构建包括上下游衔接标准和基础设施在内的产业生态环境，以便 EDA 开发商尽快开发用于 3D Si 集成设计、仿真、分析、验证、制造准备以及测试的相关软件。

从图 11.1 中还可以看到，到 2020 年，基于无凸点 W2W 键合技术（图 11.2）

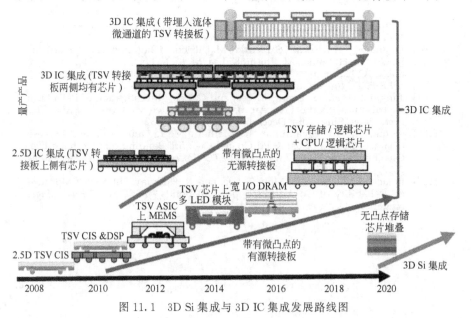

图 11.1 3D Si 集成与 3D IC 集成发展路线图

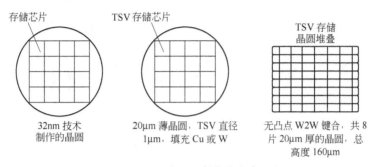

图 11.2 晶圆代工厂制作的内存立方

的 3D Si 集成技术很有希望制造出低成本、高量产、小外形尺寸的芯片堆叠存储器（内存立方）。这类产品将会由半导体晶圆代工厂生产，封测厂商得到堆叠好的无凸点晶圆后，完成晶圆凸点制作、切割、封装与测试等例行工作。

11.3 3D IC 集成发展趋势

图 1.6 曾给出了 3D IC 集成的潜在应用。到 2020 年，多数 TSV 是制作在无源转接板（2.5D IC 集成）上的，如图 11.1 所示。对于有源转接板（即器件芯片上制作有 TSV），除了在 3D MEMS[3] 和 3D LED[4] 方面的利基应用外，不得不等待产业链的形成以及自动设计能力的提高。同时，宽 I/O 存储器和 DRAM 将被图 11.3 所示的 2.5D IC 集成器件所取代：①图 1.6 和图 1.9 中的宽 I/O 存储器的 TSV SoC 逻辑芯片可以用无 TSV 的存储芯片代替；②类似地，图 1.6 和图 1.11 中的宽 I/O DRAM 的 TSV 逻辑控制器可以用无 TSV 的内存立方代替。

图 11.4 所示为采用无源 TSV 转接板的 3D IC 集成器件。可以发现，通过在 TSV 转接板上下两侧安放无 TSV 芯片，2.5D IC 集成可以成为真正的 3D IC 集成。这种情况下，芯片到芯片的互连距离较短，而且转接板尺寸较小，如图 1.23～图 1.33 所示。另外，宽 I/O 存储器（图 1.6、图 1.9）和宽 I/O DRAM（图 1.6、

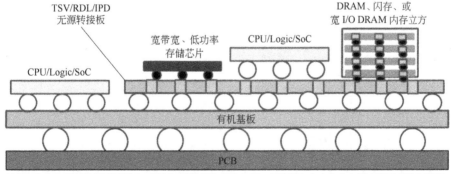

图 11.3 2.5D IC 集成的潜在应用

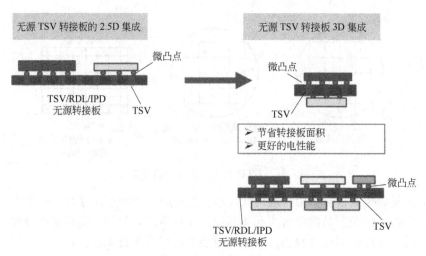

图 11.4　无源 TSV 转接板的 3D IC 集成

图 1.11）都可以采用这种布置方式。

11.4　参考文献

[1] Lau, J. H., "Evolution, Challenges, and Outlook of 3D IC/Si Integration," *Plenary Keynote at IEEE ICEP*, Nara, Japan, April 2011, pp. 1–16.
[2] Lau, J. H., "3D IC Integration and 3D Si Integration," *Plenary Keynote at IWLPC*, San Jose, CA, October 2011, pp. 1–18.
[3] Lau, J. H., C. K. Lee, C. S. Premachandran, and A. Yu, *Advanced MEMS Packaging*, McGraw-Hill, New York, 2010.
[4] Lau, J. H., R. Lee, M. Yuen, and P. Chan, "3D LED and IC Wafer Level Packaging," *Journal of Microelectronics International*, Vol. 27, Issue 2, 2010, pp. 98–105.

附录 A 量度单位换算表

科学计算和工程设计中经常遇到不同量度单位制之间的转换问题。本附录列出了目前使用的不同量度单位制的转换系数，并以列表形式给出了工程中常用单位的简写、简称，以方便读者查找。

表 A-1 量度单位换算表

长 度	
$1m = 10^{10} Å$(埃)	$1Å = 10^{-10} m$(米)
$1m = 10^9 nm$(纳米)	$1nm = 10^{-9} m$(米)
$1m = 10^6 \mu m$(微米)	$1\mu m = 10^{-6} m$(米)
$1m = 10^3 mm$(毫米)	$1mm = 10^{-3} m$(米)
$1m = 10^2 cm$(厘米)	$1cm = 10^{-2} m$(米)
$1mm = 0.0394 in$(英寸)	$1in = 25.4 mm$(毫米)
$1cm = 0.394 in$(英寸)	$1in = 2.54 cm$(厘米)
$1m = 39.4 in = 3.28 ft$(英尺)	$1ft = 12 in = 0.3048 m$(米)
$1mm = 39.37 mil$(密尔)	$1mil = 10^{-3} in = 0.0254 mm = 25.4 \mu m$
$1\mu m = 39.37 \mu in$(微英寸)	$1\mu in = 0.0254 \mu m$(微米)
面 积	
$1m^2 = 10^4 cm^2$(平方厘米)	$1cm^2 = 10^{-4} m^2$(平方米)
$1cm^2 = 10^2 mm^2$(平方毫米)	$1mm^2 = 10^{-2} cm^2$(平方厘米)
$1m^2 = 10.76 ft^2$(平方英尺)	$1ft^2 = 0.093 m^2$
$1cm^2 = 0.1550 in^2$(平方英寸)	$1in^2 = 6.452 cm^2$
体 积	
$1m^3 = 10^6 cm^3$(立方厘米)	$1cm^3 = 10^{-6} m^3$(立方米)
$1cm^3 = 10^3 mm^3$(立方毫米)	$1mm^3 = 10^{-3} cm^3$(立方厘米)
$1m^3 = 35.32 ft^3$(立方英尺)	$1ft^3 = 0.0283 m^3$(立方米)
$1cm^3 = 0.0610 in^3$(立方英寸)	$1in^3 = 16.39 cm^3$(立方厘米)

续表

质量	
$1Mg=1t=10^3 kg$(千克)	$1kg=10^{-3}t$(吨)$=10^{-3}Mg$(兆克)
$1kg=10^3 g$(克)	$1g=10^{-3}kg$(千克)
$1kg=2.205 lbm$(磅质量)	$1lbm=0.4536kg$(千克)
$1g=2.205×10^{-3} lbm$(磅质量)	$1lbm=453.6g$(克)
$1g=0.035oz$(盎司)	$1oz=28.35g$(克)
密度	
$1kg/m^3=10^{-3}g/cm^3$	$1g/cm^3=10^3 kg/m^3$
$1Mg/m^3=1t/m^3=1g/cm^3$	$1g/cm^3=1t/m^3=1Mg/m^3$
$1kg/m^3=0.0624lbm/ft^3$	$1lbm/ft^3=16.02kg/m^3$
$1g/cm^3=62.4lbm/ft^3$	$1lbm/ft^3=1.602×10^{-2}g/cm^3$
$1g/cm^3=0.0361lbm/in^3$	$1lbm/in^3=27.7g/cm^3$
力	
$1N=0.102kgf$(公斤力)	$1kgf=9.801N$(牛顿,牛)
$1N=10^5 dyn$(达因)	$1dyn=10^{-5}N=10\mu N$(微牛)
$1N=0.2248lbf$(磅力)	$1lbf=4.448N$(牛)
$1N=0.0002248k(kip)$(千磅力)	$1k=1000lbf=4.448kN$(千牛)
压力、应力、压强	
$1Pa=1N/m^2=10dyn/cm^2$	$1dyn/cm^2=0.10Pa$(帕斯卡,帕)
$1MPa=145psi$(磅力每平方英寸)	$1psi=1lbf/in^2=6.90×10^{-3}MPa$(兆帕)
$1MPa=0.102kgf/mm^2$(公斤力每平方毫米)	$1kgf/mm^2=9.807MPa$(兆帕)
$1kgf/mm^2=1422psi$(磅力每平方英寸)	$1psi=7.03×10^{-4}kgf/mm^2$(公斤力每平方毫米)
$1kPa=0.00987atm$(大气压)	$1atm=101.325kPa$(千帕)
$1kPa=0.01ba$(巴)	$1bar=100kPa=0.1MPa$(兆帕)
$1Pa=0.0075torr$(托)$=0.0075mmHg$(毫米汞柱)	$1torr=1mmHg=133.322Pa$(帕)
$1Pa=145ksi$(千磅力每平方英寸)	$1ksi=6.90MPa$(兆帕)
断裂韧性	
$1MPa·(m)^{1/2}=910psi·(in)^{1/2}$	$1psi·(in)^{1/2}=1.099×10^{-3}MPa·(m)^{1/2}$

续表

能量、功、热	
$1J=10^7 erg$(尔格)	$1erg=10^{-7}J$(焦耳,焦)
$1J=6.24\times10^{18}eV$(电子伏)	$1eV=1.602\times10^{-19}J$(焦)
$1J=0.239cal$(卡路里,卡)	$1cal=4.187J$(焦)
$1J=9.48\times10^{-4}Btu$(英热量单位)	$1Btu=1054J$(焦)
$1J=0.738ft\cdot lbf$(英尺·磅力)	$1ft\cdot lbf=1.356J$(焦)
$1eV=3.82\times10^{-20}cal$(卡)	$1cal=2.61\times10^{19}eV$(电子伏)
$1cal=3.97\times10^{-3}Btu$(英热量单位)	$1Btu=252.0cal$(卡)
功率	
$1W=1.01kgf\cdot m/s$	$1kgf\cdot m/s=9.807W$(瓦特,瓦)
$1W=1.36\times10^{-3}$马力	1 马力$=735.5W$(瓦)
$1W=0.239cal/s$(卡每秒)	$1cal/s=4.187W$(瓦)
$1W=3.414Btu/h$(英热量单位每小时)	$1Btu/h=0.293W$(瓦)
$1cal/s=14.29Btu/h$(英热量单位每小时)	$1Btu/h=0.070cal/s$(卡每秒)
$1W=10^7 erg/s$(尔格每秒)$=1J/s$(焦每秒)	$1erg/s=10^{-7}W$(瓦)
黏度	
$1Pa\cdot s=10P$(泊)	$1P=0.1Pa\cdot s$(帕·秒)
$1mPa\cdot s=1cP$(厘泊)	$1cP=10^{-3}Pa\cdot s$(帕·秒)
温度 T	
$T(K)=273.15+T(℃)$	$T(℃)=T(K)-273.15$
$T(K)=\frac{5}{9}[T(℉)-32]+273.15$	$T(℉)=\frac{9}{5}[T(K)-273.15]+32$
$T(℃)=\frac{5}{9}[T(℉)-32]$	$T(℉)=\frac{9}{5}[T(℃)+32]$
比热容	
$1J/(kg\cdot K)=2.29\times10^{-4}cal/(g\cdot K)$	$1cal/(g\cdot K)=4184J/(kg\cdot K)$
$1J/(kg\cdot K)=2.29\times10^{-4}Btu/(lbm\cdot ℉)$	$1Btu/(lbm\cdot ℉)=4184J/(kg\cdot K)$
$1cal/(g\cdot ℃)=1.0Btu(lbm\cdot ℉)$	$1Btu/(lbm\cdot ℉)=1.0cal/(g\cdot K)$
热导率	
$1W/(m\cdot K)=2.39\times10^{-3}cal/(cm\cdot s\cdot K)$	$1cal/(cm\cdot s\cdot K)=418.4W/(m\cdot K)$
$1W/(m\cdot K)=0.578Btu/(ft\cdot h\cdot ℉)$	$1Btu/(ft\cdot h\cdot ℉)=1.730W/(m\cdot K)$
$1cal/(cm\cdot s\cdot k)=241.8Btu/(ft\cdot h\cdot ℉)$	$1Btu/(ft\cdot h\cdot ℉)=4.136\times10^{-3}cal/(cm\cdot s\cdot K)$

表 A-2 单位符号及其中文名称

A,安(培)	Gb,吉(伯)	mm,毫米
Å,埃	Gy,戈(瑞)	nm,纳米
bar,巴	h,(小)时	N,牛(顿)
Btu,英热量单位	H,亨(利)	Oe,奥(斯特)
C,库(仑)	Hz,赫(兹)	psi,磅力每平方英寸
℃,摄氏度	in,英寸	P,泊
cal,卡(路里)	J,焦(耳)	Pa,帕(斯卡)
cm,厘米	K,开(尔文)	rad,拉德
cP,厘泊	kgf,千克力	s,秒
dB,分贝	kpsi,千磅每平方英寸	S,西(门子)
dyn,达因	L,升	T,特(斯拉)
erg,尔格	lbf,磅力	torr,托
eV,电子伏	lbm,磅(质量)	min,分(钟)
F,法(拉)	m,米	V,伏(特)
°F,华氏度	Mg,兆克	W,瓦(特)
ft,英尺	MPa,兆帕	Wb,韦(伯)
g,克	mil,密耳	Ω,欧(姆)

表 A-3 国际单位制中常用的词头的符号

因数	词头		符号
	英文名称	中文名称	
10^9	giga	吉	G
10^6	mega	兆	M
10^3	kilo	千	k
10^{-2}	centi	厘	c
10^{-3}	milli	毫	m
10^{-6}	micro	微	μ
10^{-9}	nano	纳	n
10^{-12}	pico	皮	p

表 A-4 常用其他符号

ppm	1×10^{-6}
ppb	1×10^{-9}

附录 B 缩 略 语 表

A
APC	Advanced Pressure Calibrator	压力校准仪
ASME	American Society of Mechanical Engineers	美国机械工程师协会
ASIC	Application Specific Integrated Circuit	专用集成电路

B
BD	Black Diamond	黑钻石
BEOL	Back End of Line	后段制程
BGA	Ball Grid Array	球栅阵列
BHE	Backside Helium	背面氦气
BISR	Build-In Self-Repair	内置自修复
BIST	Build-In Self-Test	内置自测试
BT	BismaleimideTriazine	双马来酰亚胺三嗪树脂
BU	Buildup	堆积

C
C2C	Chip to Chip	芯片到芯片键合
C2W	Chip to Wafer	芯片到晶圆键合
CFD	Computational Fluid Dynamics	计算流体动力学
CMOS	Complementary Metal Oxide Semiconductor	互补金属氧化物半导体
CMP	Chemical Mechanical Polishing	化学机械抛光
CoB	Chip on Board	板上芯片
CoC	Chip on Chip	芯片到芯片
CPU	Central Processing Unit	中央处理器
CSM	Continuous Stiffness Measurement	连续刚度测量
CSP	Chip Scale Package	芯片级封装
CT	Computed Tomography	计算机断层扫描
CTE	Coefficient of Thermal Expansion	热膨胀系数
CVD	Chemical Vapor Deposition	化学气相沉积

D
DAF	Die-Attach Film	芯片贴膜
DC	Direct Current	直流
DDR	Double Data Rate	双倍数据传输速率
DFR	Design for Reliability	可靠性设计
DI	Deionized	去离子
DoE	Design of Experiment	实验设计
DRAM	Dynamic Random Access Memory	动态随机存取存储器
DRC	Design Rule Checks	设计规则检查

DRIE	Deep Reactive Ion Etching	深反应离子刻蚀
DSC	Differential Scanning Calorimetry	差示扫描量热法
DSP	Digital Signal Processor	数字信号处理器
E		
ECTC	Electronic Components and Technology Conference	电子元件与技术会议
EDA	Electronic Design Automation	电子设计自动化
EDS	Energy Dispersive Spectrometer	能谱仪
EDX	Energy Dispersive X-ray	X射线能谱仪
EM	Electromigration	电迁移
eMMC	embedded Multimedia Card	嵌入式多媒体卡
EMS	Electronics Manufacturing Service	电子专业制造服务商
ER	Etch Rate	刻蚀速率
ERR	Energy Release Rate	能量释放率
ESC	Electrostatic Chuck	静电卡盘
eWLP	embedded Wafer Level Package	埋入式晶圆级封装
F		
FC	Flip Chip	倒装芯片
FEA	Finite-Element Analysis	有限元分析
FEOL	Front End of Line	前段制程
FIB	Focused Ion Beam	聚焦离子束
FPGA	Field Programmable Gate Array	现场可编程门阵列
FSG	Fluorinated Silicate Glass	氟硅酸玻璃
G		
GPU	Graphic Processor Unit	图像处理单元
H		
HD	High Definition	高清
HMC	Hybrid Memory Cube	混合内存立方
HRP	High-Resolution Profiler	高分辨率轮廓仪
HTS	High-Temperature-Storage Test	高温存储实验
I		
IC	Integrated Circuit	集成电路
ICP	Inductive-Coupled Plasma	感应耦合等离子体
IDM	Integrated Device Manufacturers	整合晶圆制造商
IEDM	International Electron Device Meeting	国际电子器件会议
IEEE	Institute of Electrical and Electronics Engineers	美国电气电子工程师协会
ILD	Interlayer Dielectric	中间绝缘层
IMAPS	International Microelectronics and Packaging Society	国际微电子与封装协会
IMC	Intermetallic Compounds	金属间化合物
IME	Institute of Microelectronics	新加坡微电子研究所
IMEC	Interuniversity Microelectronics Centre	比利时微电子研究中心
IMP	Ionized-Metal Plasma	电离金属等离子体
IPD	Integrated Passive Device	集成无源器件
ISSCC	International Solid-State Circuits Conference	国际固态电路会议

ITRI	Industrial Technology Research Institute	工业技术研究院(台湾)
J		
JEDEC	Joint Electronic Device Engineering Council	国际电子器件工程联合会
K		
KGD	Known-Good Die	已知合格芯片
L		
LED	Light Emitting Diode	发光二极管
LMI	Logic-Memory Interface	逻辑内存接口
LPDDR	Low Power Double Data Rate	低功耗双倍数据传输速率
LTHC	Light to Heat Conversion	光热转换
LVS	Layout Versus Schematic	电路图
M		
MC	Molding Compound	模塑料
MCP	Multichip Package	多芯片封装
MCU	Microcontroller Unit	微控制器单元
MEMS	Micro-Electro-Mechanical Systems	微机电系统
MMIC	Monolithic Microwave Integrated Circuit	单片微波集成电路
MPU	Microprocessor Unit	微处理单元
MSL	Moisture Sensitive Level	湿度敏感性等级
MTTF	Mean Time to Failure	平均寿命
MVCC	Modified Virtual Crack-Closure	修正的虚拟裂纹闭合
N		
NAND	Not-And	与非门
NEMS	Nano-Electro-Mechanical System	纳机电系统
NSF	National Science Foundation	美国国家科学基金会
O		
OEM	Original Equipment Manufacturer	原始设备制造商
OM	Optical Microscope	光学显微镜
OMEMS	Opto-Micro-Electro-Mechanical Systems	微光机电系统
OSAT	Outsourced Semiconductor Assembly and Test	外包半导体封测厂
P		
PBGA	Plastic Ball Grid Array	塑料球栅阵列
PBO	Polybenzoxazole	聚苯并噁唑
PC	Personal Computer	个人电脑
PCB	Printed Circuit Board	印刷电路板
PCT	Pressure-Cook Test	高压蒸煮实验
PECVD	Plasma Enhanced Chemical Vapor Deposition	等离子增强化学气相沉积
PEDL	Polymer-Encapsulated Dicing-line	聚合物封装切割线
PI	Polyimide	聚酰亚胺
PoP	Package on Package	封装堆叠
PVD	Physical Vapor Deposition	物理气相沉积
R		
RCP	Redistribute Chip Package	再布线芯片封装

RDL	Redistribution Layer	再分布层
RF	Radio Frequency	射频
RH	Relative Humidity	相对湿度
RoHS	Restriction of Hazardous Substances	关于限制在电子电器设备中使用某些有害成分的指令

S

SAM	Scanning Acoustic Microscope	扫描超声显微镜
SEM	Scanning Electron Microscope	扫描电子显微镜
SiP	System in Packaging	系统级封装
SLID	Solid-Liquid Interdiffusion	固-液互扩散
SME	Society of Manufacturing Engineers	美国制造工程师协会
SMT	Surface Mount Technology	表面贴装技术
SoC	System on Chip	片上系统
SoI	Silicon on Insulator	绝缘体上硅
SSD	Solid State Drive	固态驱动器

T

TAB	Tape Automated Bonding	载带自动焊
TAM	Thermomechanical Analyzer	热机械分析仪
TCB	Thermal Compression Bonding	热压键合
TCR	Temperature Coefficient of Resistance	电阻温度系数
TCT	Thermal-Cycling Test	热循环实验
TEM	Transmission Electron Microscope	透射电子显微镜
TEM	Transverse Electromagnetic	横向电磁波
TEOS	Tetraethylorthosilicate	正硅酸乙酯
TIA	Trans-Impedance Amplifier	跨阻放大器
TLP	Transient Liquid-Phase	瞬态液相
TSH	Through-SiHole	硅穿孔
TSV	Through-Silicon Vias	硅通孔
TTV	Total Thickness Variation	总厚度偏差

U

UBM	Under Bump Metallization	凸点下金属层
uHAST	unbiased Highly Accelerated Stress Test	非偏置高加速应力测试
UV	Ultraviolet	紫外线

V

VCSEL	Vertical Cavity Surface Emitting Laser	垂直腔面发射激光器

W

W2W	Wafer to Wafer	晶圆到晶圆键合
WEF	Wire-Embedded Film	内置引线薄膜
WIT	Wire-Interconnect Technology	引线互连技术
WLP	Wafer Level Package	晶圆级封装

X

XDR	X-ray Diffraction	X射线衍射

附录 C TSV 专利

在过去几年里，已有许多关于 TSV 和 3D 集成的专利被授权。本附录选取了自 2005 年以来美国的相关专利，为了方便查阅，将这些专利的专利号、名称、发明人以及签发日期一一列出。如果一些 TSV 和 3D 集成相关专利在下面没有列出，只说明在本书出版时作者不知道有这些专利存在。

8093726　Semiconductor packages having interposers, electronic products employing the same, and methods of manufacturing the same; Park, Sung-Yong; January10, 2012

8093722　System-in-package with fan-out WLCSP; Chen, Nan-Cheng; Hsu, Chih-Tai; January 10, 2012

8093711　Semiconductor device; Zudock, Frank; Meyer, Thorsten; Brunnbauer, Markus; Wolter, Andreas; January 10, 2012

8093705　Dual face package having resin insulating layer; Park, Seung Wook; Kweon, Young Do; Yuan, Jingli; Moon, Seon Hee; Hong, Ju Pyo; Lee, Jae Kwang; January 10, 2012

8093703　Semiconductor package having buried post in encapsulant and method of manufacturing the same; Kim, Pyoung-Wan; Lee, Teak-Hoon; Jang, Chul-Yong; January 10, 2012

8093696　Semiconductor device; Yoon, Kimyung; Dobritz, Stephan; Ruckmich, Stefan; January 10, 2012

8093100　Integrated circuit packaging system having through silicon via with direct interconnects and method of manufacture thereof; Choi, A Leam; Chung, Jae Han; Yang, Deok Kyung; Park, Hyung Sang; January 10, 2012

8090250　Imaging device with focus offset compensation; Lusinchi, Jean-Pierre; January 3, 2012

8076762　Variable feature interface that induces a balancedstress to prevent thin die warpage; Chandrasekaran, Arvind; Radojcic, Ratibor; December 13, 2011

8084866　Microelectronic devices and methods for filling vias in microelectronic devices; Hiatt, William M.; Kirby, Kyle K.; December 27, 2011

8084841　Systems and methods for providing high-density capacitors; Pulugurtha, Markondeya Raj; Fenner, Andreas; Malin, Anna; Goud, Dasharatham Janagama; Tummala, Rao; December 27, 2011

8080862　Systems and methods for enabling ESD protection on 3D stacked devices; Kaskoun, Kenneth; Gu, Shiqun; Nowak, Matthew; December 20, 2011

8080885　Integrated circuit packaging system with multilevel contact and method of manufacture thereof; Chow, Seng Guan (Singapore, SG); Kuan, Heap Hoe (Singapore, SG); Huang, Rui; December 20, 2011

8080445　Semiconductor device and method of forming WLP with semiconductor die embedded within penetrable encapsulant between TSV interposers; Pagaila, Reza A.; December 20, 2011

8076234　Semiconductor device and method of fabricating the same including a conductive structure is formed through at least one dielectric layer after forming avia structure; Park, Byung-Lyul; Choi, Gil-Heyun; Bang, Suk-Chul; Moon, Kwang-Jin; Lim, Dong-Chan; Jung, Deok-Young; December 7, 2011

8072079　Through-hole vias at saw streets including protrusions or recesses for interconnection; Pagaila, RezaA.; Camacho, Zigmund R.; Tay, Lionel Chien Hui; Do, Byung Tai; December 6, 2011

8072056　Apparatus for restricting moisture ingress; Mueller, Tyler (Phoenix, AZ); Batchelder, Geoffrey; Danzl, Ralph B.; Gerrish, Paul F.; Malin, Anna J.; Marrott, Trevor D.; Mattes, Michael F.; December 6, 2011

8063475　Semiconductor package system with through silicon via interposer; Choi, DaeSik (Seoul, KR);

Yang, Deok Kyung (Hanam-si, KR); Kim, Seung Won; November 22, 2011

8063469 On-chip radio frequency shield with interconnect metallization; Barth, Hans-Joachim; Koerner, Heinrich; Meyer, Thorsten; Brunnbauer, Markus; November 22, 2011

8059441 Memory array on more than one die; Taufique, Mohammed H.; Jallice, Derwin; McCauley, Donald W.; De Vale, John P.; Brekelbaum, Edward A.; Rupley, Jeffrey P., II; Loh, Gabriel H.; Black, Bryan; November 15, 2011

8067816 Techniques for placement of active and passive devices within a chip; Kim, Jonghae; Gu, Shiqun; Henderson, Brian Matthew; Toms, Thomas R.; Nowak, Matthew; November 29, 2011

8067308 Semiconductor device and method of forming an interconnect structure with TSV using encapsulant for structural support; Suthiwongsunthorn, Nathapong; Marimuthu, Pandi C.; Ku, Jae Hun; Omandam, Glenn; Goh, Hin Hwa; Heng, Kock Liang; Caparas, Jose A.; November 29, 2011

8063975 Positioning wafer lenses on electronic imagers; Butterfield, Andrew; Fabre, Sebastien; Kujanpaa, Karo; November 22, 2011

8063496 Semiconductor integrated circuit device and method of fabricating the same; Cheon, Keon-Yong; Oh, Tae-Seok; Choi, Jong-Won; Oh, Su-Young; November 22, 2011

8063424 Embedded photodetector apparatus in a 3D CMOS chip stack; Gebara, Fadi H.; Ning, Tak H.; Ouyang, Qiqing C.; Schaub, Jeremy D.; November 22, 2011

8060843 Verification of 3D integrated circuits; Wang, Chung-Hsing; Tsai, Chih Sheng; Liu, Ying-Lin; Lin, Kai-Yun; November 15, 2011

8058137 Method for fabrication of a semiconductor device and structure; Or-Bach, Zvi; Sekar, Deepak C.; Cronquist, Brian; Beinglass, Israel; de Jong, Jan Lodewijk; November 15, 2011

8058102 Package structure and manufacturing method thereof; Lin, Diann-Fang; Hu, Yu-Shan; November 15, 2011

8053902 Isolation structure for protecting dielectric layers from degradation; Chen, Ming-Fa; Lin, Sheng-Yuan; Chen, Ming-Fa (Taichung, TW); Lin, Sheng-Yuan; November 8, 2011

8053900 Through-substrate vias (TSVs) electrically connected to a bond pad design with reduced dishing effect; Yu, Chen-Hua; Chiou, Wen-Chih; Wu, Weng-Jin; November 8, 2011

8053898 Connection for off-chip electrostatic discharge protection; Marcoux, Phil P.; November 8, 2011

8049327 Through-silicon via with scalloped sidewalls; Kuo, Chen-Cheng; Chen, Chih-Hua; Chen, Ming-Fa; Chen, Chen-Shien; November 1, 2011

8049310 Semiconductor device with an interconnect element and method for manufacture; Wolter, Andreas; Hedler, Harry; Irsigler, Roland; November 1, 2011

8048761 Fabricating method for crack stop structure enhancement of integrated circuit seal ring; Yeo, Alfred; Chan, Kai Chong; November 1, 2011

8048721 Method for filling multilayer chip-stacked gaps; Hsu, Hung-Hsin; Chien, Wei-Chih; November 1, 2011

8046727 IP cores in reconfigurable three dimensional integrated circuits; Solomon, Neal; October 25, 2011

8043967 Process for through silicon via filling; Reid, Jonathan D.; Wang, Katie Qun; Wiley, Mark J.; October 25, 2011

8042082 Three dimensional memory in a system on a chip; Solomon, Neal; October 18, 2011

8039393 Semiconductor structure, method for manufacturing semiconductor structure and semiconductor package; Wang, Meng-Jen; Chen, Chien-Yu; October 18, 2011

8039386 Method for forming a through silicon via (TSV); Dao, Thuy B.; Noble, Ross E.; Triyoso, Dina H.; Dao, Thuy B. (Austin, TX); Noble, Ross E. (Austin, TX); Triyoso, Dina H.; October 18, 2011

8039356 Through silicon via lithographic alignment and registration; Herrin, Russell T.; Lindgren, Peter J.; Sprogis, Edmund J.; Stamper, Anthony K.; October 18, 2011

8035194 Semiconductor device and semiconductor package including the same; Lee, Jun-Ho; Lee, Hyung-

	Dong; Im, Hyun-Seok; October 11, 2011
8034708	Structure and process for the formation of TSVs; Kuo, Chen-Cheng; Ching, Kai-Ming; Chen, Chen-Shien; October 11, 2011
8033012	Method for fabricating a semiconductor test probe card space transformer; Hsu, Ming Cheng; Chao, Clinton Chih-Chieh; October 11, 2011
8031217	Processes and structures for IC fabrication; Sheats, Jayna; October 4, 2011
8030780	Semiconductor substrates with unitary vias and via terminals, and associated systems and methods; Kirby, Kyle K.; Parekh, Kunal R.; October 4, 2011
8030761	Mold design and semiconductor package; Kolan, Ravi Kanth; Liu, Hao; Toh, Chin Hock; October 4, 2011
8030200	Method for fabricating a semiconductor package; Eom, Yong Sung; Choi, Kwang-Seong; Bae, Hyun-Cheol; Lee, Jong-Hyun; Moon, Jong Tae; October 4, 2011
7981798	Method of manufacturing substrate; Taguchi, Yuichi; Shiraishi, Akinori; Sunohara, Masahiro; Murayama, Kei; Sakaguchi, Hideaki; Higashi, Mitsutoshi; July 19, 2011
7973415	Manufacturing process and structure of through silicon via; Kawashita, Michihiro; Yoshimura, Yasuhiro; Tanaka, Naotaka; Naito, Takahiro; Akazawa, Takashi; July 5, 2011
7973358	Coupler structure; Hanke, Andre; Nagy, Oliver; July 5, 2011
8030208	Bonding method for through-silicon-via based 3D wafer stacking; Leung, Chi Kuen Vincent; Sun, Peng; Shi, Xunqing; Chung, Chang Hwa; October 4, 2011
8030113	Thermoelectric 3D cooling; Hsu, Louis Lu-Chen; Wang, Ping-Chuan; Wei, Xiaojin; Zhu, Huilong; October 4, 2011
8026592	Through-silicon via structures including conductive protective layers; Yoon, Minseung; Kim, Namseog; Kim, Pyoungwan; Ma, Keumhee; Jo, Chajea; September 27, 2011
8026567	Thermoelectric cooler for semiconductor devices with TSV; Chang, Shih-Cheng; Pan, Hsin-Yu; September 27, 2011
8026521	Semiconductor device and structure; Or-Bach, Zvi; Sekar, Deepak C.; September 27, 2011
8020290	Processes for IC fabrication; Sheats, Jayna; September 20, 2011
8018065	Wafer-level integrated circuit package with top and bottom side, electrical connections; Lam, Ken; September 13, 2011
8017515	Semiconductor device and method of forming compliantpolymer layer between UBM and conformal dielectric layer/RDL for stress relief; Marimuthu, Pandi C.; Suthiwongsunthorn, Nathapong; Huang, Shuangwu; September 13, 2011
8017497	Method for manufacturing semiconductor; Oi, Hideo; September 13, 2011
8017451	Electronic modules and methods for forming the same; Racz, Livia M.; Tepolt, Gary B.; Thompson, Jeffrey C.; Langdo, Thomas A.; Mueller, Andrew J.; September 13, 2011
8014166	Stacking integrated circuits containing serializer and deserializer blocks using through silicon via; Yazdani, Farhang; September 6, 2011
8012808	Integrated microchannels for 3D through silicon architectures; Shi, Wei; Lu, Daoqiang; Bai, Yiqun; Zhou, Qing A.; He, Jianqqi; September 6, 2011
8008764	Bridges for interconnecting interposers in multichip integrated circuits; Joseph, Douglas James; Knickerbocker, John Ulrich; August 30, 2011
8008195	Method for manufacturing semiconductor device; Koike, Osamu; Kadogawa, Yutaka; August 30, 2011
8008192	Conductive interconnect structures and formation methods using supercritical fluids; Sulfridge, Marc; August 30, 2011
8005326	Optical clock signal distribution using through silicon vias; Chang, Shih-Cheng; Lin, Jin-Lien; Hsu, Kuo-Ching; Ching, Kai-Ming; Wu, Jiun Yi; Chen, Yen-Huei; August 23, 2011
7998862	Method for fabricating semiconductor device; Park, Kunsik; Baek, Kyu-Ha; Do, Lee-Mi; Kim,

Dong-Pyo; Park, Ji Man; August 16, 2011

7994041　Method of manufacturing stacked semiconductor package using improved technique of forming through via; Lim, Kwon-Seob; Kang, Hyun Seo; August 9, 2011

7990174　Circuit for calibrating impedance and semiconductor apparatus using the same; Park, Nak Kyu; August 2, 2011

7989318　Method for stacking semiconductor dies; Yang, Ku-Feng; Wu, Weng-Jin; Chiou, Wen-Chih; Yu, Chen-Hua; August 2, 2011

7989226　Clocking architecture in stacked and bonded dice; Peng, Mark Shane; August 2, 2011

7986582　Method of operating a memory apparatus, memory device and memory apparatus; Ruckerbauer, Hermann; Sichert, Christian; July 26, 2011

7986042　Method for fabrication of a semiconductor device and structure; Or-Bach, Zvi; Cronquist, Brian; Beinglass, Israel; de Jong, Jan Lodewijk; Sekar, Deepak C.; July 26, 2011

7977962　Apparatus and methods for through substrate via test; Hargan, Ebrahim H.; Bunker, Layne; Dimitriu, Dragos; King, Gregory; July 12, 2011

7977781　Semiconductor device; Ito, Kiyoto; Saen, Makoto; Kuroda, Yuki; July 12, 2011

7973416　Thru silicon enabled die stacking scheme; Chauhan, Satyendra Singh; July 5, 2011

7973413　Through-substrate via for semiconductor device; Kuo, Chen-Cheng; Chen-Shien, Chen; Ching, Kai-Ming; Chen, Chih-Hua; July 5, 2011

7973411　Microfeature workpieces having conductive interconnect structures formed by chemically reactive processes, and associated systems and methods; Borthakur, Swarnal; July 5, 2011

7972969　Method and apparatus for thinning a substrate; Yang, Ku-Feng; Chiou, Wen-Chih; Wu, Weng-Jin; Zuo, Kewei; July 5, 2011

7972905　Packaged electronic device having metal comprising self-healing die attach material; Wainerdi, James C.; Tellkamp, John P.; July 5, 2011

7972902　Method of manufacturing a wafer including providing electrical conductors isolated from circuitry; Youn, Sunpil; Lee, Seok-Chan; July 5, 2011

7969193　Differential sensing and TSV timing control scheme for 3D-IC; Wu, Wei-Cheng; Chen, Yen-Huei; Chang, Meng-Fan; June 28, 2011

7969013　Through silicon via with dummy structure and method for forming the same; Chen, Chih-Hua; Chen, Chen-Shien; Kuo, Chen-Cheng; Shen, Wen-Wei; June 28, 2011

7969009　Through silicon via bridge interconnect; Chandrasekaran, Arvind; June 28, 2011

7964916　Method for fabrication of a semiconductor device and structure; Or-Bach, Zvi; Cronquist, Brian; Beinglass, Israel; de Jong, Jan L.; Sekar, Deepak C.; June 21, 2011

7964502　Multilayered through via; Dao, Thuy B.; Vuong, Chanh M.; June 21, 2011

7960773　Capacitor device and method for manufacturing the same; Chang, Shu-Ming; Chiang, Chia-Wen; June 14, 2011

7960282　Method of manufacture an integrated circuit system with through silicon via; Yelehanka, Pradeep Ramachandramurthy; Tan, Denise; Lek, ChungMeng; Thiam, Thomas; Lam, Jeffrey C.; Hsia, Liang-Choo; June 14, 2011

7960242　Method for fabrication of a semiconductor device and structure; Or-Bach, Zvi; Cronquist, Brian; Beinglass, Israel; de Jong, Jan Lodewijk; Sekar, Deepak C.; June 14, 2011

7958627　Method of attaching an electronic device to an MLCC having a curved surface; Randall, Michael S.; Wayne, Chris; McConnell, John; June 14, 2011

7957209　Method of operating a memory apparatus, memory device and memory apparatus; Ruckerbauer, Hermann; June 7, 2011

7957173　Composite memory having a bridging device for connecting discrete memory devices to a system; Kim, Jin-Ki; June 7, 2011

7956443　Through-wafer interconnects for photo imager and memory wafers; Akram, Salman; Watkins,

	Charles M.; Hiatt, Mark; Hembree, David R.; Wark, James M.; Farnworth, Warren M.; Tuttle, Mark E.; Rigg, SidneyB.; Oliver, Steven D.; Kirby, Kyle K.; Wood, Alan G.; Velicky, Lu; June 7, 2011
7956442	Backside connection to TSVs having redistribution lines; Hsu, Kuo-Ching; Chen, Chen-Shien; June 7, 2011
7955895	Structure and method for stacked wafer fabrication; Yang, Ku-Feng; Chiou, Wen-Chih; Wu, Weng-Jin; Tu, Hung-Jung; June 7, 2011
7952904	Three-dimensional memory-based three-dimensional memory module; Zhang, Guobiao; May 31, 2011
7952903	Multimedia three-dimensional memory module (M3DMM) system; Zhang, Guobiao; May 31, 2011
7952478	Capacitance-based microchip exploitation detection; Bartley, Gerald K.; Becker, Darryl J.; Dahlen, Paul E.; Germann, Philip R.; Maki, Andrew B.; Maxson, MarkO.; May 31, 2011
7952176	Integrated circuit packaging system and method of manufacture thereof; Pagaila, Reza Argenty; Do, Byung Tai; Chua, Linda Pei Ee; May 31, 2011
7952171	Die stacking with an annular via having a recessed socket; Pratt, Dave; May 31, 2011
7951647	Performing die-to-wafer stacking by filling gaps between dies; Yang, Ku-Feng; Chiou, Wen-Chih; Wu, Weng-Jin; Sung, Ming-Chung; May 31, 2011
7948064	System on a chip with on-chip RF shield; Barth, Hans-Joachim (Munich, DE); Hanke, Andre; Jenei, Snezana; Nagy, Oliver; Morinaga, Jiro; Adler, Bernd; Koerner, Heinrich; May 24, 2011
7944038	Semiconductor package having an antenna on the molding compound thereof; Chiu, Chi-Tsung; Lee, Pao-Nan; May 17, 2011
7943513	Conductive through connection and forming method thereof; Lin, Shian-Jyh; May 17, 2011
7943473	Minimum cost method for forming high density passive capacitors for replacement of discrete board capacitors using a minimum cost 3D wafer-to-wafer modular integration scheme; Ellul, Joseph Paul; Tran, Khanh; Bergemont, Albert; May 17, 2011
7940074	Data transmission circuit and semiconductor apparatus including the same; Ku, Young Jun; May 10, 2011
7939941	Formation of through via before contact processing; Chiou, Wen-Chih; Yu, Chen-Hua; Wu, Weng-Jin; May 10, 2011
7939369	3D integration structure and method using bonded metal planes; Farooq, Mukta G.; Iyer, Subramanian S; May 10, 2011
7936622	Defective bit scheme for multilayer integrated memory device; Li, Hai; Chen, Yiran; Setiadi, Dadi; Liu, Harry Hongyue; Lee, Brian; May 3, 2011
7936052	On-chip RF shields with backside redistribution lines; Barth, Hans-Joachim; Pohl, Jens; Beer, Gottfried; Koerner, Heinrich; May 3, 2011
7923370	Method for stacking serially-connected integrated circuits and multichip device made from same; Pyeon, Hong Beom; April 12, 2011
7923290	Integrated circuit packaging system having dual sided connection and method of manufacture thereof; Ko, Chan Hoon (Ichon si, KR); Park, Soo-San (Seoul, KR); Kim, Young Chul; April 12, 2011
7933428	Microphone apparatus; Sawada, Tatsuhiro; April 26, 2011
7932608	Through-silicon via formed with a post passivation interconnect structure; Tseng, Ming-Hong; Jao, Sheng Huang; April 26, 2011
7930664	Programmable through silicon via; Feng, Kai Di; Hsu, Louis Lu-Chen; Wang, Ping-Chuan; Yang, Zhijian; April 19, 2011
7928563	3D ICs with microfluidic interconnects and methods of constructing same; Bakir, Muhannad S.; Sekar, Deepak; Dang, Bing; King, Calvin, Jr.; Meindl, James D.; April 19, 2011
7928534	Bond pad connection to redistribution lines having tapered profiles; Hsu, Kuo-Ching; Chen, Chen-Shien; Huang, Hon-Lin; April 19, 2011
7916511	Semiconductor memory device including plurality of memory chips; Park, Ki-Tae; March 29, 2011

7915736　Microfeature workpieces and methods for forming interconnects in microfeature workpieces; Kirby, Kyle K. ; Hiatt, William M. ; Stocks, Richard L. ; March 29, 2011

7913000　Stacked semiconductor memory device with compound read buffer; Chung, Hoe-ju; March 22, 2011

7910473　Through-silicon via with air gap; Chen, Ming-Fa; March 22, 2011

7906857　Molded integrated circuit package and method of forming a molded integrated circuit package; Hoang, Lan H. ; Chaware, Raghunandan; Yip, Laurene; March 15, 2011

7906431　Semiconductor device fabrication method; Mistuhashi, Toshiro; March 15, 2011

7904273　In-line depth measurement for thru silicon via; Liu, Qizhi; Wang, Ping-Chuan; Watson, Kimball M. ; Yang, Zhijian J. ; March 8, 2011

7902674　Three-dimensional die-stacking package structure; Chang, Hsiang-Hung; Chang, Shu-Ming; March 8, 2011

7902643　Microfeature workpieces having interconnects and conductive, back planes, and associated systems and methods; Tuttle, Mark E. ; March 8, 2011

7902069　Small area, robust silicon via structure and process; Andry, Paul S. ; Cotte, John M. ; Knickerbocker, John Ulrich; Tsang, Cornelia K; March 8, 2011

7900519　Microfluidic measuring tool to measure through silicon via depth; Chandrasekaran, Arvind; March 8, 2011

7894199　Hybrid package; Chang, Li-Tien; February 22, 2011

7893529　Thermoelectric 3D cooling; Hsu, Louis Lu-Chen; Wang, Ping-Chuan; Wei, Xiaojin; Zhu, Huilong; February 22, 2011

7893526　Semiconductor package apparatus; Mun, Sung-ho (Suwon-si, KR); Kang, Sun-won (Seoul, KR); Baek, Seung-duk; February 22, 2011

7888806　Electrical connections for multichip modules; Lee, Seok-Chan; Kim, Min-Woo; February 15, 2011

7888668　Phase change memory; Kuo, Chien-Li; Wu, Kuei-Sheng; Lin, Yung-Chang; February 15, 2011

7884625　Capacitance structures for defeating microchip tampering; Bartley, Gerald K. (Rochester, MN); Becker, Darryl J. (Rochester, MN); Dahlen, Paul E. (Rochester, MN); Germann, Philip R. (Oronoco, MN); Maki, Andrew B. (Rochester, MN); Maxson, Mark; February 8, 2011

7884016　Liner materials and related processes for 3D integration; Sprey, Hessel; Nakano, Akinori; February 8, 2011

7884015　Methods for forming interconnects in microelectronic workpieces and microelectronic workpieces formed using such methods; Sulfridge, Marc; February 8, 2011

7880494　Accurate capacitance measurement for ultra-largescale integrated circuits; Doong, Yih-Yuh; Chang, Keh-Jeng; Mii, Yuh-Jier; Liu, Sally; Hung, Lien Jung; Chang, Victor Chih Yuan; February 1, 2011

7875948　Backside illuminated image sensor; Hynecek, Jaroslav (Lake Oswego, OR); Forbes, Leonard (LakeOswego, OR); Haddad, Homayoon (Lake Oswego, OR); Joy, Thomas; January 25, 2011

7872332　Interconnect structures for stacked dies, including penetrating structures for through-silicon vias, and associated systems and methods; Fay, Owen R. (Meridian, ID); Farnworth, Warren M. (Nampa, ID); Hembree, David R. ; January 18, 2011

7867910　Method of accessing semiconductor circuits from the backside using ion-beam and gas-etch; Scrudato, Carmelo F. ; Gu, George Y. ; Hahn, Loren L. ; Herschbein, Steven B. ; January 11, 2011

7867821　Integrated circuit package system with through semiconductor vias and method of manufacture thereof; Chin, Chee Keong; January 11, 2011

7863721　Method and apparatus for wafer level integration using tapered vias; Suthiwongsunthorn, Nathapong; Marimuthu, Pandi Chelvam; January 4, 2011

7863187　Microfeature workpieces and methods for forming interconnects in microfeature workpieces; Hiatt, William M. ; Dando, Ross S. ; January 4, 2011

7863106　Silicon interposer testing for three-dimensional chipstack; Christo, Michael Anthony; Maldonado,

	JulioAlejandro; Weekly, Roger Donell; Zhou, Tingdong; January 4, 2011
7859099	Integrated circuit packaging system having through silicon via with direct interconnects and method of manufacture thereof; Choi, A Leam; Chung, Jae Han; Yang, Deok Kyung; Park, Hyung Sang; December 28, 2010
7858512	Semiconductor with bottom-side wrap-around flange contact; Marcoux, Phil P.; December 28, 2010
7851893	Semiconductor device and method of connecting a shielding layer to ground through conductive vias; Kim, Seung Won; Yang, Dae Wook; December 14, 201
7851342	In-situ formation of conductive filling material in through-silicon via; Xu, Dingying; Eitan, Amram; December 14, 2010
7848153	High-speed memory architecture; Bruennert, Michael (Munich, DE); Gregorius, Peter; Braun, Georg; Gaertner, Andreas; Ruckerbauer, Hermann; Alexander, George William; Stecker, Johannes; December 7, 2010
7847383	Multichip package for reducing parasitic load of pin; So, Byung-Se; Lee, Dong-Ho; December 7, 2010
7847379	Lightweight and compact through-silicon via stack package with excellent electrical connections and method for manufacturing the same; Chung; QwanHo December 7, 2010
7843072	Semiconductor package having through holes; Park, Sung Su; Kim, Jin Young; Jin, Jeong Gi; November 30, 2010
7843064	Structure and process for the formation of TSVs; Kuo, Chen-Cheng; Ching, Kai-Ming; Chen-Shien, Chen; November 30, 2010
7843052	Semiconductor devices and fabrication methods thereof; Yoo, Min; Lee, Ki Wook; Lee, Min Jae; November 30, 2010
7842548	Fixture for P-through silicon via assembly; Lee, Chien-Hsiun; Chen, Chen-Shien; Lii, Mirng-Ji; Karta, Tjandra Winata; November 30, 2010
7839163	Programmable through silicon via; Feng, Kai Di; Hsu, Louis Lu-Chen; Wang, Ping-Chuan; Yang, Zhijian; November 23, 2010
7838975	Flip-chip package with fan-out WLCSP; Chen, Nan-Cheng; November 23, 2010
7838967	Semiconductor chip having TSV (through silicon via) and stacked assembly including the chips; Chen, Ming-Yao; November 23, 2010
7838337	Semiconductor device and method of formingan interposer package with through silicon vias; Marimuthu, Pandi Chelvam (Singapore, SG); Suthiwongsunthorn, Nathapong (Singapore, SG); Shim, Il Kwon (Singapore, SG); Heng, Kock Liang; November 23, 2010
7834440	Semiconductor device with stacked memory and processor LSIs; Ito, Kiyoto; Saen, Makoto; Kuroda, Yuki; November 16, 2010
7830692	Multichip memory device with stacked memory chips, method of stacking memory chips, and method of controlling operation of multichip package memory; Chung, Hoe-ju; Lee, Jung-bae; Kang, Uk-song; November 9, 2010
7830018	Partitioned through-layer via and associated systems and methods; Lee, Teck Kheng; November 9, 2010
7829976	Microelectronic devices and methods for forming interconnects in microelectronic devices; Kirby, Kyle K.; Akram, Salman; Hembree, David R.; Rigg, Sidney B.; Farnworth, Warren M.; Hiatt, William M.; November 9, 2010
7825517	Method for packaging semiconductor dies having through-silicon vias; Su, Chao-Yuan; November 2, 2010
7825024	Method of forming through-silicon vias; Lin, Chuan-Yi; Lee, Song-Bor; Huang, Ching-Kun; Lin, Sheng-Yuan; November 2, 2010
7821107	Die stacking with an annular via having a recessed socket; Pratt, Dave; October 26, 2010
7816945	3D chip-stack with fuse-type through silicon via; Feng, Kai Di; Hsu, Louis Lu-Chen; Wang, Ping-

	Chuan; Yang, Zhijian; October 19, 2010
7816776	Stacked semiconductor device and method of forming serial path thereof; Choi, Young-Don; October 19, 2010
7816227	Tapered through-silicon via structure; Chen, Chen Shien; Kuo, Chen Cheng; Ching, Kai-Ming; Chen, Chih-Hua; October 19, 2010.
7816181	Method of underfilling semiconductor die in a die stack and semiconductor device formed thereby; Bhagath, Shrikar; Takiar, Hem; October 19, 2010
7813043	Lens assembly and method of manufacture; Lusinchi, Jean-Pierre; Kui, Xiao-Yun; October 12, 2010
7812459	Three-dimensional integrated circuits with protection layers; Yu, Chen-Hua; Chiou, Wen-Chih; Wu, Weng-Jin; Tu, Hung-Jung; Yang, Ku-Feng; October 12, 2010
7812449	Integrated circuit package system with redistribution layer; Kuan, Heap Hoe; Chow, Seng Guan; Huang, Rui; October 12, 2010
7812446	Semiconductor device; Kurita, Yoichiro; October 12, 2010
7812426	TSV-enabled twisted pair; Peng, Mark Shane; Chao, Clinton; Hsu, Chao-Shun; October 12, 2010
7808105	Semiconductor package and fabricating method thereof; Paek, Jong Sik; October 5, 2010
7803714	Semiconductor through silicon vias of variable size and method of formation; Ramiah, Chandrasekaram; Sanders, Paul W.; September 28, 2010
7799678	Method for forming a through silicon via layout; Kropewnicki, Thomas J.; Chatterjee, Ritwik; Junker, Kurt H.; September 21, 2010
7799613	Integrated module for data processing system; Dang, Bing; Knickerbocker, John U.; Tsang, Cornelia K.; September 21, 2010
7796446	Memory dies for flexible use and method for configuringmemory dies; Ruckerbauer, Hermann; Bruennert, Michael; Menczigar, Ullrich; Mueller, Christian; Tontisirin, Sitt; Braun, Georg; Savignac, Dominique; September 14, 2010
7795735	Methods for forming single dies with multilayer interconnect structures and structures formed therefrom; Hsu, Chao-Shun; Tang, Chen-Yao; Chao, Clinton; Peng, Mark Shane; September 14, 2010
7795650	Method and apparatus for backside illuminated image sensors using capacitively coupled readout integrated circuits; Eminoglu, Selim; Lauxtermann, Stefan C.; September 14, 2010
7795139	Method for manufacturing semiconductor package; Han, Kwon Whan; Park, Chang Jun; Suh, Min Suk; Kim, Seong Cheol; Kim, Sung Min; Yang, SeungTaek; Lee, Seung Hyun; Kim, Jong Hoon; Lee, Ha Na; September 14, 2010
7795134	Conductive interconnect structures and formation methods using supercritical fluids; Sulfridge, Marc; September 14, 2010
7791919	Semiconductor memory device capable of identifying a plurality of memory chips stacked in the same package; Shimizu, Yuui; September 7, 2010
7791175	Method for stacking serially connected integrated circuits and multichip device made from same; Pyeon, Hong Beom; September 7, 2010
7786008	Integrated circuit packaging system having through silicon vias with partial depth metal fill regions and method of manufacture thereof; Do, Byung Tai; Chow, Seng Guan; Yoon, Seung Uk; August 31, 2010
7777330	High bandwidth cache-to-processing unit communicationin a multiple processor/cache system; Pelley, Perry H.; McShane, Michael B.; August 17, 2010
7776741	Process for through silicon via filing; Reid, Jonathan D.; Wang, Katie Qun; Willey, Mark J.; August 17, 2010
7775119	Media-compatible electrically isolated pressure sensor for high temperature applications; Suminto, James Tjanmeng; Yunus, Mohammad; August 17, 2010
7772880	Reprogrammable three-dimensional intelligent system on a chip; Solomon, Neal; August 10, 2010
7772868	Accurate capacitance measurement for ultra-large scale integrated circuits; Doong, Yih-Yuh; Chang,

	Keh-Jeng; Mii, Yuh-Jier; Liu, Sally; Hung, Lien Jung; Chang, Victor Chih Yuan; August 10, 2010
7772124	Method of manufacturing a through-silicon-via on chip passive MMW bandpass filter; Bavisi, Amit; Ding, Hanyi; Wang, Guoan; Woods, Wayne H., Jr.; Xu; Jiansheng; August 10, 2010
7772081	Semiconductor device and method of forming high frequency circuit structure and method thereof; Lin, Yaojian; Fang, Jianmin; Chen, Kang; Cao, Haijing; August 10, 2010
7760144	Antennas integrated in semiconductor chips; Chang, Chung-Long; Lu, David Ding-Chung; Chung, Shine; July 20, 2010
7759800	Microelectronics devices, having vias, and packaged microelectronic devices having vias; Rigg, Sidney B.; Watkins, Charles M.; Kirby, Kyle K.; Benson, Peter A.; Akram, Salman; July 20, 2010
7759165	Nanospring; Bajaj, Rajeev; July 20, 2010
7755173	Series-shunt switch with thermal terminal; Mondi, Anthony Paul; Bukowski, Joseph Gerard; July 13, 2010
7750459	Integrated module for data processing system; Dang, Bing; Knickerbocker, John Ulrich; Tsang, CorneliaKang-I; July 6, 2010
7749899	Microelectronic workpieces and methods and systems for forming interconnects in microelectronic workpieces; Clark, Douglas; Oliver, Steven D.; Kirby, Kyle K.; Dando, Ross S.; July 6, 2010
7741156	Semiconductor device and method of forming through vias with reflowed conductive material; Pagaila, Reza A.; Chua, Linda Pei Ee; Do, Byung Tai; June 22, 2010
7741148	Semiconductor device and method of forming an interconnect, structure for 3D devices using encapsulant for structural support; Marimuthu, Pandi C.; Suthiwongsunthorn, Nathapong; Heng, Kock Liang; June 22, 2010
7738249	Circuitized substrate with internal cooling structure and electrical assembly utilizing same; Chan, Benson; Egitto, Frank D.; Lin, How T.; Magnuson, Roy H.; Markovich, Voya R.; Thomas, David L.; June 15, 2010
7701252	Stacked die network-on-chip for FPGA; Chow, Francis Man-Chit; Patel, Rakesh H.; Pistorius, Erhard Joachim; April 20, 201
7701244	False connection for defeating microchip exploitation; Bartley, Gerald K.; Becker, Darryl J.; Dahlen, Paul E.; Germann, Philip R.; Maki, Andrew B.; Maxson, Mark O.; Sheets, John E., II; April 20, 2010
7698470	Integrated circuit, chip stack and data processingsystem; Ruckerbauer, Hermann; Savignac, Dominique; April 13, 2010
7692946	Memory array on more than one die; Taufique, Mohammed H.; Jallice, Derwin; McCauley, Donald W.; De Vale, John P.; Brekelbaum, Edward A.; Rupley, Jeffrey P., II; Loh, Gabriel H.; Black, Bryan; April 6, 2010
7692448	Reprogrammable three-dimensional field programmable gate arrays; Solomon, Neal; April 6, 2010
7691748	Through-silicon via and method for forming thesame; Han, Kwon Whan; April 6, 2010
7692310	Forming a hybrid device; Park, Chang-Min; Ramanathan, Shriram; April 6, 2010
7692278	Stacked-die packages with silicon vias and surface activated bonding; Periaman, Shanggar; Ooi, Kooi Chi; Cheah, Bok Eng; April 6, 2010
7691748	Through-silicon via and method for forming thesame; Han, Kwon Whan; April 6, 2010
7683459	Bonding method for through-silicon-via based 3D wafer stacking; Ma, Wei; Shi, Xunqing; Chung, Chang Hwa; March 23, 2010
7683458	Through-wafer interconnects for photoimager and memory wafers; Akram, Salman; Watkins, Charles M.; Hiatt, William M.; Hembree, David R.; Wark; James M.; Farnworth; Warren M.; Tuttle; Mark E., Rigg; Sidney B., Oliver; Steven D., Kirby; Kyle K., Wood; Alan G. Velicky, Lu; March 23, 2010
7670950	Copper metallization of through silicon via; Richardson, Thomas B.; Zhang, Yun; Wang, Chen;

	Paneccasio, Vincent, Jr.; Wang, Cai; Lin, Xuan; Hurtubise, Richard; Abys, Joseph A.; March 2, 2010
7666768	Through-die metal vias with a dispersed phase of graphitic structures of carbon for reduced thermal expansion and increased electrical conductance; Raravikar, Nachiket R.; Suh, Daewoong; Arana, Leonel; Matayabas, James C., Jr.; February 23, 2010
7666711	Semiconductor device and method of forming double-sided through vias in saw streets; Pagaila, Reza A.; Do, Byung Tai; February 23, 2010
7656031	Stackable semiconductor package having metal pin within through hole of package; Chen, Cheng-Chung; Wang, Chia-Chung; Tan, Chin Hock; Lin, Charles W. C.; February 2, 2010
7638867	Microelectronic package having solder-filled through-vias; Xu, Dingying; Hackitt, Dale A.; December 29, 2009
7633165	Introducing a metal layer between SiN and TiN toimprove CBD contact resistance for P-TSV; Hsu, Kuo-Ching; Chen, Chen-Shien; Su, Boe; Huang, Hon-Lin; December 15, 2009
7629249	Microfeature workpieces having conductive interconnect structures formed by chemically reactive processes, and associated systems and methods; Borthakur, Swarnal; December 8, 2009
7622377	Microfeature workpiece substrates having through substrate vias, and associated methods of formation; Lee, Teck Kheng; Lim, Andrew Chong Pei; November 24, 2009
7598617	Stack package utilizing through vias and re-distribution lines; Lee, Seung Hyun; Suh, Min Suk; October 6, 2009
7598523	Test structures for stacking dies having through-silicon vias; Luo, Wen-Liang; Kuo, Yung-Liang; Cheng, HsuMi; October 6, 2009
7592697	Microelectronic package and method of cooling same; Arana, Leonel R.; Newman, Michael W.; Chang, Je-Young; September 22, 2009
7589008	Methods for forming interconnects in microelectronic workpieces and microelectronic workpieces formed using such methods; Kirby, Kyle K.; September 15, 2009
7576435	Low-cost and ultrafine integrated circuit packaging technique; Chao, Clinton; August 18, 2009
7564115	Tapered through-silicon via structure; Chen, Chen-Shien; Kuo, Chen-Cheng; Ching, Kai-Ming; Chen, Chih-Hua; July 21, 2009
7541203	Conductive adhesive for thinned silicon wafers with through silicon vias; Knickerbocker, John U.; June 2, 2009
7531453	Microelectronic devices and methods for forming interconnects in microelectronic devices; Kirby, Kyle K.; Akram, Salman; Hembree, David R.; Rigg, Sidney B.; Farnworth, Warren M.; Hiatt, William M.; May 12, 2009
7528006	Integrated circuit die containing particle-filled through-silicon metal vias with reduced thermal expansion; Arana, Leonel; Newman, Michael; Natekar, Devendra; May 5, 2009
7514116	Horizontal carbon nanotubes by vertical growth and rolling; Natekar, Devendra; Tomita, Yoshihiro; Hwang, Chi-Won; April 7, 2009
7494846	Design techniques for stacking identical memory dies; Hsu, Chao-Shun; Liu, Louis; Chao, Clinton; Peng, Mark Shane; February 24, 200
7446420	Through silicon via chip stack package capable of facilitating chip selection during device operation; Kim, Jong Hoon; November 4, 2008
7435913	Slanted vias for electrical circuits on circuit boards and other substrates; Chong, Chin Hui (Singapore, SG); Lee, Choon Kuan; October 14, 2008
7432592	Integrated microchannels for 3D through silicon architectures; Shi, Wei; Lu, Daoqiang; Bai, Yiqun; Zhou, Qing A.; He, Jiangqi; October 7, 2008
7427803	Electromagnetic shielding using through-silicon vias; Chao, Clinton; Hsu, Chao-Shun; Peng, Mark Shane; Lu, Szu Wei; Karta, Tjandra Winata; September 23, 2008
7425499	Methods for forming interconnects in vias and microelectronic workpieces including such

	interconnects; Oliver, Steven D. ; Kirby, Kyle K. ; Hiatt, William M. ; September 16, 2008
7413979	Methods for forming vias in microelectronic devices, and methods for packaging microelectronic devices; Rigg, Sidney B. ; Watkins, Charles M. ; Kirby, Kyle K. ; Benson, Peter A. ; Akram, Salman; August 19, 2008
7410884	3D integrated circuits using thick metal for backside connections and offset bumps; Ramanathan, Shriram; Kim, Sarah E. ; Morrow, Patrick R. ; August 12, 2008
7400033	Package on package design to improve functionality and efficiency; Cheah, Bok Eng (Penang, MY); Periaman, Shanggar (Penang, MY); Ooi, Kooi Chi; July 15, 2008
7317256	Electronic packaging including die with through silicon via; Williams, Christina K. ; Thomas, Rainer E. ; January 8, 2008
7241675	Attachment of integrated circuit structures and other substrates to substrates with vias; Savastiouk, Sergey; Kao, Sam; July 10, 2007
7111149	Method and apparatus for generating a device ID for stacked devices; Eilert, Sean S. ; September 19, 2006
7081408	Method of creating a tapered via using a receding mask and resulting structure; Lane, Ralph L. ; Hill, Charles D. ; July 25, 2006
6924551	Through silicon via, folded flex microelectronic package; Rumer, Christopher L. ; Zarbock, Edward A. ; August 2, 2005

附录 D 推荐阅读材料

D.1 TSV、3D 集成与可靠性

1. Moore, G., "Cramming More Components Onto Integrated Circuits," *Electronics*, Vol. 38, No. 8, April 19, 1965.
2. Lau, J. H., *Reliability of RoHS-Compliant 2D and 3D IC Interconnects*, McGraw-Hill, New York, 2011.
3. Lau, J. H., C. K. Lee, C. S. Premachandran, and A. Yu, *Advanced MEMS Packaging*, McGraw-Hill, New York, 2010.
4. Lau, J. H., "Overview and Outlook of TSV and 3D Integrations," *Journal of Microelectronics International*, Vol. 28, No. 2, 2011, pp. 8–22.
5. Lau, J. H., M. S. Zhang, and S. W. R. Lee, "Embedded 3D Hybrid IC Integration System-in-Package (SiP) for Opto-Electronic Interconnects in Organic Substrates," *ASME Transactions, Journal of Electronic Packaging*, Vol. 133, September 2011, pp. 031010: 1-7.
6. Lau, J. H., "Critical Issues of 3D IC Integrations," *IMAPS Transactions, Journal of Microelectronics and Electronic Packaging*, First Quarter, 2010, pp. 35–43.
7. Lau, J. H., "Design and Process of 3D MEMS Packaging," *IMAPS Transactions, Journal of Microelectronics and Electronic Packaging*, First Quarter, 2010, pp. 10–15.
8. Lau, J. H., R. Lee, M. Yuen, M., and P. Chan, "3D LED and IC Wafer Level Packaging," *Journal of Microelectronics International*, Vol. 27, No. 2, 2010, pp. 98–105.
9. Lau, J. H., "State-of-the-art and Trends in 3D Integration," *Chip Scale Review*, March–April 2010, pp. 22–28.
10. Lau, J. H., "TSV Manufacturing Yield and Hidden Costs for 3D IC Integration," *IEEE/ECTC Proceedings*, Las Vegas, NV, June 2010, pp. 1031–1041.
11. Lau, J. H., Y. S. Chan, and R. S. W. Lee, "3D IC Integration with TSV Interposers for High-Performance Applications," *Chip Scale Review*, September–October 2010, pp. 26–29.
12. Yu, A., J. H. Lau, S. Ho, A. Kumar, Y. Wai, D. Yu, M. Jong, V. Kripesh, D. Pinjala, and D. Kwong, "Study of 15-μm-Pitch Solder Microbumps for 3D IC Integration," *IEEE/ECTC Proceedings*, San Diego, CA, May 2009, pp. 6–10.
13. Yu, A., J. H. Lau, S. Ho, A. Kumar, H. Yin, J. Ching, V. Kripesh, D. Pinjala, S. Chen, C. Chan, C. Chao, C. Chiu, M. Huang, and C. Chen, "Three Dimensional Interconnects with High Aspect Ratio TSVs and Fine Pitch Solder Microbumps," *IEEE/ECTC Proceedings*, San Diego, CA, May 2009, pp. 350–354. Also, *IEEE Transactions in Advanced Packaging* (accepted).
14. Yu, A., A. Kumar, S. Ho, H. Yin, J. H. Lau, J. Ching, V. Kripesh, D. Pinjala, S. Chen, C. Chan, C. Chao, C. Chiu, M. Huang, and C. Chen, "Development of Fine Pitch Solder Microbumps for 3D Chip Stacking," *IEEE/EPTC Proceedings*, Singapore, December 2008, pp. 387–392. Also, *IEEE Transactions in Advanced Packaging* (accepted).
15. Yu, A., N. Khan, G. Archit, D. Pinjalal, K. Toh, V. Kripesh, S. Yoon, and J. H. Lau, "Development of Silicon Carriers with Embedded Thermal Solutions for High Power 3D Package," *IEEE Transactions on Components and Packaging Technology*, Vol. 32, No. 3, September 2009, pp. 566–571.
16. Tang, G., O. Navas, D. Pinjala, J. H. Lau, A. Yu, and V. Kripesh, "Integrated Liquid Cooling Systems for 3D Stacked TSV Modules," *IEEE Transactions on*

Components and Packaging Technologies, Vol. 33, No. 1, 2010, pp. 184–195.
17. Chen, K., C. Premachandran, K. Choi, C. Ong, X. Ling, A. Khairyanto, B. Ratmin, P. Myo, and J. H. Lau, "C2W Bonding Method for MEMS Applications," *IEEE/ECTC Proceedings*, December 2008, pp. 1283–1287.
18. Premachandran, C. S., J. H. Lau, X. Ling, A. Khairyanto, K. Chen, and Myo Ei Pa Pa, "A Novel, Wafer-Level Stacking Method for Low-Chip Yield and Non-Uniform, Chip-Size Wafers for MEMS and 3D SIP Applications," *IEEE/ECTC Proceedings*, Orlando, FL, May 27–30, 2008, pp. 314–318.
19. Chai, T., X. Zhang, J. H. Lau, C. Selvanayagam, K. Biswas, S. Liu, D. Pinjala, G. Tang, Y. Ong, S. Vempati, E. Wai, H. Li, B. Liao, N. Ranganathan, V. Kripesh, J. Sun, J. Doricko, and C. Vath, "Development of Large Die Fine-Pitch Cu/Low-k FCBGA Package with Through Silicon Via (TSV) Interposer," *IEEE/ECTC Proceedings*, May 2009, pp. 305–312. Also, *IEEE Transactions on CPMT*, Vol. 1, No. 5, 2011, pp. 660-672.
20. Hoe, G., G. Tang, P. Damaruganath, C. Chong, J. H. Lau, X. Zhang, and K. Vaidyanathan, "Effect of TSV Interposer on the Thermal Performance of FCBGA Package," *IEEE/ECTC Proceedings*, Singapore, December 2009, pp. 778–786.
21. Choi, W. O., C. S. Premachandran, S. Ong, Ling, X., E. Liao, K. Ahmad, B. Ratmin, K. Chen, P. Thaw, and J. H. Lau, "Development of Novel Intermetallic Joints Using Thin Film Indium Based Solder by Low Temperature Bonding Technology for 3D IC Stacking," *IEEE/ECTC Proceedings*, San Diego, CA, May 2009, pp. 333–338.
22. Kumar, A., X. Zhang, Q. Zhang, M. Jong, G. Huang, V. Kripesh, C. Lee, J. H. Lau, D. Kwong, V. Sundaram, R. Tummula, and M. Georg, "Evaluation of Stresses in Thin Device Wafer Using Piezoresistive Stress Sensor," *IEEE Proceedingsof EPTC, December 2008*, , pp. 1270–1276. Also, *IEEE Transactions in Components and Packaging Technologies* Vol. 1, No. 6, June 2011, pp. 841-850.
23. Khan, N., L. Yu, P. Tan, S. Ho, N. Su, H. Wai, K. Vaidyanathan, D. Pinjala, J. H. Lau, T. Chuan, "3D Packaging with Through Silicon Via (TSV) for Electrical and Fluidic Interconnections," *IEEE/ECTC Proceedings*, San Diego, CA, May 2009, pp. 1153–1158.
24. Sekhar, V. N., S. Lu, A. Kumar, T. C. Chai, V. Lee, S. Wang, X. Zhang, C. S. Premchandran, V. Kripesh, and J. H. Lau, "Effect of Wafer Back Grinding on the Mechanical Behavior of Multilayered Low-k for 3D-Stack Packaging Applications," *IEEE/ECTC Proceedings*, Orlando, FL, May 27–30, 2008, pp. 1517–1524. Also, *IEEE Transactions on Components and Packaging Technologies* (in press).
25. Khan, N., V. Rao, S. Lim, S. Ho, V. Lee, X. Zhang, R. Yang, E. Liao, Ranganathan, T. Chai, V. Kripesh, and J. H. Lau, "Development of 3D Silicon Module with TSV for System in Packaging," *IEEE/ECTC Proceedings*, Orlando, FL, May 27–30, 2008, pp. 550–555.
26. Ho, S., S. Yoon, Q. Zhou, K. Pasad, V. Kripesh and J. H. Lau, "High RF Performance TSV for Silicon Carrier for High Frequency Application," *IEEE/ECTC Proceedings*, Orlando, FL, May 27–30, 2008, pp. 1956–1962.
27. Lim, S., V. Rao, H. Yin, W. Ching, V. Kripesh, C. Lee, J. H. Lau, J. Milla, and A. Fenner, "Process Development and Reliability of Microbumps," *IEEE/ECTC Proceedings*, December 2008, pp. 367–372. Also, *IEEE Transactions in Components and Packaging Technology*, Vol. 33, No. 4, December 2010, pp. 747-753.
28. Selvanayagam, C., J. H. Lau, X. Zhang, S. Seah, K. Vaidyanathan, and T. Chai, "Nonlinear Thermal Stress/Strain Analysis of Copper Filled TSV (Through Silicon Via) and Their Flip-Chip Microbumps," *IEEE/ECTC Proceedings*, Orlando, FL, May 27–30, 2008, pp. 1073–1081. Also, *IEEE Transactions on Advanced Packaging*, Vol. 32, No. 4, November 2009, pp. 720–728.
29. Zhang, X., A. Kumar, Q. X. Zhang, Y. Y. Ong, S. W. Ho, C. H. Khong, V. Kripesh, J. H. Lau, D.-L. Kwong, V. Sundaram, Rao R. Tummula, and G. Meyer-Berg, "Application of Piezoresistive Stress Sensors in Ultra Thin Device Handling and Characterization," *Journal of Sensors & Actuators: A. Physical*, Vol. 156, November 2009, pp. 2–7.
30. Lau, J. H., T. G. Yue, G. Y. Y. Hoe, X.W. Zhang, C. T. Chong, P. Damaruganath, and K. Vaidyanathan, "Effects of TSV (Through Silicon Via) Interposer/

Chip on the Thermal Performances of 3D IC Packaging," ASME Paper no. IPACK2009–89380.
31. Lau, J. H., "State-of-the-Art and Trends in Through-Silicon Via (TSV) and 3D Integrations," ASME Paper no. IMECE2010-37783.
32. Lau, J. H., and G. Tang, "Thermal Management of 3D IC Integration with TSV (Through Silicon Via)," *IEEE/ECTC Proceedings*, San Diego, May 2009, pp. 635–640.
33. Carson, F., K. Ishibashi, S. Yoon, P. Marimuthu, and D. Shariff, "Development of Super Thin TSV PoP," *Proceedings of IEEE CPMT Symposium Japan*, August 2010, pp. 7–10.
34. Kohara, S., K. Sakuma, Y. Takahashi, T. Aoki, K. Sueoka, K. Matsumoto, P. Andry, C. Tsang, E. Sprogis, J. Knickerbocker, and Y. Orii, "Thermal Stress Analysis of 3D Die Stacks with Low-Volume Interconnections," *Proceedings of IEEE CPMT Symposium Japan*, August 2010, pp. 165–168.
35. Sekiguchi, M., H. Numata, N. Sato, T. Shirakawa, M. Matsuo, H. Yoshikawa, M. Yanagida, H. Nakayoshi, and K. Takahashi, "Novel Low Cost Integration of Through Chip Interconnection and Application to CMOS Image Sensor," *IEEE/ECTC Proceedings*, San Diego, CA, May 2006, pp. 1367–1374.
36. Takahashi, K., and M. Sekiguchi, "Through Silicon Via and 3D Wafer/Chip Stacking Technology," *IEEE Proceedings of Symposium on VLSI Circuits Digest of Technical Papers*, 2006, pp. 89–92.
37. Juergen, M., K. Wolf, A. Zoschke, R. Klumpp, M. Wieland, L. Klein, A. Nebrich, I. Heinig, W. Limansyah, O. Weber, O. Ehrmann, and H. Reichl, "3D Integration of Image Sensor SiP Using TSV Silicon Interposer," *IEEE/ECTC Proceedings*, December 2009, pp. 795–800.
38. Limansyah, I., M. J. Wolf, A. Klumpp, K.Zoschke, R. Wieland, M. Klein, H. Oppermann, L. Nebrich, A. Heinig, A. Pechlaner, H. Reichl, and W. Weber, "3D Image Sensor SiP with TSV Silicon Interposer," *IEEE/ECTC Proceedings*, May 2009, pp. 1430–1436.
39. Garrou, P., C. Bower, and P. Ramm, *3D Integration: Technology and Applications*, Wiley, Hoboken, NJ, 2009.
40. Ramm, P., M. Wolf, A. Klumpp, R. Wieland, B. Wunderle, B. Michel, and H. Reichl, "Through Silicon Via Technology: Processes and Reliability for Wafer-Level 3D System Integration," *IEEE/ECTC Proceedings*, Orlando, FL, May 2008, pp. 847–852.
41. Andry, P. S., C. K. Tsang, B. C. Webb, E. J. Sprogis, S. L. Wright, B. Bang, and D. G. Manzer, "Fabrication and Characterization of Robust Through-Silicon Vias for Silicon-Carrier Applications," *IBM Journal of Research and Development*, Vol. 52, No. 6, 2008, pp. 571–581.
42. Knickerbocker, J. U., P. S. Andry, B. Dang, R. R. Horton, C. S. Patel, R. J. Polastre, K. Sakuma, E. S. Sprogis, C. K. Tsang, B. C. Webb, and S. L. Wright, "3D Silicon Integration," *IEEE/ECTC Proceedings*, May 2008, pp. 538–543.
43. Kumagai, K., Y. Yoneda, H. Izumino, H. Shimojo, M. Sunohara, and T. Kurihara, "A Silicon Interposer BGA Package with Cu-Filled TSV and Multilayer Cu-Plating Interconnection," *IEEE/ECTC Proceedings*, Orlando, FL, May 2008, pp. 571–576.
44. Sunohara, M., T. Tokunaga, T. Kurihara, and M. Higashi, "Silicon Interposer with TSVs (Through Silicon Vias) and Fine Multilayer Wiring," *IEEE/ECTC Proceedings*, Orlando, FL, May 2008, pp. 847–852.
45. Lee, H. S., Y.-S. Choi, E. Song, K. Choi, T. Cho, and S. Kang, "Power Delivery Network Design for 3D SIP Integrated over Silicon Interposer Platform," *IEEE/ECTC Proceedings*, Reno, NV, May 2007, pp. 1193–1198.
46. Matsuo, M., N. Hayasaka, and K. Okumura, "Silicon Interposer Technology for High-Density Package," *IEEE/ECTC Proceedings*, May 2000, pp. 1455–1459.
47. Wong, E., J. Minz, and S. K. Lim, "Effective Thermal Via and Decoupling Capacitor Insertion for 3D System-on-Package," *IEEE/ECTC Proceedings*, San Diego, CA, May 2006, pp. 1795–1801.
48. Kang, U., H. Chung, S. Heo, D. Park, H. Lee, J. Kim, S. Ahn, S. Cha, J. Ahn, D. Kwon, J. Lee, H. Joo, W. Kim, D. Jang, N. Kim, J. Choi, T. Chung, J. Yoo, J. Choi, C. Kim, and Y. Jun, "8-Gb 3D DDR3 DRAM Using Through-Silicon-Via Technology," *IEEE Journal of Solid-State Circuits*, Vol. 45, No. 1, January 2010, pp. 111–119.
49. Lau, J. H., Y. Chan, and R. Lee, "Thermal-Enhanced and Cost-Effective 3D IC

Integration with TSV (Through-Silicon Via) Interposers for High-Performance Applications," ASME Paper no. IMECE2010-40975.
50. Shi, X., P. Sun, Y. Tsui, P. Law, S. Yau, C. Leung, Y. Liu, C. Chung, S. Ma, M. Miao, and Y. Jin, "Development of CMOS-Process-Compatible Interconnect Technology for 3D Stacking of NAND Flash Memory Chips," *IEEE/ECTC Proceedings*, Las Vegas, NV, June 2010, pp. 74–78.
51. Kikuchi, K., C. Ueda, K. Takemura, O. Shimada, T. Gomyo, Y. Takeuchi, T. Ookubo, K. Baba, M. Aoyagi, T. Sudo, and K. Otsuka, "Low-Impedance Evaluation of Power Distribution Network for Decoupling Capacitor Embedded Interposers of 3D Integrated LSI System," *IEEE/ECTC Proceedings*, Las Vegas, NV, June 2010, pp. 1455–1460.
52. Sridharan, V., S. Min, V. Sundaram, V. Sukumaran, S. Hwang, H. Chan, F. Liu, C. Nopper, and R. Tummala, "Design and Fabrication of Bandpass Filters in Glass Interposer with Through-Package-Vias (TPV)," *IEEE/ECTC Proceedings*, Las Vegas, NV, June 2010, pp. 530–535.
53. Sukumaran, V., Q. Chen, F. Liu, N. Kumbhat, T. Bandyopadhyay, H. Chan, S. Min, C. Nopper, V. Sundaram, and R. Tummala, "Through-Package-Via Formation and Metallization of Glass Interposers," *IEEE/ECTC Proceedings*, Las Vegas, NV, June 2010, pp. 557–563.
54. Sakuma, K., K. Sueoka1, S. Kohara, K. Matsumoto, H. Noma, T. Aoki, Y. Oyama, H. Nishiwaki, P. S. Andry, C. K. Tsang, J. Knickerbocker, and Y. Orii, "IMC Bonding for 3D Interconnection," *IEEE/ECTC Proceedings*, Las Vegas, NV, June 2010, pp. 864–871.
55. Doany, F., B. Lee, C. Schow, C. Tsang, C. Baks, Y. Kwark, R. John, J. Knickerbocker, and J. Kash, "Terabit/s-Class 24-Channel Bidirectional Optical Transceiver Module Based on TSV Si Carrier for Board-Level Interconnects," *IEEE/ECTC Proceedings*, Las Vegas, NV, June 2010, pp. 58–65.
56. Khan, N., D. Wee, O. Chiew, C. Sharmani, L. Lim, H. Li, and S. Vasarala, "Three Chips Stacking with Low Volume Solder Using Single Re-Flow Process," *IEEE/ECTC Proceedings*, Las Vegas, NV, June 2010, pp. 884–888.
57. Trigg, A., L. Yu, X. Zhang, C. Chong, C. Kuo, N. Khan, and D. Yu, "Design and Fabrication of a Reliability Test Chip for 3D-TSV," *IEEE/ECTC Proceedings*, Las Vegas, NV, June 2010, pp. 79–83.
58. Agarwal, R., W. Zhang, P. Limaye, R. Labie, B. Dimcic, A. Phommahaxay, and P. Soussan, "Cu/Sn Microbumps Interconnect for 3D TSV Chip Stacking," *IEEE/ECTC Proceedings*, Las Vegas, NV, June 2010, pp. 858–863.
59. Töpper, M., I. Ndip, R. Erxleben, L. Brusberg, N. Nissen, H. Schröder, H. Yamamoto, G. Todt, and H. Reichl, "3D Thin Film Interposer Based on TGV (Through Glass Vias): An Alternative to Si-Interposer," *IEEE/ECTC Proceedings*, Las Vegas, NV, June 2010, pp. 66–73.
60. Bouchoucha, M., P. Chausse, D. Henry, and N. Sillon, "Process Solutions and Polymer Materials for 3D-WLP Through Silicon Via Filling," *IEEE/ECTC Proceedings*, Las Vegas, NV, June 2010, pp. 1696–1698.
61. Liu, H., K. Wang, K. Aasmundtveit, and N. Hoivik, "Intermetallic Cu3Sn as Oxidation Barrier for Fluxless Cu-Sn Bonding," *IEEE/ECTC Proceedings*, Las Vegas, NV, June 2010, pp. 853–857.
62. Kang, I., G. Jung, B. Jeon, J. Yoo, and S. Jeong, "Wafer Level Embedded System in Package (WL-eSiP) for Mobile Applications," *IEEE/ECTC Proceedings*, Las Vegas, NV, June 2010, pp. 309–315.
63. Reed, J., M. Lueck, C. Gregory, A. Huffman, J. Lannon, Jr., and D. Temple, "High Density Interconnect at 10-μm Pitch with Mechanically Keyed Cu/Sn-Cu and Cu-Cu Bonding for 3D Integration," *IEEE/ECTC Proceedings*, Las Vegas, NV, June 2010, pp. 846–852.
64. Kraft, J., F. Schrank, J. Teva, J. Siegert, G. Koppitsch, C. Cassidy, E. Wachmann, F. Altmann, S. Brand, C. Schmidt, and M. Petzold, "3D Sensor Application with Open Through Silicon Via Technology," *IEEE ECTC Proceedings*, Orlando, FL, June 2011, pp. 560–566.
65. Wang, Y., and T. Suga, "Influence of Bonding Atmosphere on Low-Temperature Wafer Bonding," *IEEE/ECTC Proceedings*, Las Vegas, NV, June 2010, pp. 435–439.
66. Au, K., S. Kriangsak, X. Zhang, W. Zhu, and C. Toh, "3D Chip Stacking and Reliability Using TSV-Micro C4 Solder Interconnection," *IEEE/ECTC Proceedings*, Las Vegas, NV, June 2010, pp. 1376–1384.

67. Hsieh, M. C., S. Wu, C. Wu, J. H. Lau, R. Tain, and R. Lo, "Investigation of Energy Release Rate for Through Silicon Vias (TSVs) in 3D IC integration," *EuroSimE Proceedings*, April 2011, pp. 1/7–7/7.
68. Hsin, Y. C., C. Chen, J. H. Lau, P. Tzeng, S. Shen, Y. Hsu, S. Chen, C. Wn, J. Chen, T. Ku, and M. Kao, "Effects of Etch Rate on Scallop of Through-Silicon Vias (TSVs) in 200-mm and 300-mm Wafers," *IEEE ECTC Proceedings*, Orlando, FL, June 2011, pp. 1130–1135.
69. Chen, J. C., P. J. Tzeng, S. C. Chen, C. Y. Wu, J. H. Lau, C. C. Chen, C. H. Lin, Y. C. Hsin, T. K. Ku, and M. J. Kao, "Impact of Slurry in Cu CMP (Chemical Mechanical Polishing) on Cu Topography of Through Silicon Vias (TSVs), Re-distributed Layers, and Cu Exposure," *IEEE/ECTC Proceedings*, Orlando, FL, June 2011, pp. 1389–1394.
70. Chien, J., Y. Chao, J. H. Lau, M. Dai, R. Tain, M. Dai, P. Tzeng, C. Lin, Y. Hsin, S. Chen, J. Chen, C. Chen, C. Ho, R. Lo, T. Ku, and M. Kao, "A Thermal Performance Measurement Method for Blind Through Silicon Vias (TSVs) in a 300-mm Wafer," *IEEE ECTC Proceedings*, Orlando, FL, June 2011, pp. 1204–1210.
71. Tsai, W., H. Chang, C. Chien, J. H. Lau, H. Fu, C. W. Chiang, T. Y. Kuo, Y. H. Chen, R. Lo, and M. J. Kao, "How to Select Adhesive Materials for Temporary Bonding and De-Bonding of Thin-Wafer Handling in 3D IC Integration," *IEEE/ECTC Proceedings*, Orlando, FL, June 2011, pp. 989–998.
72. Lee, C. K., T. Chang, Y. Huang, H. Fu, J. H. Huang, Z. Hsiao, J. H. Lau, C. T. Ko, R. Cheng, K. Kao, Y. Lu, R. Lo, and M. J. Kao, "Characterization and Reliability Assessment of Solder Microbumps and Assembly for 3D IC Integration," *IEEE ECTC Proceedings*, Orlando, FL, June 2011, pp. 1468–1474.
73. Lin, Y., C. Zhan, J. Juang, J. H. Lau, T. Chen, R. Lo, M. Kao, T. Tian, and K. N. Tu, "Electromigration in Ni/Sn Intermetallic Micro Bump Joint for 3D IC Chip Stacking," *IEEE/ECTC Proceedings*, Orlando, FL, June 2011, pp. 351–357.
74. Zhan, C., J. Juang, Y. Lin, Y. Huang, K. Kao, T. Yang, S. Lu, J. H. Lau, T. Chen, R. Lo, and M. J. Kao, "Development of Fluxless Chip-on-Wafer Bonding Process for 3D chip Stacking with 30μm Pitch Lead-Free Solder Micro Bump Interconnection and Reliability Characterization," *IEEE/ECTC Proceedings*, Orlando, FL, June 2011, pp. 14–21.
75. Huang, S., T. Chang, R. Cheng, J. Chang, C. Fan, C. Zhan, J. H. Lau, T. Chen, R. Lo, and M. Kao, "Failure Mechanism of 20-μm Pitch Micro Joint Within a Chip Stacking Architecture," *IEEE/ECTC Proceedings*, Orlando, FL, June 2011, pp. 886–892.
76. Zhang, X., R. Rajoo, C. Selvanayagam, A. Kumar, V. Rao, N. Khan, V. Kripesh, J. H. Lau, D.-L. Kwong, V. Sundaram, and R. Tummula, "Application of Piezoresistive Stress Sensor in Wafer Bumping and Drop Impact Test of Embedded Ultra Thin Device," *IEEE/ECTC Proceedings*, Orlando, FL, June 2011, pp. 1276–1282.
77. Dorsey, P., "Xilinx Stacked Silicon Interconnect Technology Delivers Breakthrough FPGA Capacity, Bandwidth, and Power Efficiency," Xilinx white paper: Virtex-7 FPGAs, WP380, October 27, 2010, pp. 1–10.
78. Banijamali, B., S. Ramalingam, K. Nagarajan, and R. Chaware, "Advanced Reliability Study of TSV Interposers and Interconnects for the 28-nm Technology FPGA," *IEEE/ECTC Proceedings*, Orlando, FL, June 2011, pp. 285–290.
79. Banijamali, B., S. Ramalingam, N. Kim, and R. Wyland, "Ceramics vs. Low CTE Organic Packaging of TSV Silicon Interposers," *IEEE/ECTC Proceedings*, Orlando, FL, June 2011, pp. 573–576.
80. Kim, N., D. Wu, D. Kim, A. Rahman, and P. Wu, "Interposer Design Optimization for High Frequency Signal Transmission in Passive and Active Interposer Using Through Silicon Via (TSV)," *IEEE/ECTC Proceedings*, Orlando, FL, June 2011, pp. 1160–1167.
81. Khan, N., V. Rao, S. Lim, S. Ho, V. Lee, X. Zhang, R. Yang, E. Liao, Ranganathan, T. Chai, V. Kripesh, and J. H. Lau, "Development of 3D Silicon Module with TSV for System in Packaging," *IEEE Transactions on Components, Packaging and Manufacturing Technology*, Vol. 33, No. 1, March 2010, pp. 3–9.
82. Lau, J. H., H. C. Chien, and R. Tain, "TSV Interposers with Embedded Microchannels for 3D IC and Multiple High-Power LEDs Integration SiP," ASME paper no. InterPACK2011-52204, July 2011.
83. Lau, J. H., and X. Zhang, "Effects of TSV Interposer on the Reliability of 3D

IC Integration SiP," ASME paper no. InterPACK2011-52205, July 2011.
84. Lau, J. H., "The Most Cost-Effective Integrator (TSV Interposer) for 3D IC Integration SiP," ASME paper no. InterPACK2011-52189, July 2011.
85. Tummula, R., and M. Swaminathan, *System-on-Package: Miniaturization of the Entire System*, McGraw-Hill, New York, 2008.
86. Lau, J. H., M. Dai, Y. Chao, W. Li, S. Wu, J. Hung, M. Hsieh, J. Chien, R. Tain, C. Tzeng, K. Lin, E. Hsin, C. Chen, M. Chen, C. Wu, J. Chen, J. Chien, C. Chiang, Z. Lin, L. Wu, H. Chang, W. Tsai, C. Lee, T. Chang, C. Ko, T. Chen, S. Sheu, S. Wu, Y. Chen, R. Lo, T. Ku, M. Kao, F. Hsieh, and D. Hu, "Feasibility Study of a 3D IC Integration System-in-Packaging (SiP)," *IEEE/ICEP Proceedings*, Nara, Japan, April 13, 2011, pp. 210–216.
87. Lau, J. H., "Evolution, Outlook, and Challenges of 3D IC/Si Integration," *IEEE/ICEP Proceedings* (Keynote), Nara, Japan, April 13, 2011, pp. 1–17.
88. Chai, T., X. Zhang, J. H. Lau, C. Selvanayagam, K. Biswas, S. Liu, D. Pinjala, G. Tang, Y. Ong, S. Vempati, E. Wai, H. Li, B. Liao, N. Ranganathan, V. Kripesh, J. Sun, J. Doricko, and C. Vath, "Development of Large Die Fine-Pitch Cu/Low-k FCBGA Package with Through Silicon Via (TSV) Interposer," *IEEE Transactions on Components, Packaging and Manufacturing Technology*, Vol. 1, No. 5, 2011, pp. 660–672.
89. Kannan, S., S. Evana, A. Gupta, B. Kim, and L. Li, "3D Copper Based TSV for 60-GHz Applications," *IEEE/ECTC Proceedings*, Orlando, FL, June 2011, pp. 1168–1175.
90. Spiller, S., F. Molina, J. Wolf, J. Grafe, A. Schenke, D. Toennies, M. Hennemeyer, T. Tabuchi, and H. Auer, "Processing of Ultrathin 300-mm Wafers with Carrierless Technology," *IEEE/ECTC Proceedings*, Orlando, FL, June 2011, pp. 984–988.
91. Choi, K., K. Sung, H. Bae, J. Moon, and Y. Eom, "Bumping and Stacking Processes for 3D IC using Fluxfree Polymer," *IEEE/ECTC Proceedings*, Orlando, FL, June 2011, pp. 1746–1751.
92. Sukumaran, V., T. Bandyopadhyay, Q. Chen, N. Kumbhat, F. Liu, R. Pucha, Y. Sato, M. Watanabe, K. Kitaoka, M. Ono, Y. Suzuki, C. Karoui, C. Nopper, M. Swaminathan, V. Sundaram, and R. Tummala, "Design, Fabrication and Characterization of Low-Cost Glass Interposers with Fine-Pitch Through-Package-Vias," *IEEE/ECTC Proceedings*, Orlando, FL, June 2011, pp. 583–588.
93. Zhang, Y. C. King, Jr., J. Zaveri, Y. Kim, V. Sahu, Y. Joshi, and M. Bakir, "Coupled Electrical and Thermal 3D IC Centric Microfluidic Heat Sink Design and Technology," *IEEE/ECTC Proceedings*, Orlando, FL, June 2011, pp. 2037–2044.
94. Parekh, M., P. Thadesar, and M. Bakir, "Electrical, Optical and Fluidic Through-Silicon Vias for Silicon Interposer Applications," *IEEE/ECTC Proceedings*, Orlando, FL, June 2011, pp. 1992–1998.
95. Chen, Q., T. Bandyopadhyay, Y. Suzuki, F. Liu, V. Sundaram, R. Pucha, M. Swaminathan, and R. Tummala, "Design and Demonstration of Low Cost, Panel-Based Polycrystalline Silicon Interposer with Through-Package-Vias (TPVs)," *IEEE/ECTC Proceedings*, Orlando, FL, June 2011, pp. 855–860.
96. Liu, X., Q. Chen, V. Sundaram, M. Simmons-Matthews4, K. Wachtler, R. Tummala, and S. Sitaraman, "Thermo-Mechanical Behavior of Through Silicon Vias in a 3D Integrated Package with Inter-Chip Microbumps," *IEEE/ECTC Proceedings*, Orlando, FL, June 2011, pp. 1190–1195.
97. Mitra1, J., M. Jung, S. Ryu, R. Huang, S. Lim, and D. Pan, "A Fast Simulation Framework for Full-chip Thermo-Mechanical Stress and Reliability Analysis of Through-Silicon-Via-Based 3D ICs," *IEEE/ECTC Proceedings*, Orlando, FL, June 2011, pp. 746–753.
98. Zhang, R., S. Lee, D. Xiao, and H. Chen, "LED Packaging Using Silicon Substrate with Cavities for Phosphor Printing and Copper-filled TSVs for 3D Interconnection," *IEEE/ECTC Proceedings*, Orlando, FL, June 2011, pp. 1616–1621.
99. Zhou, Z., C. Liu, X. Wang, X. Luo, and S. Liu, "Integrated Process for Silicon Wafer Thinning," *IEEE/ECTC Proceedings*, Orlando, FL, June 2011, pp. 1811–1814.
100. Wan, Z., X. Luo, and S. Liu, "Effect of Blind Hole Depth and Shape of Solder Joint on the Reliability of Through Silicon Via (TSV)," *IEEE/ECTC Proceedings*, Orlando, FL, June 2011, pp. 1657–1661.

101. Chen, Z., S. Zhou, Z. Lv, C. Liu, X. Chen, X. Jia, K. Zeng, B. Song, F. Zhu, M. Chen, X. Wang, H. Zhang, and S. Liu, "Expert Advisor for Integrated Virtual Manufacturing and Reliability for TSV/SiP Based Modules," *IEEE/ECTC Proceedings*, Orlando, FL, June 2011, pp. 1183–1189.
102. Lee, G., Y. Kim, S. Jeon, K. Byun, and D. Kwon, "Interfacial Reliability and Micropartial Stress Analysis Between TSV and CPB Through NIT and MSA," *IEEE/ECTC Proceedings*, Orlando, FL, June 2011, pp. 1436–1443.
103. Nimura, M., J. Mizuno, K. Sakuma, and S. Shoji, "Solder/Adhesive Bonding Using Simple Planarization Technique for 3D Integration," *IEEE/ECTC Proceedings*, Orlando, FL, June 2011, pp. 1147–1152.
104. Maria, J., B. Dang, S. Wright, C. Tsang, P. Andry, R. Polastre, Y. Liu, L. Wiggins and J. Knickerbocker, "3D Chip Stacking with 50-μm Pitch Lead-Free Micro-C4 Interconnections," *IEEE/ECTC Proceedings*, Orlando, FL, June 2011, pp. 268–273.
105. Sakuma, K., K. Toriyama, H. Noma, K. Sueoka, N. Unami, J. Mizuno, S. Shoji, and Y. Orii, "Fluxless Bonding for Fine-Pitch and Low-Volume Solder 3D Interconnections," *IEEE/ECTC Proceedings*, Orlando, FL, June 2011, pp. 7–13.
106. Andry, P., B. Dang, J. Knickerbocker, K. Tamura, and N. Taneichi, "Low-Profile 3D Silicon-on-Silicon Multi-chip Assembly," *IEEE/ECTC Proceedings*, Orlando, FL, June 2011, pp. 553–559.
107. Doany, F., C. Schow, B. Lee, R. Budd, C. Baks, R. Dangel, R. John, F. Libsch, and J. Kash, "Terabit/sec-Class Board-Level Optical Interconnects Through Polymer Waveguides Using 24-Channel Bidirectional Transceiver Modules," *IEEE/ECTC Proceedings*, Orlando, FL, June 2011, pp. 790–797.
108. Che, F., T. Chai, S. Lim, R. Rajoo, and X. Zhang, "Design and Reliability Analysis of Pyramidal Shape 3-Layer Stacked TSV Die Package," *IEEE/ECTC Proceedings*, Orlando, FL, June 2011, pp. 1428–1435.
109. Lee, J., D. Fernandez, M. Paing, Y. Yeo, and S. Gao, "Novel Chip Stacking Process for 3D Integration," *IEEE/ECTC Proceedings*, Orlando, FL, June 2011, pp. 1939–1943.
110. Au, K., J. Beleran, Y. Yang, Y. Zhang, S. Kriangsak, P. Wilson, Y. Drake, C. Toh, and C. Surasit, "Thru Silicon Via Stacking & Numerical Characterization for Multi-Die Interconnections using Full Array and Very Fine Pitch Micro C4 Bumps," *IEEE/ECTC Proceedings*, Orlando, FL, June 2011, pp. 296–303.
111. Lu, K., S. Ryu, Q. Zhao, K. Hummler, J. Im, R. Huang, and P. Ho, "Temperature-Dependent Thermal Stress Determination for Through-Silicon-Vias (TSVs) by Combining Bending Beam Technique with Finite Element Analysis," *IEEE/ECTC Proceedings*, Orlando, FL, June 2011, pp. 1475–1480.
112. Mitra, J., M. Jung, S. Ryu, R. Huang, S. Lim, and D. Pan, "A Fast Simulation Framework for Full-Chip Thermo-Mechanical Stress and Reliability Analysis of Through-Silicon-Via-Based 3D ICs," *IEEE/ECTC Proceedings*, Orlando, FL, June 2011, pp. 746–753.
113. Lin, Y., C. Hsieh, C. Yu, C. Tung, and D. Yu, "Study of the Thermo-Mechanical Behavior of Glass Interposer for Flip Chip Packaging Applications," *IEEE/ECTC Proceedings*, Orlando, FL, June 2011, pp. 634–638.
114. Lin, T., R. Wang, M. Chen, C. Chiu, S. Chen, T. Yeh, L. Lin, S. Hou, J. Lin, K. Chen, S. Jeng, and D. Yu, "Electromigration Study of Micro Bumps at Si/Si Interface in 3DIC Package for 28-nm Technology and Beyond," *IEEE/ECTC Proceedings*, Orlando, FL, June 2011, pp. 346–350.
115. Wei, C., C. Yu, C. Tung, R. Huang, C. Hsieh, C. Chiu1, H. Hsiao, Y. Chang, C. Lin, Y. Liang, C. Chen, T. Yeh, L. Lin, and D. Yu, "Comparison of the Electromigration Behaviors Between Micro-Bumps and C4 Solder Bumps," *IEEE/ECTC Proceedings*, Orlando, FL, June 2011, pp. 706–710.
116. Bouchoucha, M., P. Chausse, S. Moreau, L. Chapelon, N. Sillon, and O. Thomas, "Reliability Study of 3D-WLP Through Silicon Via with Innovative Polymer Filling Integration," *IEEE/ECTC Proceedings*, Orlando, FL, June 2011, pp. 567–572.
117. Shariff, D., P. Marimuthu, K. Hsiao, L. Asoy, C. Yee, A. Oo, K. Buchanan1, K. Crook, T. Wilby, and S. Burgess, "Integration of Fine-Pitched Through-Silicon Vias and Integrated Passive Devices," *IEEE/ECTC Proceedings*, Orlando, FL, June 2011, pp. 844–848.
118. Malta, D., C. Gregory, M. Lueck, D. Temple, M. Krause, F. Altmann, M. Petzold, M. Weatherspoon, and J. Miller, "Characterization of Thermo-Mechanical Stress and Reliability Issues for Cu-Filled TSVs," *IEEE/ECTC Proceedings*, Orlando, FL, June 2011, pp. 1815–1821.

119. Ramkumar, S., H. Venugopalan, and K. Khanna, "Novel Anisotropic Conductive Adhesive for 3D Stacking and Lead-Free PCB Packaging: A Review," *IEEE/ECTC Proceedings*, Orlando, FL, June 2011, pp. 246–254.
120. Oprins, H., V. Cherman, B. Vandevelde, C. Torregiani, M. Stucchi, G. Van der Plas, P. Marchal, and E. Beyne, "Characterization of the Thermal Impact of Cu-Cu Bonds Achieved Using TSVs on Hot Spot Dissipation in 3D Stacked ICs," *IEEE/ECTC Proceedings*, Orlando, FL, June 2011, pp. 861–868.
121. Tsukada, A., R. Sato, S. Sekine, R. Kimura, K. Kishi, Y. Sato, Y. Iwata, and H. Murata, "Study on TSV with New Filling Method and Alloy for Advanced 3D-SiP," *IEEE/ECTC Proceedings*, Orlando, FL, June 2011, pp. 861–868.
122. Yoon, S., K. Ishibashi, S. Dzafir, M. Prashant, P. Marimuthu, and F. Carson, "Development of Super Thin TSV PoP," *IEEE/ECTC Proceedings*, Orlando, FL, June 2011, pp. 274–278.
123. Meinshausen, L., K. Weide-Zaage, and M. Petzold, "Electro- and Thermomigration in Micro Bump Interconnects for 3D Integration," *IEEE/ECTC Proceedings*, Orlando, FL, June 2011, pp. 1444–1451.
124. Ko, Y., H. Fujii, Y. Sato, C. Lee, and S. Yoo, "Advanced Solder TSV Filling Technology Developed with Vacuum and Wave Soldering," *IEEE/ECTC Proceedings*, Orlando, FL, June 2011, pp. 2091–2095.
125. Jung, M., Y. Song, T. Yim, and J. Lee, "Evaluation of Additives and Current Mode on Copper Via Fill," *IEEE/ECTC Proceedings*, Orlando, FL, June 2011, pp. 1908–1912.
126. Choi, Y., J. Shin, and K. Paik, "A Study on the 3D-TSV Interconnection Using Wafer-Level Non-Conductive Adhesives (NCAs)," *IEEE/ECTC Proceedings*, Orlando, FL, June 2011, pp. 1126–1129.
127. Zoschke, K., J. Wolf, C. Lopper, I. Kuna, N. Jürgensen, V. Glaw, K. Samulewicz, J. Röder, M. Wilke, O. Wünsch, M. Klein, M. Suchodoletz, H. Oppermann, T. Braun, R. Wieland, and O. Ehrmann, "TSV Based Silicon Interposer Technology for Wafer Level Fabrication of 3D SiP Modules," *IEEE/ECTC Proceedings*, Orlando, FL, June 2011, pp. 836–843.
128. Brusberg, L., N. Schlepple, and H. Schröder, "Chip-to-Chip Communication by Optical Routing Inside a Thin Glass Substrate," *IEEE/ECTC Proceedings*, Orlando, FL, June 2011, pp. 805–812.
129. Schröder, H., L. Brusberg, N. Arndt-Staufenbiel, J. Hofmann, and S. Marx, "Glass Panel Processing for Electrical and Optical Packaging," *IEEE/ECTC Proceedings*, Orlando, FL, June 2011, pp. 625–633.
130. Redolfi, A., D. Velenis, S. Thangaraju, P. Nolmans, P. Jaenen, M. Kostermans, U. Baier, E. Van Besien, H. Dekkers, T. Witters, N. Jourdan, A. Van Ammel, K. Vandersmissen, S. Rodet, H. Philipsen, A. Radisic, N. Heylen, Y. Travaly, B. Swinnen, and E. Beyne, "Implementation of an Industry Compliant, SiN 50-μm Via-Middle TSV Technology on 300-mm Wafers," *IEEE/ECTC Proceedings*, Orlando, FL, June 2011, pp. 1384–1388.
131. Jourdain, A., T. Buisson, A. Phommahaxay, A. Redolfi, S. Thangaraju, Y. Travaly, E. Beyne, and B. Swinnen, "Integration of TSVs, Wafer Thinning and Backside Passivation on Full 300-mm CMOS Wafers for 3D Applications," *IEEE/ECTC Proceedings*, Orlando, FL, June 2011, pp. 1122–1125.
132. Pham, N., V. Cherman, B. Vandevelde, P. Limaye, N. Tutunjyan, R. Jansen, N. Van Hoovels, D. Tezcan, P. Soussan, E. Beyneand, and H. Tilmans, "Zero-Level Packaging for (RF-)MEMS Implementing TSVs and Metal Bonding," *IEEE/ECTC Proceedings*, Orlando, FL, June 2011, pp. 1588–1595.
133. Halder, S., A. Jourdain, M. Claes, I. Wolf, Y. Travaly, E. Beyne, B. Swinnen, V. Pepper, P. Guittet, G. Savage, and L. Markwort, "Metrology and Inspection for Process Control During Bonding and Thinning of Stacked Wafers for Manufacturing 3D SICs," *IEEE/ECTC Proceedings*, Orlando, FL, June 2011, pp. 999–1002.
134. Vos, J., A. Jourdain, M. Erismis, W. Zhang, K. De Munck, A. La Manna, D. Tezcan, and P. Soussan, "High Density 20-μm Pitch CuSn Microbump Process for High-End 3D Applications," *IEEE/ECTC Proceedings*, Orlando, FL, June 2011, pp. 27–31.
135. Zhang, W., B. Dimcic, P. Limaye, A. Manna, P. Soussan, and E. Beyne, "Ni/Cu/Sn Bumping Scheme for Fine-Pitch Micro-Bump Connections," *IEEE/ECTC Proceedings*, Orlando, FL, June 2011, pp. 109–113.

136. Che, F., H. Li, X. Zhang, S. Gao, and K. Teo, "Wafer Level Warpage Modeling Methodology and Characterization of TSV Wafers," *IEEE/ECTC Proceedings*, Orlando, FL, June 2011, pp. 1196–1203.
137. Kwon, W., J. Lee, V. Lee, J. Seetoh, Y. Yeo, Y. Khoo, N. Ranganathan, K. Teo, and S. Gao, "Novel Thinning/Backside Passivation for Substrate Coupling Depression of 3D IC," *IEEE/ECTC Proceedings*, Orlando, FL, June 2011, pp. 1395–1399.
138. Lee, J., V. Lee, J. Seetoh, S. Thew, Y. Yeo, H. Li, K. Teo, and S. Gao, "Advanced Wafer Thinning and Handling for Through Silicon Via Technology," *IEEE/ECTC Proceedings*, Orlando, FL, June 2011, pp. 1852–1857.
139. Oprins, H., V. Cherman, B. Vandevelde, C. Torregiani, M. Stucchi, G. Van der Plas, P. Marchal, and E. Beyne, "Characterization of the Thermal Impact of Cu-Cu Bonds Achieved Using TSVs on Hot Spot Dissipation in 3D Stacked ICs," *IEEE/ECTC Proceedings*, Orlando, FL, June 2011, pp. 861–868.
140. Peng, L., H. Li, D. Lim, S. Gao, and C. Tan, "Thermal Reliability of Fine Pitch Cu-Cu Bonding with Self-Assembled Monolayer (SAM) Passivation for Wafer-on-Wafer 3D-Stacking," *IEEE/ECTC Proceedings*, Orlando, FL, June 2011, pp. 22–26.
141. Xie, L., W. Choi, C. Premachandran, C. Selvanayagam, K. Bai, Y. Zeng, S. Ong, E. Liao, A. Khairyanto, V. Sekhar, and S. Thew, "Design, Simulation and Process Optimization of AuInSn Low Temperature TLP Bonding for 3D IC Stacking," *IEEE/ECTC Proceedings*, Orlando, FL, June 2011, pp. 279–284.
142. Akasaka, Y., "Three-Dimensional IC Trends," *Proceedings of the IEEE*, Vol. 74, No. 12, December 1986, pp. 1703–1714.
143. Akasaka, Y., and Nishimura, T., "Concept and Basic Technologies for 3D IC Structure," *IEEE Proceedings of International Electron Devices Meetings*, Vol. 32, 1986, pp. 488–491.
144. Chen, K., S. Lee, P. Andry, C. Tsang, A. Topop, Y. Lin, Y., J. Lu, A. Young, M. Ieong, and W. Haensch, "Structure, Design and Process Control for Cu Bonded Interconnects in 3D Integrated Circuits," *IEEE Proceedings of International Electron Devices Meeting (IEDM 2006)*, San Francisco, CA, December 11–13, 2006, pp. 367–370.
145. Liu, F., R. Yu, A. Young, J. Doyle, X. Wang, L. Shi, K. Chen, X. Li, D. Dipaola, D. Brown, C. Ryan, J. Hagan, K. Wong, M. Lu, X. Gu, N. Klymko, E. Perfecto, A. Merryman, K. Kelly, S. Purushothaman, S. Koester, R. Wisnieff, and W. Haensch, "A 300-mm Wafer-Level Three-Dimensional Integration Scheme Using Tungsten Through-Silicon Via and Hybrid Cu-Adhesive Bonding," *IEEE Proceedings of IEDM*, December 2008, pp. 1–4.
146. Yu, R., F. Liu, R. Polastre, K. Chen, X. Liu, L. Shi, E. Perfecto, N. Klymko, M. Chace, T. Shaw, D. Dimilia, E. Kinser, A. Young, S. Purushothaman, S. Koester, and W. Haensch, "Reliability of a 300-mm-Compatible 3DI Technology Based on Hybrid Cu-Adhesive Wafer Bonding," *Proceedings of Symposium on VLSI Technology Digest of Technical Papers*, 2009, pp. 170–171.
147. Shigetou, A., T. Itoh, K. Sawada, and T. Suga, "Bumpless Interconnect of 6-μm Pitch Cu Electrodes at Room Temperature," *IEEE Proceedings of ECTC*, Lake Buena Vista, FL, May 27–30, 2008, pp. 1405–1409.
148. Tsukamoto, K., E. Higurashi, and T. Suga, "Evaluation of Surface Microroughness for Surface Activated Bonding," *Proceedings of IEEE CPMT Symposium Japan*, August 2010, pp. 147–150.
149. Kondou, R., C. Wang, and T. Suga, "Room-Temperature Si-Si and Si-SiN Wafer Bonding," *Proceedings of IEEE CPMT Symposium Japan*, August 2010, pp. 161–164.
150. Shigetou, A., T. Itoh, M. Matsuo, N. Hayasaka, K. Okumura, and T. Suga, "Bumpless Interconnect Through Ultrafine Cu Electrodes by Mans of Surface-Activated Bonding (SAB) Method," *IEEE Transaction on Advanced Packaging*, Vol. 29, No. 2, May 2006, p. 226.
151. Wang, C., and T. Suga, "A Novel Moire Fringe Assisted Method for Nanoprecision Alignment in Wafer Bonding," *IEEE/ECTC Proceedings*, San Diego, CA, May 25–29, 2009, pp. 872–878.
152. Wang, C., and T. Suga, "Moire Method for Nanoprecision Wafer-to-Wafer Alignment: Theory, Simulation and Application," *IEEE Proceedings of International Conference on Electronic Packaging Technology and High-Density Packaging*, August 2009, pp. 219–224.

153. Higurashi, E., D. Chino, T. Suga, and R. Sawada, "Au-Au Surface-Activated Bonding and Its Application to Optical Microsensors with 3D Structure," *IEEE Journal of Selected Topic in Quantum Electronics*, Vol. 15, No. 5, September–October 2009, pp. 1500–1505.
154. Burns, J., B. Aull, C. Keast, C. Chen, C. Chen, C. Keast, J. Knecht, V. Suntharalingam, K. Warner, P. Wyatt, and D. Yost, "A Wafer-Scale 3D Circuit Integration Technology," *IEEE Transactions on Electron Devices*, Vol. 53, No. 10, October 2006, pp. 2507–2516.
155. Chen, C., K. Warner, D. Yost, J. Knecht, V. Suntharalingam, C. Chen, J. Burns, and C. Keast, "Sealing Three-Dimensional SOI Integrated-Circuit Technology," *IEEE Proceedings of International SOI Conference*, 2007, pp. 87–88.
156. Chen, C., C. Chen, D. Yost, J. Knecht, P. Wyatt, J. Burns, K. Warner, P. Gouker, P. Healey, B. Wheeler, and C. Keast, "Three-Dimensional Integration of Silicon-on-Insulator RF Amplifier," *Electronics Letters*, Vol. 44, No. 12, June 2008, pp. 1–2.
157. Chen, C., C. Chen, D. Yost, J. Knecht, P. Wyatt, J. Burns, K. Warner, P. Gouker, P. Healey, B. Wheeler, and C. Keast, "Wafer-Scale 3D Integration of Silicon-on-Insulator RF Amplifiers," *IEEE Proceedings of Silicon Monolithic IC in RF Systems*, 2009, pp. 1–4.
158. Chen, C., C. Chen, P. Wyatt, P. Gouker, J. Burns, J. Knecht, D. Yost, P. Healey, and C. Keast, "Effects of Through-Box Vias on SOI MOSFETs," *IEEE Proceedings of VLSI Technology, Systems and Applications*, 2008, pp. 1–2.
159. Chen, C., C. Chen, J. Burns, D. Yost, K. Warner, J. Knecht, D. Shibles, and C. Keast, "Thermal Effects of Three-Dimensional Integrated Circuit Stacks," *IEEE Proceedings of International SOI Conference*, 2007, pp. 91–92.
160. Aull, B., J. Burns, C. Chen, B. Felton, H. Hanson, C. Keast, J. Knecht, A. Loomis, M. Renzi, A. Soares, V. Suntharalingam, K. Warner, D. Wolfson, D. Yost, and D. Young, "Laser Radar Imager Based on 3D Integration of Geiger-Mode Avalanche Photodiodes with Two SOI Timing Circuit Layers," *IEEE Proceedings of International Solid-State Circuits Conference*, 2006, p. 16.9.
161. Chatterjee, R., M. Fayolle, P. Leduc, S. Pozder, B. Jones, E. Acosta, B. Charlet, T. Enot, M. Heitzmann, M. Zussy, A. Roman, O. Louveau, S. Maitreqean, D. Louis, N. Kernevez, N. Sillon, G. Passemard, V. Pol, V. Mathew, S. Garcia, T. Sparks, and Z. Huang, "Three-Dimensional Chip Stacking Using a Wafer-to-Wafer Integration," *IEEE Proceedings of IITC*, 2007, pp. 81–83.
162. Ledus, P., F. Crecy, M. Fayolle, M. Fayolle, B. Charlet, T. Enot, M. Zussy, B. Jones, J. Barbe, N. Kernevez, N. Sillon, S. Maitreqean, D. Louis, and G. Passemard, "Challenges for 3D IC Integration: Bonding Quality and Thermal Management," *IEEE Proceedings of IITC*, 2007, pp. 210–212.
163. Poupon, G., N. Sillon, D. Henry, C. Gillot, A. Mathewson, L. Cioccio, B. Charlet, P. Leduc, M. Vinet, and P. Batude, "System on Wafer: A New Silicon Concept in Sip," *Proceedings of the IEEE*, Vol. 97, No. 1, January 2009, pp. 60–69.
164. Fujimoto, K., N. Maeda, H. Kitada, Y. Kim, A. Kawai, K. Arai, T. Nakamura, K. Suzuki, and T. Ohba, "Development of Multi-Stack Process on Wafer-on-Wafer (WoW)," *Proceedings of IEEE CPMT Symposium Japan*, August 2010, pp. 157–160.
165. Chen, Q., D. Zhang, Z. Wang, L. Liu, and J. Lu, "Chip-to-Wafer (C2W) 3D Integration with Well-Controlled Template Alignment and Wafer-Level Bonding," *IEEE/ECTC Proceedings*, Orlando, FL, June 2011, pp. 1–6.
166. Healy, M., and S. Lim, "Power Delivery System Architecture for Many-Tier 3D Systems," *IEEE/ECTC Proceedings*, Las Vegas, NV, June 2010, pp. 1682–1688.
167. Liu, F., X. Gu, K. A. Jenkins, E. A. Cartier, Y. Liu, P. Song, and S. J. Koester, "Electrical Characterization of 3D Through-Silicon-Vias," *IEEE/ECTC Proceedings*, Las Vegas, NV, June 2010, pp. 1100–1105.
168. Gu, X., B. Wu, M. Ritter, and L. Tsang, "Efficient Full-Wave Modeling of High Density TSVs for 3D Integration," *IEEE/ECTC Proceedings*, Las Vegas, NV, June 2010, pp. 663–666.
169. Okoro, C., R. Agarwal, P. Limaye, B. Vandevelde, D. Vandepitte, and E. Beynel, "Insertion Bonding: A Novel Cu-Cu Bonding Approach for 3D Integration," *IEEE/ECTC Proceedings*, Las Vegas, NV, June 2010, pp. 1370–1375.
170. Huyghebaert, C., J. Olmen, O. Chukwudi, J. Coenen, A. Jourdain, M. Cauwenberghe, R. Agarwahl, R., A. Phommahaxay, M. Stucchi, and P. Soussan., "Enabling 10-μm Pitch Hybrid Cu-Cu IC Stacking with Through

171. Pak, J., J. Cho, J. Kim, J. Lee, H. Lee, K. Park, and J. Kim, "Slow Wave and Dielectric Quasi-TEM Modes of Metal-Insulator-Semiconductor (MIS) Structure Through Silicon Via (TSV) in Signal Propagation and Power Delivery in 3D Chip Package," *IEEE/ECTC Proceedings*, Las Vegas, NV, June 2010, pp. 667–672.
172. Cioccioa, L., P. Gueguena, E. Grouillera, L. Vandrouxa, V. Delayea, M. Rivoireb, J. Lugandb, and L. Claveliera, "Vertical Metal Interconnect Thanks to Tungsten Direct Bonding," *IEEE/ECTC Proceedings*, Las Vegas, NV, June 2010, 1359–1363.
173. Gueguena, P., L. Cioccioa, P. Morfoulib, M. Zussya, J. Dechampa, L. Ballya, and L. Claveliera, "Copper Direct Bonding: An Innovative 3D Interconnect," *IEEE/ECTC Proceedings*, Las Vegas, NV, June 2010, pp. 878–883.
174. Taibi, R., L. Ciocciob, C. Chappaz, L. Chapelon, P. Gueguenb, J. Dechampb, R. Fortunierc, and L. Clavelierb, "Full Characterization of Cu/Cu Direct Bonding for 3D Integration," *IEEE/ECTC Proceedings*, Las Vegas, NV, June 2010, pp. 219–225.
175. Lim, D., J. Wei, C. Ng, and C. Tan, "Low Temperature Bump-Less Cu-Cu Bonding Enhancement with Self-Assembled Monolayer (SAM) Passivation for 3D Integration," *IEEE/ECTC Proceedings*, Las Vegas, NV, June 2010, pp. 1364–1369.
176. Onkaraiah, S., and C. Tan, "Mitigating Heat Dissipation and Thermo-Mechanical Stress Challenges in 3D IC Using Thermal Through Silicon Via (TTSV)," *IEEE/ECTC Proceedings*, Las Vegas, NV, June 2010, pp. 411–416.
177. Campbell, D., "Process Characterization Vehicles for 3D Integration," *IEEE/ECTC Proceedings*, Las Vegas, NV, June 2010, pp. 1112–1116.
178. Bieck, F., S. Spiller, F. Molina, M. Töpper, C. Lopper, I. Kuna, T. Seng, and T. Tabuchi, "Carrierless Design for Handling and Processing of Ultrathin Wafers," *IEEE/ECTC Proceedings*, Las Vegas, NV, June 2010, pp. 316–322.
179. Itabashi, T., and M. Zussman, "High Temperature Resistant Bonding Solutions Enabling Thin Wafer Processing (Characterization of Polyimide Base Temporary Bonding Adhesive for Thinned Wafer Handling)," *IEEE/ECTC Proceedings*, Las Vegas, NV, June 2010, pp. 1877–1880.
180. Lee, G., H. Son, J. Hong, K. Byun, and D. Kwon, "Quantification of Micropartial Residual Stress for Mechanical Characterization of TSV Through Nanoinstrumented Indentation Testing," *IEEE/ECTC Proceedings*, Las Vegas, NV, June 2010, pp. 200–205.
181. Dang, B., P. Andry, C. Tsang, J. Maria, R. Polastre, R. Trzcinski, A. Prabhakar, and J. Knickerbocker, "CMOS Compatible Thin Wafer Processing Using Temporary Mechanical Wafer, Adhesive and Laser Release of Thin Chips/Wafers for 3D Integration," *IEEE/ECTC Proceedings*, Las Vegas, NV, June 2010, pp. 1393–1398.
182. Kang, U., H. Chung, S. Heo, D. Park, H. Lee, J. Kim, S. Ahn, S. Cha, J. Ahn, D. Kwon, J. Lee, H. Joo, W. Kim, D. Jang, N. Kim, J. Choi, T. Chung, J. Yoo, J. Choi, C. Kim, and Y. Jun, "8-Gb 3D DDR3 DRAM Using Through-Silicon-Via Technology," *IEEE Journal of Solid-State Circuits*, Vol. 45, No. 1, January 2010, pp. 111–119.
183. Farooq, M. G., T. L. Graves-Abe, W. F. Landers, C. Kothandaraman, B. A. Himmel, P. S. Andry, C. K. Tsang, E. Sprogis, R. P. Volant, K. S. Petrarca, K. R. Winstel, J. M. Safran, T. D. Sullivan, F. Chen, M. J. Shapiro, R. Hannon, R. Liptak, D. Berger, and S. S. Iyer, "3D Copper TSV Integration, Testing and Reliability," *Proceedings of IEEE IEDM*, Washington, DC, December 2011, pp. 7.1.1–7.1.4.
184. Dorsey, P., "Xilinx Stacked Silicon Interconnect Technology Delivers Breakthrough FPGA Capacity, Bandwidth, and Power Efficiency," Xilinx white paper: Virtex-7 FPGAs, WP380, October 27, 2010, pp. 1–10.
185. Banijamali, B., S. Ramalingam, K. Nagarajan, and R. Chaware, "Advanced Reliability Study of TSV Interposers and Interconnects for the 28-nm Technology FPGA," *IEEE/ECTC Proceedings*, Orlando, FL, June 2011, pp. 285–290.
186. Kim, N., D. Wu, D. Kim, A. Rahman, and P. Wu, "Interposer Design Optimization for High Frequency Signal Transmission in Passive and Active

Interposer using Through Silicon Via (TSV)," *IEEE/ECTC Proceedings*, Orlando, FL, June 2011, pp. 1160–1167.
187. Lau, J. H., M. Dai, Y. Chao, W. Li, S. Wu, J. Hung, M. Hsieh, J. Chien, R. Tain, C. Tzeng, K. Lin, E. Hsin, C. Chen, M. Chen, C. Wu, J. Chen, J. Chien, C. Chiang, Z. Lin, L. Wu, H. Chang, W. Tsai, C. Lee, T. Chang, C. Ko, T. Chen, S. Sheu, S. Wu, Y. Chen, R. Lo, T. Ku, M. Kao, F. Hsieh, and D. Hu, "Feasibility Study of a 3D IC Integration System-in-Packaging (SiP)," *IEEE/ICEP Proceedings*, Nara, Japan, April 13, 2011, pp. 210–216.
188. Lau, J. H., C.-J. Zhan, P.-J. Tzeng, C.-K. Lee, M.-J. Dai, H.-C. Chien, Y.-L. Chao, W. Li, S.-T. Wu, J.-F. Hung, R.-M. Tain, C.-H. Lin, Y.-C. Hsin, C.-C. Chen, S.-C. Chen, C.-Y. Wu, J.-C. Chen, C.-H. Chien, C.-W. Chiang, H. Chang, W.-L. Tsai, R.-S. Cheng, S.-Y. Huang, Y.-M. Lin, T.-C. Chang, C.-D. Ko, T.-H. Chen, S.-S. Sheu, S.-H. Wu, Y.-H. Chen, W.-C. Lo, T.-K. Ku, M.-J. Kao, and D.-Q. Hu, "Feasibility Study of a 3D IC Integration System-in-Packaging (SiP) from a 300-mm Multi-Project Wafer (MPW)," *Proceedings of IMAPS International Conference*, Long Beach, CA, October 2011, pp. 446–454.
189. Zhan, C.-J., J. H. Lau, P.-J. Tzeng, C.-K. Lee, M.-J. Dai, H.-C. Chien, Y.-L. Chao, W. Li, S.-T. Wu, J.-F. Hung, R.-M. Tain, C.-H. Lin, Y.-C. Hsin, C.-C. Chen, S.-C. Chen, C.-Y. Wu, J.-C. Chen, C.-H. Chien, C.-W. Chiang, H. Chang, W.-L. Tsai, R.-S. Cheng, S.-Y. Huang, Y.-M. Lin, T.-C. Chang, C.-D. Ko, T.-H. Chen, S.-S. Sheu, S.-H. Wu, Y.-H. Chen, W.-C. Lo, T.-K. Ku, M.-J. Kao, and D.-Q. Hu, "Assembly Process and Reliability Assessment of TSV/RDL/IPD Interposer with Multi-Chip-Stacking for 3D IC Integration SiP," *IEEE/ECTC Proceedings* (in press).
190. Sheu, S., Z. Lin, J. Hung, J. H. Lau, P. Chen, S. Wu, K. Su, C. Lin, S. Lai, T. Ku, W. Lo, and M. Kao, "An Electrical Testing Method for Blind Through Silicon Vias (TSVs) for 3D IC Integration," *Proceedings of IMAPS International Conference*, Long Beach, CA, October 2011, pp. 208–214.
191. Wu, C., S. Chen, P. Tzeng, J. H. Lau, Y. Hsu, J. Chen, Y. Hsin, C. Chen, S. Shen, C. Lin, T. Ku, and M. Kao, "Oxide Liner, Barrier and Seed Layers, and Cu-Plating of Blind Through Silicon Vias (TSVs) on 300-mm Wafers for 3D IC Integration," *Proceedings of IMAPS International Conference*, Long Beach, CA, October 2011, pp. 1–7.
192. Chien, H., J. H. Lau, Y. Chao, R. Tain, M. Dai, S. Wu, W. Lo, and M. Kao, "Thermal Performance of 3D IC Integration with Through-Silicon Via (TSV)," *Proceedings of IMAPS International Conference*, Long Beach, CA, October 2011, pp. 25–32.
193. Chang, H. H., J. H. Lau, W. L. Tsai, C. H. Chien, P. J. Tzeng, C. J. Zhan, C. K. Lee, M. J. Dai, H. C. Fu, C. W. Chiang, T. Y. Kuo, Y. H. Chen, W. C. Lo, T. K. Ku, and M. J. Kao, "Thin Wafer Handling of 300-mm Wafer for 3D IC Integration," *Proceedings of IMAPS International Conference*, Long Beach, CA, October 2011, pp. 202–207.
194. Lau, J. H., P.-J. Tzeng, C.-K. Lee, C.-J. Zhan, M.-J. Dai, Li Li, C.-T. Ko, S.-W. Chen, H. Fu, Y. Lee, Z. Hsiao, J. Huang, W. Tsai, P. Chang, S. Chung, Y. Hsu, S.-C. Chen, Y.-H. Chen, T.-H. Chen, W.-C. Lo, T.-K. Ku, M.-J. Kao, J. Xue, and M. Brillhart, "Wafer Bumping and Characterizations of Fine-Pitch Lead-Free Solder Microbumps on 12-inch (300-mm) Wafer for 3D IC Integration," *Proceedings of IMAPS International Conference*, Long Beach, CA, October 2011, pp. 650–656.
195. Hsin, Y. C., C. Chen, J. H. Lau, P. Tzeng, S. Shen, Y. Hsu, S. Chen, C. Wn, J. Chen, T. Ku, and M. Kao, "Effects of Etch Rate on Scallop of Through-Silicon Vias (TSVs) in 200-mm and 300-mm Wafers," *IEEE/ECTC Proceedings*, Orlando, FL, June 2011, pp. 1130–1135.
196. Chen, J. C., J. H. Lau, P. J. Tzeng, S. C. Chen, C. Y. Wu, C. C. Chen, C. H. Lin, Y. C. Hsin, T. K. Ku, and M. J. Kao, "Impact of Slurry in Cu CMP (Chemical Mechanical Polishing) on Cu Topography of Through Silicon Vias (TSVs), Re-distributed Layers, and Cu Exposure," *IEEE/ECTC Proceedings*, Orlando, FL, June 2011, pp. 1389–1394. Also, *IEEE Transactions on CPMT* (in press).
197. Chien, J., Y. Chao, J. H. Lau, M. Dai, R. Tain, M. Dai, P. Tzeng, C. Lin, Y. Hsin, S. Chen, J. Chen, C. Chen, C. Ho, R. Lo, T. Ku, and M. Kao, "A Thermal Performance Measurement Method for Blind Through Silicon Vias (TSVs) in a 300-mm Wafer," *IEEE/ECTC Proceedings*, Orlando, FL, June 2011, pp. 1204–1210.

198. Tsai, W., H. H. Chang, C. H. Chien, J. H. Lau, H. C. Fu, C. W. Chiang, T. Y. Kuo, Y. H. Chen, R. Lo, and M. J. Kao, "How to Select Adhesive Materials for Temporary Bonding and De-Bonding of Thin-Wafer Handling in 3D IC Integration?" *IEEE/ECTC Proceedings*, Orlando, FL, June 2011, pp. 989–998.
199. Chang, H., J. Huang, C. Chiang, Z. Hsiao, H. Fu, C. Chien, Y. Chen, W. Lo, and K. Chiang, "Process Integration and Reliability Test for 3D Chip Stacking with Thin Wafer Handling Technology," *IEEE/ECTC Proceedings*, Orlando, FL, June 2011, pp. 304–311.
200. Lee, C. K., T. C. Chang, Y. Huang, H. Fu, J. H. Huang, Z. Hsiao, J. H. Lau, C. T. Ko, R. Cheng, K. Kao, Y. Lu, R. Lo, and M. J. Kao, "Characterization and Reliability Assessment of Solder Microbumps and Assembly for 3D IC Integration," *IEEE/ECTC Proceedings*, Orlando, FL, June 2011, pp. 1468–1474.
201. Zhan, C., J. Juang, Y. Lin, Y. Huang, K. Kao, T. Yang, S. Lu, J. H. Lau, T. Chen, R. Lo, and M. J. Kao, "Development of Fluxless Chip-on-Wafer Bonding Process for 3D Chip Stacking with 30-μm Pitch Lead-Free Solder Micro Bump Interconnection and Reliability Characterization," *IEEE/ECTC Proceedings*, Orlando, FL, June 2011, pp. 14–21.
202. Cheng, R., K. Kao, J. Chang, Y. Hung, T. Yang, Y. Huang, S. Chen, T. Chang, Q. Hunag, R. Guino, G. Hoang, J. Bai, and K. Becker, "Achievement of Low Temperature Chip Stacking by a Pre-Applied Underfill Material," *IEEE/ECTC Proceedings*, Orlando, FL, June 2011, pp. 1858–1863.
203. Huang, S., T. Chang, R. Cheng, J. Chang, C. Fan, C. Zhan, J. H. Lau, T. Chen, R. Lo, and M. Kao, "Failure Mechanism of 20-μm Pitch Micro Joint Within a Chip Stacking Architecture," *IEEE/ECTC Proceedings*, Orlando, FL, June 2011, pp. 886–892.
204. Lin, Y., C. Zhan, J. Juang, J. H. Lau, T. Chen, R. Lo, M. Kao, T. Tian, and K. N. Tu, "Electromigration in Ni/Sn Intermetallic Micro Bump Joint for 3D IC Chip Stacking," *IEEE/ECTC Proceedings*, Orlando, FL, June 2011, pp. 351–357.
205. Pang, X., T. T. Chua, H. Y. Li, E. B. Liao, W. S. Lee, and F. X. Che, "Characterization and Management of Wafer Stress for Various Pattern Densities in 3D Integration Technology," *IEEE/ECTC Proceedings*, Las Vegas, NV, June 2010, pp. 1866–1869.
206. Zoschke, K., M. Wegner, M. Wilke, N. Jürgensen, C. Lopper, I. Kuna, V. Glaw, J. Röderl, O. Wünsch1, M. J. Wolf, O. Ehrmann, and H. Reichl, "Evaluation of Thin Wafer Processing Using a Temporary Wafer Handling System as Key Technology for 3D System Integration," *IEEE/ECTC Proceedings*, Las Vegas, NV, June 2010, pp. 1385–1392.
207. Charbonnier, J., R. Hida, D. Henry, S. Cheramy, P. Chausse, M. Neyret, O. Hajji, G. Garnier, C. Brunet-Manquat, P. Haumesser, L. Vandroux, R. Anciant, N. Sillon, A. Farcy, M. Rousseau, J. Cuzzocrea, G. Druais, and E. Saugier, "Development and Characterisation of a 3D Technology Including TSV and Cu Pillars for High Frequency Applications," *IEEE/ECTC Proceedings*, Las Vegas, NV, June 2010, pp. 1077–1082.
208. Sun, Y., X. Li, J. Gandhi, S. Luo, and T. Jiang, "Adhesion Improvement for Polymer Dielectric to Electrolytic-Plated Copper," *IEEE/ECTC Proceedings*, Las Vegas, NV, June 2010, pp. 1106–1111.
209. Kawano, M., N. Takahashi, M. Komuro, and S. Matsui, "Low-Cost TSV Process Using Electroless Ni Plating for 3D Stacked DRAM," *IEEE/ECTC Proceedings*, Las Vegas, NV, June 2010, pp. 1094–1099.
210. Malta, D., C. Gregory, D. Temple, T. Knutson, C. Wang, T. Richardson, Y. Zhang, and R. Rhoades, "Integrated Process for Defect-Free Copper Plating and Chemical-Mechanical Polishing of Through-Silicon Vias for 3D Interconnects," *IEEE/ECTC Proceedings*, Las Vegas, NV, June 2010, pp. 1769–1775.
211. Campbell, D., "Yield Modeling of 3D Integrated Wafer Scale Assemblies," *IEEE/ECTC Proceedings*, Las Vegas, NV, June 2010, pp. 1935–1938.
212. Archard, D., K. Giles, A. Price, S. Burgess, and K. Buchanan, "Low Temperature PECVD of Dielectric Films for TSV Applications," *IEEE/ECTC Proceedings*, Las Vegas, NV, June 2010, pp. 764–768.
213. Shigetou, A., and T. Suga, "Modified Diffusion Bonding for Both Cu and SiO_2 at 150°C in Ambient Air," *IEEE/ECTC Proceedings*, Las Vegas, NV, June 2010, pp. 872–877.
214. Amagai, M., and Y. Suzuki, "TSV Stress Testing and Modeling," *IEEE/ECTC*

Proceedings, Las Vegas, NV, June 2010, pp. 1273–1280.
215. Gupta, A., S. Kannan, B. Kim, F. Mohammed, and B. Ahn, "Development of Novel Carbon Nanotube TSV Technology," *IEEE/ECTC Proceedings*, Las Vegas, NV, June 2010, pp. 1699–1702.
216. Kannan, S., A. Gupta, B. Kim, F. Mohammed, and B. Ahn, "Analysis of Carbon Nanotube Based Through Silicon Vias," *IEEE/ECTC Proceedings*, Las Vegas, NV, June 2010, pp. 51–57.
217. Lu, K., S. Ryu, Q. Zhao, X. Zhang, J. Im, R. Huang, and P. Ho, "Thermal Stress Induced Delamination of Through Silicon Vias in 3D Interconnects," *IEEE/ECTC Proceedings*, Las Vegas, NV, June 2010, pp. 40–45.
218. Miyazaki, C., H. Shimamoto, T. Uematsu, Y. Abe, K. Kitaichi, T. Morifuji, and S. Yasunaga, "Development of High Accuracy Wafer Thinning and Pickup Technology for Thin Wafer (Die)," *Proceedings of IEEE CPMT Symposium Japan*, August 2010, pp. 139–142.
219. Cho, J., K. Yoon, J. Pak, J. Kim, J. Lee, H. Lee, K. Park, and J. Kim, "Guard Ring Effect for Through Silicon Via (TSV) Noise Coupling Reduction," *Proceedings of IEEE CPMT Symposium Japan*, August 2010, pp. 151–154.
220. Nonake, T., K. Fujimaru, A. Shimada, N. Asahi, Y. Tatsuta, H. Niwa, and Y. Tachibana, "Wafer and/or Chip Bonding Adhesives for 3D Package," *Proceedings of IEEE CPMT Symposium Japan*, August 2010, pp. 169–172.
221. Lau, J. H., "Heart and Soul of 3D IC Integration," posted at 3D InCites Conference, June 29, 2010; http://www.semineedle.com/posting/34277.
222. Lau, J. H., "Who Invented the TSV and When?" posted at 3D InCites Conference, April 24, 2010; http://www.semineedle.com/posting/31171.
223. Gat, A., L. Gerzberg, J. Gibbons, T. Mages, J. Peng, and J. Hong, "CW Laser of Polycrystalline Silicon: Crystalline Structure and Electrical Properties," *Applied Physics Letters*, Vol. 33, No. 8, October 1978, pp. 775–778.
224. Lau, J. H., *Flip Chip Technology*, McGraw-Hill, New York, 1996.
225. Lau, J. H., *Low Cost Flip Chip Technologies*, McGraw-Hill, New York, 2000.
226. Hu, G., H. Kalyanam, S. Krishnamoorthy, and L. Polka, "Package Technology to Address the Memory Bandwidth Challenge for Tera-Scale Computing," *INTEL Technology Journal*, Vol. 11, 2007, pp. 197–206.
227. Ong, Y., S. Ho, K. Vaidyanathan, V. Sekhar, M. Jong, S. Long, V. Lee, C. Leong, V. Rao, J. Ong, X. Ong, X. Zhang, Y. Seung, J. H. Lau, Y. Lim, D. Yeo, K. Chan, Z. Yanfeng, J. Tan, and D. Sohn. "Design, Assembly and Reliability of Large Die and Fine-Pitch Cu/Low-*k* Flip Chip Package," *Journal of Microelectronics Reliability*, Vol. 50, 2010, pp. 986–994.
228. Lau, J. H., and R. S. W. Lee, *Microvias for Low-Cost, High-Density Interconnects*, McGraw-Hill, New York, 2001.
229. Lau, J. H., *Ball-Grid Array Technology*, McGraw-Hill, New York, 1995.
230. Lau, J. H., T. Tseng, and D. Cheng, "Heat Spreader with a Placement Recess and Bottom Saw-Teeth for Connection to Ground Planes on a This Two-Side Single-Core BGA Substrate," U.S. Patent No. 6,057,601; granted May 2, 2000; filed November 27, 1998.
231. Lau, J. H., and K. L. Chen, "Thermal and Mechanical Evaluations of a Cost-Effective Plastic Ball Grid Array Package," *ASME Transactions, Journal of Electronic Packaging*, Vol. 119, September 1997, pp. 208–212.
232. Lau, J. H., and T. Chen, "Cooling Assessment and Distribution of Heat Dissipation of a Cavity Down Plastic Ball Grid Array Package: NuBGA," *IMAPS Transactions, International Journal of Microelectronics & Electronic Packaging*, Vol. 21, No. 1, 1998, pp. 20–28.
233. Lau, J. H., and T. Chou, "Electrical Design of a Cost-Effective Thermal Enhanced Plastic Ball Grid Array Package: NuBGA," *IEEE Transactions on CPMT, Part B*, Vol. 21, No. 1, February 1998, pp. 35–42.
234. Lau, J. H., T. Chen, and R. Lee, "Effect of Heat Spreader Sizes on the Thermal Performance of Large Cavity-Down Plastic Ball Grid Array Packages," *ASME Transactions, Journal of Electronic Packaging*, Vol. 121, No. 4, 1999, pp. 242–248.
235. Lau, J. H., "Design, Manufacturing, and Testing of a Novel Plastic Ball Grid Array Package," *Journal of Electronics Manufacturing*, Vol. 9, No. 4, December 1999, pp. 283–291.
236. Lau, J. H., and T. Chen, "Low-Cost Thermal and Electrical Enhanced Plastic Ball Grid Array Package: NuBGA," *Microelectronics International*, 1999.
237. Lau, J. H., "Solder Joint Reliability of Flip Chip and Plastic Ball Grid Array

Assemblies Under Thermal, Mechanical, and Vibration Conditions," *IEEE Transaction on CPMT, Part B*, Vol. 19, No. 4, November 1996, pp. 728–735.
238. Lau, J. H., with R. Lee, "Design for Plastic Ball Grid Array Solder Joint Reliability," *Journal of the Institute of Interconnection Technology*, Vol. 23, No. 2, January 1997, pp. 11–13.
239. Lau, J. H., with W. Jung and Y. Pao, "Nonlinear Analysis of Full-Matrix and Perimeter Plastic Ball Grid Array Solder Joints," *ASME Transactions, Journal of Electronic Packaging*, Vol. 119, September 1997, pp. 163–170.
240. Lau, J. H., with R. Lee, "Solder Joint Reliability of Cavity-Down Plastic Ball Grid Array Assemblies," *Journal of Soldering and Surface Mount Technology*, Vol. 10, No. 1, February 1998, pp. 26–31.
241. Lau, J. H., *Solder Joint Reliability: Theory and Applications*, Van Nostrand Reinhold, New York, 1991.

D. 2   3D MEMS 与 IC 集成

1. Madou, M. J., *Fundamentals of Microfabrication: The Science of Miniaturization*, CRC Press, Boca Raton, FL, 2002.
2. Nguyen, C., "MEMS Technology for Timing and Frequency Control," *IEEE Trans. Ultrason. Ferroelect. Freq. Contr.* 54:251–270, 2007.
3. Gad-el-Hak, M. *The Mems Handbook*, CRC Press, Boca Raton, FL, 2002.
4. Hacker, J. B., R. E. Mihailovich, M. Kim, and J. F. DeNatale, "A Ka-Band 3-Bit rf MEMS True-Time-Delay Network," *IEEE Trans. Microwave Theory Tech.* 51:305–308, 2003.
5. Menz, W., J. Mohr, and O. Paul, *Microsystem Technology*, Wiley-VCH, Hoboken, NJ, 2001.
6. Goldsmith, C. L., Z. Yao, S. Eshelman, and D. Denniston, "Performance of Low-Loss RF MEMS Capacitive Switches," *IEEE Microwave Wireless Compon. Lett.* 8:269–271, 1998.
7. Senturia, S. D., *Microsystem Design*, Springer, New York, 2000.
8. Rebeiz, G. M., *RF MEMS: Theory, Design and Technology*, Wiley, Hoboken, NJ, 2003.
9. Anagnostou, D. E., C. G. Christodoulou, G. Tzeremes, T. S. Liao, and P. K. L. Yu, "Fractal Antennas with RF-MEMS Switches for Multiple Frequency Applications," *Proceedings of the IEEE APS/URSI International Symposium*, Vol. 2, San Antonio, TX, June 2002, pp. 22–25.
10. Yano, M., F. Yamagishi, and T. Tsuda, "Optical MEMS for Photonic Switching Compact and Stable Optical Crossconnect Switches for Simple, Fast, and Flexible Wavelength Applications in Recent Photonic Networks," *J. Selected Topics Quantum Elect.* 11:383–394, 2005.
11. Anagnostou, D. E., G. Zheng, M. Chryssomallis, J. C. Lyke, G. E. Ponchak, J. Papapolymerou, and C. G. Christodoulou, "Design, Fabrication and Measurements of a Self-Similar Re-Configurable Antenna with RF-MEMS Switches," *IEEE Trans. Antennas Propagat.* 54:422–432, 2006.
12. Liu, A. Q., and X. M. Zhang, "A Review of MEMS External-Cavity Tunable Lasers," *J. Micromech. Microeng.* 17:R1–R13, 2007.
13. Huff, G. H., and J. T. Bernhard, "Integration of Packaged rf MEMS Switches with Radiation Pattern Reconfigurable Square Spiral Microstrip Antennas," *IEEE Trans. Antennas Propagat.* 54:464–469, 2006.
14. Kingsley, N., D. E. Anagnostou, M. Tentzeris, and J. Papapolymerou, "RF MEMS Sequentially Reconfigurable Sierpinski Antenna on a Flexible Organic Substrate with Novel DC-Biasing Technique," *IEEE/ASME J. Microelectromech. Syst.* 16:1185–1192, 2007.
15. Van Caekenberghe, K., and K. Sarabandi, "A 2-Bit Ka-Band RF MEMS Frequency Tunable Slot Antenna," *IEEE Antennas Wireless Propagat. Lett.* 7:179–182, 2008.
16. Nguyen, C., "MEMS Technology for Timing and Frequency Control," *IEEE Trans. Ultrason. Ferroelect. Freq. Contr.* 54:251–270, 2007.
17. Young, R. M., J. D. Adam, C. R. Vale, T. T. Braggins, S. V. Krishnaswamy, C.

E. Milton, D. W. Bever, L. G. Chorosinski, Li-Shu Chen, D. E. Crockett, C. B. Freidhoff, S. H. Talisa, E. Capelle, R. Tranchini, J. R. Fende, J. M. Lorthioir, and A. R. Tories, "Low-Loss Bandpass RF Filter Using MEMS Capacitance Switches to Achieve a One-Octave Tuning Range and Independently Variable Bandwidth," *IEEE MTT-S Int. Microwave Symp. Digest* 3:1781–1784, 2003.

18. Tan, G. L., R. E. Mihailovich, J. B. Hacker, J. F. DeNatale, and G. M. Rebeiz, "Low-Loss 2- and 4-bit TTD MEMS Phase Shifters Based on SP4T Switches," *IEEE Trans. Microwave Theory Tech.* 51:297–304, 2003.

19. Hacker, J. B., R. E. Mihailovich, M. Kim, and J. F. DeNatale, "A Ka-Band 3-Bit RF MEMS True-Time-Delay Network," *IEEE Trans. Microwave Theory Tech.* 51:305–308, 2003.

20. Ford, J. E., K. W. Goossen, J. A. Walker, D. T. Neilson, D. M. Tennant, S. Y. Park, and J. W. Sulhoff, "Interference-Based Micromechanical Spectral Equalizers," *IEEE J. Selected Topics Quantum Elect.* 10:579–587, 2004.

21. Nordquist, C. D., C. W. Dyck, G. M. Kraus, I. C. Reines, L. Goldsmith, D. Cowan, T. A. Plut, F. Austin, IV, P. S. Finnegan, M. H. Ballance, and T. Sullivan, "A DC to 10 GHz 6-Bit RF MEMS Time Delay Circuit," *IEEE Microwave Wireless Compon. Lett.* 16:305–307, 2006.

22. Perruisseau-Carrier, J., R. Fritschi, P. Crespo-Valero, and A. K. Skrivervik, "Modeling of Periodic Distributed MEMS Application to the Design of Variable True-Time-Delay Lines," *IEEE Trans. Microwave Theory Tech.* 54:383–392, 2006.

23. Kim, C.-H., N. Park, and Y.-K. Kim, "MEMS Reflective Type Variable Optical Attenuator Using Off-Axis Misalignment," *Proc. IEEE/LEOS Int. Conf. Opt. MEMS*, Lugano, Switzerland, 2002, pp. 55–56.

24. Lakshminarayanan, B., and T. M. Weller, "Design and Modeling of 4-Bit Slow wave MEMS Phase Shifters," *IEEE Trans. Microwave Theory Tech.* 54:120–127, 2006.

25. Lakshminarayanan, B., and T. M. Weller, "Optimization and Implementation of Impedance-Matched True-Time-Delay Phase Shifters on Quartz Substrate," *IEEE Trans. Microwave Theory Tech.* 55:335–342, 2007.

26. Van Caekenberghe, K., and T. Vaha-Heikkila, "An Analog RF MEMS Slotline True-Time-Delay Phase Shifter," *IEEE Trans. Microwave Theory Tech.* 56:2151–2159, 2008.

27. Maciel, J. J., J. F. Slocum, J. K. Smith, and J. Turtle, "MEMS Electronically Steerable Antennas for Fire Control Radars," *IEEE Aerosp. Electron. Syst. Mag.*, November 2007, pp. 17–20.

28. Yeow, T.-W., K. L. E. Law, and A. Goldenberg, "MEMS Optical Switches," *IEEE Commun. Mag.* 39:158–163, 2001.

29. Herrick, K. J., G. Jerinic, R. P. Molfino, S. M. Lardizabal, and B. Pillans, "S-Ku Band Intelligent Amplifier Microsystem," *Proc. SPIE* 6232, May 2006.

30. Neukermans, A., and R. Ramaswami, "MEMS Technology for Optical Networking," *IEEE Commun. Mag.* 39:62–69, 2001.

31. Pranonsatit, S., A. S. Holmes, I. D. Robertson, and S. Lucyszyn, "Single-Pole Eight-Throw RF MEMS Rotary Switch," *IEEE/ASME J. Microelectromech. Syst.* 15:1735–1744, 2006.

32. Lin, L. Y., and E. L. Goldstein, "Opportunities and Challenges for MEMS in Lightwave Communications," *IEEE J. Selected Topics Quantum Elect.* 8:163–172, 2002.

33. Vaha-Heikkila, T., K. Van Caekenberghe, J. Varis, J. Tuovinen, and G. M. Rebeiz, "RF MEMS Impedance Tuners for 6–24-GHz Applications," *Wiley Int. J. RF Microwave Computer-Aided Eng.* 17: 265–278, 2007.

34. Syms, R. A., and D. F. Moore, "Optical MEMS for Telecoms," *Materials Today* 5:26–35, 2002.

35. Schoebel, J., T. Buck, M. Reimann, M. Ulm, M. Schneider, A. Jourdain, G. Carchon, and H. Tilmans, "Design Considerations and Technology Assessment of Phased Array Antenna Systems with RF MEMS for Automotive Radar Applications," *IEEE Trans. Microwave Theory Tech.* 53:1968–1975, 2005.

36. Wu, M. C., O. Solgaard, and J. E. Ford, "Optical MEMS for Lightwave Communication," *J. Lightwave Technol.* 24:4433–4454, 2006.

37. Mailloux, R. J. *Phased Array Antenna Handbook*, Artech House, New York, 2005.

38. Hoffmann. M., and E. Voges, "Bulk Silicon Micromachining for MEMS in Optical Communication Systems," *J. Micromech. Microeng.* 12:349–360, 2002.

39. Jung, C., M. Lee, G. P. Li, and F. D. Flaviis, "Reconfigurable Scan-Beam Singlearm Spiral Antenna Integrated with RF MEMS Switches," *IEEE Trans. Antennas Propagat.* 54:455–463, 2006.
40. Chang-Hasnain, C. J. "Tunable VCSEL," *J. Selected Topics Quantum Elect.* 6: 978–987, 2000.
41. Lijie, J. Z., and D. Uttamchandani, "Integrated Self-Assembling and Holding Technique Applied to a 3D MEMS Variable Optical Attenuator," *IEEE J. Microelectromech. Syst.* 13:83–90, 2004.
42. Wikipedia, "Microelectromechanical Systems," available at http://en.wikipedia.org/wiki/MEMS; accessed April 1, 2008.
43. Yole Development, "World MEMS Markets: The 2006–2012 MEMS Market Database," 2008.
44. Lau, J. H., C. P. Wong, N. C. Lee, and R. Lee, *Electronics Manufacturing with Lead-Free, Halogen-Free, and Adhesive Materials*, McGraw-Hill, New York, 2003.
45. Lau, J. H., and R. Lee, *Microvias for Low-Cost, High-Density Interconnects*, McGraw-Hill, New York, 2001.
46. Lau, J. H., *Low-Cost Flip-Chip Technologies for DCA, WLCSP, and PBGA Assemblies*, McGraw-Hill, New York, 2000.
47. Lau, J. H., and R. Lee, *Chip Scale Package Design: Materials, Process, Reliability, and Applications*, McGraw-Hill, New York, 1999.
48. Lau, J. H., C. P. Wong, J. Prince, and W. Nakayama, *Electronic Packaging: Design, Materials, Process, and Reliability*, McGraw-Hill, New York, 1998.
49. Lau, J. H., and Y. Pao, *Solder Joint Reliability of BGA, CSP, Flip Chip, and Fine Pitch SMT Assemblies*, McGraw-Hill, New York, 1997.
50. Lau, J. H., *Flip Chip Technologies*, McGraw-Hill, New York, 1996.
51. Lau, J. H., *Ball Grid Array Technology*, McGraw-Hill, New York, 1995.
52. Lau, J. H., *Chip on Board Technologies for Multichip Modules*, Van Nostrand Reinhold, New York, 1994.
53. Lau, J. H., *Handbook of Fine Pitch Surface Mount Technology*, Van Nostrand Reinhold, New York, 1994.
54. Lau, J. H., *Thermal Stress and Strain in Microelectronics Packaging*, Van Nostrand Reinhold, New York, 1993.
55. Lau, J. H., *Handbook of Tape Automated Bonding*, Van Nostrand Reinhold, New York, 1992.
56. Lau, J. H., *Solder Joint Reliability: Theory and Applications*, Van Nostrand Reinhold, New York, 1991.
57. Mohamed Gad-el-Hak. *The MEMS Handbook*, CRC Press, Boca Raton, FL, 2002.
58. Van Caekenberghe, K., and K. Sarabandi, "A 2-Bit Ka-Band RF MEMS Frequency Tunable Slot Antenna," *IEEE Antennas and Wireless Propagat. Lett.* 7:179–182, 2008.
59. Premachandran, C. S., M. Chew, W. Choi, A. Khairyanto, K. Chen, J. Singh, S. Wang, Y. Xu, N. Chen, C. Sheppard, M. Olivo, and J. H. Lau, "Influence of Optical Probe Packaging on a 3D MEMS Scanning Micro Mirror for Optical Coherence Tomography (OCT) Applications," *IEEE Proceedings of Electronic Components and Technology Conference*, Orlando, FL, May 27–30, 2008, pp. 829–833.
60. Premachandran, C. S., J. H. Lau, X. Ling, A. Khairyanto, K. Chen, and Myo Ei Pa Pa, "A Novel, Wafer-Level Stacking Method for Low-Chip Yield and Non-Uniform, Chip-Size Wafers for MEMS and 3D SiP Applications," *IEEE Proceedings of Electronic Components and Technology Conference*, Orlando, FL, May 27–30, 2008, pp. 314–318.
61. Lau, J. H., "3D MEMS Packaging," *IMAPS Proceedings*, San Jose, CA, November 2009.
62. Chen, K., C. Premachandran, K. Choi, C. Ong, X. Ling, A. Ratmin, M. Pa, and J. H. Lau, "C2W Low Temperature Bonding Method for MEMS Applictions," *IEEE Proceedings of Electronics Packaging Technology Conference*, Singapore, December 2008, pp. 1–7.
63. Klumpp, A., "Vertical System Integration by Using Inter-Chip Vias and Solid-Liquid Interdiffusion Bonding," *Jpn. J. Appl. Phys.* 43:L829–L830, 2004.
64. Chen, K., "Microstructure Examination of Copper Wafer Bonding," *J. Electron. Mater.* 30:331–335, 2001.
65. Morrow, P. R., "Three-Dimensional Wafer Stacking via Cu-Cu Bonding Integrated with 65-nm Strained-Si/Low-*k* CMOS Technology," *IEEE Electron.*

Device Lett. 27:335–337, 2006.
66. Shimbo, M., K. Furukawa, K. Fukuda, and K. Tanzawa, "Silicon-to-Silicon Direct Bonding Method," *J. Appl. Phys.* 60:2987–2989, 1986.
67. Made, R., C. L. Gan, L. Yan, A. Yu, S. U. Yoon, J. H. Lau, and C. Lee, "Study of Low Temperature Thermocompression Bonding in Ag-In Solder for Packaging Applications," *J. Electron. Mater.* 38:365–371, 2009.
68. Yan, L.-L., C.-K. Lee, D.-Q. Yu, A.-B. Yu, W.-K. Choi, J. H. Lau, and S.-U. Yoon, "A Hermetic Seal Using Composite Thin Solder In/Sn as Intermediate Layer and Its Interdiffusion Reaction with Cu," *J. Electron. Mater.* 38:200–207, 2009.
69. Yan, L.-L., V. Lee, D. Yu, W. K. Choi, A. Yu, S.-U. Yoon, and J. H. Lau, "A Hermetic Chip to Chip Bonding at Low Temperature with Cu/In/Sn/Cu Joint," *IEEE/ECTC Proceedings,* Orlando, FL, May 2008, pp. 1844–1848.
70. Yu, A., C. Lee, L. Yan, R. Made, C. Gan, Q. Zhang, S. Yoon, and J. H. Lau, "Development of Wafer Level Packaged Scanning Micromirrors," *Proc. Photon. West* 6887:1–9, 2008.
71. Lee, C., A. Yu, L. Yan, H. Wang, J. Han, Q. Zhang, and J. H. Lau, "Characterization of Intermediate In/Ag Layers of Low Temperature Fluxless Solder Based Wafer Bonding for MEMS Packaging," *J. Sensors Actuators* (in press).
72. Yu, D.-Q., C. Lee, L. L. Yan, W. K. Choi, A. Yu, and J. H. Lau, "The Role of Ni Buffer Layer on High-Yield, Low-Temperature Hermetic Wafer Bonding Using In/Sn/Cu Metallization," *Appl. Phys. Lett.* 94 (3):034105 – 034105-3, 2009.
73. Yu, D. Q., L. L. Yan, C. Lee, W. K. Choi, S. U. Yoon, and J. H. Lau, "Study on High Yield Wafer to Wafer Bonding Using In/Sn and Cu Metallization," *Proceedings of the Eurosensors Conference,* Dresden, Germany, 2008, pp. 1242–1245.
74. Yu, D., C. Lee, and J. H. Lau, "The Role of Ni Buffer Layer Between InSn Solder and Cu Metallization for Hermetic Wafer Bonding," *Proceedings of the International Conference on Electronics Materials and Packaging,* Taipei, Taiwan, October 22–24, 2008, pp. 335–338.
75. Yu, D., L. Yan, C. Lee, W. Choi, M. Thew, C. Foo, and J. H. Lau, "Wafer Level Hermetic Bonding Using Sn/In and Cu/Ti/Au Metallization," *IEEE/ECTC Proceeding,* Singapore, December 2008, pp. 1–6.
76. Choi, W., D. Yu, C. Lee, L. Yan, A. Yu, S. Yoon, J. H. Lau, M. Cho, Y. Jo, and H. Lee, "Development of Low Temperature Bonding Using In-Based Solders," *IEEE/ECTC Proceedings,* Orlando, FL, May 2008, pp. 1294–1299.
77. Choi, W., C. Premachandran, C. Ong, X. Ling, E. Liao, A. Khairyanto, K. Chen, K., P. Thaw, and J. H. Lau, "Development of Novel Intermetallic Joints Using Thin Film Indium Based Solder by Low Temperature Bonding Technology for 3D IC Stacking," *IEEE/ECTC Proceedings,* San Diego, CA, May 2009, pp. 333–338.
78. Chen, K., C. Premachandran, K. Choi, C. Ong, X. Ling, A. Khairyanto, B. Ratmin, P. Myo, and J. H. Lau, "C2W Bonding Method for MEMS Applications, *IEEE/ECTC Proceedings,* Singapore, December 2008, pp. 1283–1287.
79. Premachandran, C. S., J. H. Lau, X. Ling, A. Khairyanto, K. Chen, and Myo Ei Pa Pa. "A Novel, Wafer-Level Stacking Method for Low-Chip Yield and Non-Uniform, Chip-Size Wafers for MEMS and 3D SIP Applications," *IEEE Proceedings of Electronic Components and Technology Conference,* Orlando, FL, May 27–30, 2008, pp. 314–318.
80. Simic, V., and Z. Marinkovic, "Room Temperature Interactions in Ag–Metals Thin Film Couples." *Thin Solid Films* 61:149–160, 1979.
81. Lin, J.-C., "Solid-Liquid Interdiffusion Bonding Between In-Coated Silver Thick Films," *Thin Solid Films* 61:212–221, 1979.
82. Roy, R., and S. K. Sen, "The Kinetics of Formation of Intermetallics in Ag/In Thin Film Couples," *Thin Solid Films* 61:303–318, 1979.
83. Chuang, R. W., and C. C. Lee, "Silver-Indium Joints Produced at Low Temperature for High Temperature Devices," *IEEE Trans. Components Packag. Technol.* A25:453–458, 2002.
84. Chuang, R. W., and C. C. Lee, "High-Temperature Non-Eutectic Indiumtin Joints Fabricated by a Fluxless Process," *Thin Solid Films* 414:175–179, 2002.
85. Lee, C. C., and R. W. Chuang, "Fluxless Non-eutectic Joints Fabricated using gold-tin multilayer composite," *IEEE Trans. Components Packag. Technol.* A26:416–422, 2003.
86. Humpston, G., and D. Jacobson, *Principles of Soldering and Brazing,* ASM

International, Materials Park, MD, 1993, pp 128–132.
87. Chuang, T., H. Lin, and C. Tsao, "Intermetallic Compounds Formed During Diffusion Soldering of Au/Cu/Al$_2$O$_3$ and Cu/Ti/Si with Sn/In Interlayer," *J. Electron. Mater.* 35:1566–1570, 2006.
88. Lee, C., and S. Choe, "Fluxless In-Sn Bonding Process at 140°C," *Mater. Sci. Eng.* A333:45–50, 2002.
89. Lee, C. "Wafer Bonding by Low-Temperature Soldering," *Sensors & Actuators* 85:330–334, 2000.
90. Tazzoli, A., M. Rinaldi, and G. Piazza, "Ovenized High Frequency Oscillators Based on Aluminum Nitride Contour-Mode MEMS Resonators," *IEEE/IEDM Proceedings*, December 2011, pp. 20.2.1–20.2.4.
91. Hwang, E., A. Driscoll, and S. Bhave, "Platform for JFET-Based Sensing of RF MEMS Resonators in CMOS Technology," *IEEE/IEDM Proceedings*, December 2011, pp. 20.4.1–20.4.4.
92. Liu, H., C. Lee1, T. Kobayashi, C. Tay, and C. Quan, "A MEMS-Based Wideband Piezoelectric Energy Harvester System Using Mechanical Stoppers," *IEEE/IEDM Proceedings*, December 2011, pp. 29.6.1–29.6.4.
93. Vigna, B., "Tri-Axial MEMS Gyroscopes and Six Degree-Of-Freedom Motion Sensors," *IEEE/IEDM Proceedings*, December 2011, pp. 29.1.1–29.1.4.
94. Elfrink, R., S. Matova, C. de Nooijer, M. Jambunathan, M. Goedbloed, J. van de Molengraft, V. Pop, R.J.M. Vullers, M. Renaud, and R. van Schaijk, "Shock Induced Energy Harvesting with a MEMS Harvester for Automotive Applications," *IEEE/IEDM Proceedings*, December 2011, pp. 29.5.1–29.5.4.
95. Lin, Y., T. Riekkinen, W. Li, E. Alon, and C. Nguyen, "A Metal Micromechanical Resonant Switch for On-Chip Power Applications," *IEEE/IEDM Proceedings*, December 2011, pp. 20.6.1–20.6.4.
96. Sdeghi, M., R. Peterson, and K. Najafi, "Micro-Hydraulic Structure for High Performance Bio-Mimetic Air Flow Sensor Arrays," *IEEE/IEDM Proceedings*, December 2011, pp. 29.4.1–29.4.4.
97. Bhugra, H., Y. Wang, W. Pan, D. Lei, and S. Lee, "High Performance pMEMS Oscillators: The Next Generation Frequency," *IEEE/IEDM Proceedings*, December 2011, pp. 20.1.1–20.1.4.
98. Vullers, R., R. Schaijk, M. Goedbloed, R. Elfrink, Z. Wang, and C. Hoof, "Process Challenges of MEMS Harvesters and Their Effect on Harvester Performance," *IEEE/IEDM Proceedings*, December 2011, pp. 10.2.1–10.2.4.
99. Lau, J. H., *Reliability of RoHS-Compliant 2D and 3D IC Interconnects*, McGraw-Hill, New York, 2011.
100. Lai, J. H., C. Lee, C. Premachandran, and A. Yu, *Advanced MEMS Packaging*, McGraw-Hill, New York, 2010.

D.3 半导体 IC 封装

1. Lau, J. H., *Through-Silicon Vias for 3D Integration*, McGraw-Hill, New York, 2013.
2. Lau, J. H., *Reliability of RoHS-Compliant 2D and 3D IC Interconnects*, McGraw-Hill, New York, 2011.
3. Lau, J. H., C. K. Lee, C. S. Premachandran, and A. Yu, *Advanced MEMS Packaging*, McGraw-Hill, New York, 2010.
4. Lau, J. H., C. P. Wong, N. C. Lee, and S. W. Lee, *Electronics Manufacturing with Lead-Free, Halogen-Free and Conductive-Adhesive Materials*, McGraw-Hill, New York, 2003.
5. Lau, J. H, and S. W. Ricky Lee, *Microvias for Low Cost, High Density Interconnects*, McGraw-Hill, New York, 2001.
6. Lau, J. H., *Low-Cost Flip-Chip Technologies*, McGraw-Hill, New York, 2000.
7. Lau, J. H., and S. W. R. Lee, *Chip-Scale Package*, McGraw-Hill, New York, 1999.
8. Lau, J. H., C. P. Wong, J. Prince, and W. Nakayama, *Electronic Packaging: Design, Materials, Process, and Reliability*, McGraw-Hill, New York, 1998.
9. Lau, J. H. and Y.-H. Pao, *Solder Joint Reliability of BGA, CSP, Flip Chip and Fine*

Pitch SMT Assemblies, McGraw-Hill, New York, 1997.
10. Lau, J. H. (ed.), *Flip Chip Technologies*, McGraw-Hill, New York, 1996.
11. Lau, J. H. (ed.), *Ball Grid Array Technology*, McGraw-Hill, New York, 1995.
12. Lau, J. H. (ed.), *Chip-on-Board: Technologies for Multichip Modules*, Van Nostrand Reinhold, New York, 1994.
13. Lau, J. H. (ed.), *Handbook of Fine Pitch Surface Mount Technology*, Van Nostrand Reinhold, New York, 1994.
14. Frear, D., H. Morgan, S. Burchett, and J. H. Lau (eds.), *The Mechanics of Solder Alloy Interconnects*, Van Nostrand Reinhold, New York, 1994.
15. Lau, J. H. (ed.), *Thermal Stress and Strain in Microelectronics Packaging*, Van Nostrand Reinhold, New York, 1993.
16. Lau, J. H. (ed.), *Handbook of Tape Automated Bonding*, Van Nostrand Reinhold, New York, 1992.
17. Lau, J. H. (ed.), *Solder Joint Reliability: Theory and Applications*, Van Nostrand Reinhold, New York, 1991.
18. Tummala, R. R., and Madhavan Swaminathan, *System on Package: Miniaturization of the Entire System*, McGraw-Hill, New York, 2008.
19. Tummala, R. R., *Fundamentals of Microsystems Packaging*, McGraw-Hill, New York, 2001.
20. Tummala, R. R., and E. J. Rymaszewski (eds.), *Microelectronics Packaging Handbook*, Van Nostrand Reinhold, New York, 1989.
21. Tummala, R. E., E. J. Rymasewski, and A. G. Klopfenstein (eds.), *Microelectronics Packaging Handbook: Semiconductor Packaging*, Part II, 2nd ed., Chapman & Hall, New York, 1997.
22. Tummala, R. E., E. J. Rymasewski, and A. G. Klopfenstein (eds.), *Microelectronics Packaging Handbook: Subsystem Packaging*, Part III, 2nd ed., Chapman & Hall, New York, 1997.
23. Wong, C. P., *Polymers for Electronic and Photonic Application*, Academic Press, New York, 1993.
24. Lu, D., and C. P. Womg (eds.), *Materials for Advanced Packaging*, Springer, New York, 2008.
25. Suhir, E., Y. C. Lee, and C. P. Wong (eds.), *Micro- and Opto-Electronic Materials and Structures: Physics, Mechanics, Design, Reliability, Packaging*, Springer, New York, 2007.
26. Garrou, P., C. Bower, and P. Ramm (eds.), *Handbook of 3D Integration*, Wiley, New York, 2008.
27. Wu, B., A. Kumar, and S. Ramaswami (eds.), *3D IC Stacking Technology*, McGraw-Hill, New York, 2011.
28. Xie, Y., J. Cong, and S. Sapatnekar (eds.), *Three-Dimensional Integrated Circuit Design*, Springer, New York, 2010.
29. Flick, E. W., *Adhesives, Sealants and Coatings for the Electronics Industry*, 2nd ed., Noyes Publications, Park Ridge, NJ, 1992.
30. Bar-Cohen, A., and A. D. Kraus (eds.), *Advances in Thermal Modeling of Electronic Components and Systems*, Vol. 1, Hemisphere Publishing Corporation, New York, 1988.
31. Bar-Cohen, A., and A. D. Kraus (eds.), *Advances in Thermal Modeling of Electronic Components and Systems*, Vol. 2, ASME Press, New York, 1990.
32. Bar-Cohen, A., and A. D. Kraus (eds.), *Advances in Thermal Modeling of Electronic Components and Systems*, Vol. 3, ASME Press, New York, 1993.
33. Bar-Cohen, A., and A. D. Kraus (eds.), *Advances in Thermal Modeling of Electronic Components and Systems*, Vol. 4, ASME Press, New York, 1998.
34. Shockley, W., *Electrons and Holes in Semiconductors: With Applications to Transistor Electronics*, Van Nostrand Co., New York, 1950.
35. Terano, T., K. Asai, and M. Sugeno, *Applied Fuzzy Systems*, Academic Press Professional, Cambridge, MA, 1989.
36. Hwang, J. S., *Ball Grid Array and Fine Pitch Peripheral Interconnections: A Handbook of Technology and Applications for Microelectronics/Electronics Manufacturing*, Electrochemical Publications, Ayr, Scotland, 1995.
37. Rahn, A., *The Basics of Soldering*, Wiley, New York, 1993.
38. Moore, T. M., and R. G. McKenna (eds.), *Characterization of Integrated Circuit Packaging Materials*, Butterworth-Heinemann, Boston, MA, 1993.
39. Bakoglu, H. B., *Circuits, Interconnections, and Packaging for VSLI*, Addison-Wesley, New York, 1990.

40. Solberg, V., *Design Guidelines for Surface Mount and Fine Pitch Technology*, 2nd ed., McGraw-Hill, New York, 1995.
41. Parker, C. B. (ed.), *Dictionary of Scientific and Technical Terms*, 5th ed., McGraw-Hill, New York, 1994.
42. Kavanagh, P., *Downsizing for Client/Server Applications*, Academic Press Professional, Cambridge, MA, 1995.
43. ASM International Handbook Committee, *Electronic Materials Handbook:* Vol. I, *Packaging,* ASM, Materials Park, OH, 1989.
44. Landers, T. L., W. D. Brown, E. W. Fant, E. M. Malstrom, and N. M. Schmitt, *Electronics Manufacturing Processes*, Prentiss-Hall, Englewood Cliffs, NJ, 1994.
45. Baker, D., D. C. Koehler, W. O. Fleckenstein, C. E. Roden, and R. Sabia, *Physical Design of Electronic Systems:* Vol. III: Integrated Device and Connection Technology, Prentice-Hall, Englewood Cliffs, NJ, 1971.
46. Keyser, C. A., *Materials: Science in Engineering*, 2nd ed., Merrill Publishing Co., Columbus, OH, 1968.
47. Harper, C. A., and R. M. Sampson, *Electronic Materials and Processes Handbook*, 2nd ed., McGraw-Hill, New York, 1994.
48. Harper, C. A. (ed.), *Electronic Packaging and Interconnection Handbook*, McGraw-Hill, New York, 1991.
49. Klir, G. J., and T. A Folger, *Fuzzy Sets, Uncertainty, and Information*, Prentice-Hall, Englewood Cliffs, NJ, 1988.
50. Morris, J. E. (ed.), *Electronics Packaging Forum*, Vol. 1, Van Nostrand Reinhold, New York, 1990.
51. Dieter, G. E., *Engineering Design: A Materials and Processing Approach*, McGraw-Hill, New York, 1991.
52. Morris, J. E. (ed.), *Electronics Packaging Forum*, Vol. 2, Van Nostrand Reinhold, New York, 1991.
53. Ruoff, A. L., *Introduction to Materials Science*, Prentice-Hall, Englewood Cliffs, NJ, 1972.
54. Morris, J. E. (ed.), *Electronics Packaging Forum: Multichip Module Technology Issues*, IEEC Press, New York, 1994.
55. Tittel, E., and M. Robbins, *E-mail Essentials*, Academic Press Professional, Cambridge, MA, 1994.
56. *Engineered Materials Handbook*: Vol. 1, *Composites*, ASM International, New York, 1987.
57. *Engineered Materials Handbook:* Vol. 2, *Engineering Plastics*, ASM International, New York, 1988.
58. *Engineered Materials Handbook:* Vol. 3, *Adhesives and Sealants*, ASM International, New York, 1990.
59. *Engineered Materials Handbook:* Vol. 4, *Ceramics and Glasses*, ASM International, New York, 1991.
60. Marcoux, P. P., *Fine Pitch Surface Mount Technology: Quality, Design, and Manufacturing Techniques*, Van Nostrand Reinhold, New York, 1992.
61. Fjelstad, J., *Flexible Circuit Technology*, Silicon Valley Publishers Group, Campbell, CA 1994.
62. Stearns, T. H., *Flexible Printed Circuitry*, McGraw-Hill, New York, 1996.
63. McNeill, F. M., and E. Thro, *Fuzzy Logic: A Practical Approach*, Academic Press Professional, Cambridge, MA, 1994.
64. Cox, E., *The Fuzzy Systems Handbook: A Practitioner's Guide to Building, Using, and Maintaining Fuzzy Systems*, Academic Press Professional, Cambridge, MA, 1994.
65. *General Requirements for Implementation of Statistical Process Control*, ANSI/IPC-PC-90, IPC, Lincolnwood, IL, October 1990.
66. *Guidelines for Multichip Module Technology Utilization*, IPC-MC-790, IPC, Lincolnwood, IL, July 1992.
67. Pecht, M. (ed.), *Handbook of Electronic Package Design*, Marcel Dekker, New York, 1991.
68. Matisoff, B. S., *Handbook of Electronics Packaging Design and Engineering*, 2nd ed., Van Nostrand Reinhold, New York, 1990.
69. Harper, C. A. (ed.), *Handbook of Plastics, Elastomers, and Composites*, 2nd ed., McGraw-Hill, New York, 2002.
70. Johnson, H. W., and M. Graham, *High-Speed Digital Design: A Handbook of Black Magic*, Prentice-Hall, Englewood Cliffs, NJ, 1993.

71. Licari, J. J., and L. R. Enlow, *Hybrid Microcircuit Technology Handbook: Materials, Processes, Design, Testing and Production*, Noyes Publications, Park Ridge, NJ, 1988.
72. Sergent, J. E., and C. A. Harper (eds.), *Hybrid Microelectronics Handbook*, 2nd ed., McGraw-Hill, Inc., New York, 1995.
73. Pecht, M., *Integrated Circuit, Hybrid, and Multichip Module Package Design Guidelines: A Focus on Reliability*, Wiley, New York, 1994.
74. O'Mara, W. C., *Liquid Crystal Flat Panel Display: Manufacturing Science and Technology*, Van Nostrand Reinhold, New York, 1993.
75. Kirschman, R. K. (ed.), *Low-Temperature Electronics*, IEEE Press, New York, 1986.
76. *Mastering and Implementing BGA Technology*, AEIC, New York, 1995.
77. Yost, F. G., F. M. Hosking, and D. R. Frear, *The Mechanics of Solder Alloy Wetting and Spreading*, Van Nostrand Reinhold, New York, 1993.
78. Grovenor, C. R. M., *Microelectronic Materials*, Institute of Physics Publishing, Washington, DC, 1994.
79. Tewksbury, S. K. (ed.), *Microelectronic System Interconnections: Performance and Modeling*, IEEE Press, New York, 1994.
80. Landzberg, A. H. (ed.), *Microelectronics Manufacturing Diagnostics Handbook*, Van Nostrand Reinhold, New York, 1993.
81. Motorola Semiconductor Products, Inc., *Microprocessor Applications Manual*, McGraw-Hill, New York, 1975.
82. Lowenheim, F. (ed.), *Modern Electroplating*, 3rd ed., Wiley, New York, 1974.
83. Doane, D. A., and Franzon, P. D., *Multichip Module Technologies and Alternatives: The Basics*, Van Nostrand Reinhold, New York, 1993.
84. Johnson, R. W., R. K. F. Teng, and J. W. Balde (eds.), *Multichip Modules: Systems Advantages, Major Constructions, and Materials Technologies*. IEEE Press, New York, 1991.
85. Rao, G. K., *Multilevel Interconnect Technology*, McGraw-Hill, New York, 1993.
86. Tittel, E., and M. Robbins, *Network Design Essentials: Everything You Need to Know*, Academic Press Professional, Cambridge, MA, 1994.
87. Simon, A. R., *Network Re-Engineering: Foundations of Enterprise Computing*, Academic Press Professional, Cambridge, MA, 1994.
88. Beranek, L. L. (ed.), *Noise and Vibration Control*, rev. ed., McGraw-Hill, New York, 1971.
89. Montgomery, S. L., *Object-Oriented Information Engineering: Analysis, Design, and Implementation*, Academic Press Professional, Cambridge, MA, 1994.
90. Simon, A. R., and T. Wheeler, *Open Systems Handbook*, 2nd ed., Academic Press Professional, Cambridge, MA., 1995.
91. Dally, J. W., *Packaging of Electronic Systems: A Mechanical Engineering Approach*, McGraw-Hill, New York, 1990.
92. Neugebauer, C. A., A. F. Yerman, R. O. Carlson, J. F. Burgess, H. F. Webster, and H. H. Glascock, *The Packaging of Power Semiconductor Devices*, Electrocomponent Science Monographs, Vol. 7, Gordon and Breach Science Publishers, New York, 1986.
93. Hannemann, R. J., A. D. Kraus, and M. Pecht (eds.), *Physical Architecture of VLSI Systems*, Wiley, New York, 1994.
94. Pecht, M., *Placement and Routing of Electronic Modules*, Marcel Dekker, New York, 1993.
95. Pecht, M. G., L. T. Nguyen, and E. B. Hakim (eds.), *Plastic-Encapsulated Microelectronics: Materials, Processes, Quality, Reliability, and Applications*, Wiley, New York, 1995.
96. Manzione, L. T., *Plastic Packaging of Microelectronic Devices*, Van Nostrand Reinhold, New York, 1990.
97. Seraphim, D. P., R. Lasky, and C.-Y. Li, *Principles of Electronic Packaging*, McGraw-Hill, New York, 1989.
98. Coombs, C. F., Jr., *Printed Circuits Handbook*, 4th ed., McGraw-Hill, New York, 1996.
99. Stevens, R. T., *Quick Reference to Computer Graphics Terms*, Academic Press Professional, Cambridge, MA, 1993.
100. *Reliability Assessment of Wafer Scale Integration Using Finite Element Analysis*, Rome Laboratory, Griffiss AFB, New York, October 1991.
101. Harman, George G., *Reliability and Yield Problems of Wire Bonding in Microelectronics*, International Society for Hybrid Microelectronics,

Washington, DC, 1991.
102. *Rome Laboratory Reliability Engineer's Toolkit: An Application Oriented Guide for the Practicing Reliability Engineer*, Systems Reliability Division, AFMC, Griffiss AFB, New York, 1993.
103. Frear, D. R., W. B. Jones, and K. R. Kinsman (ed.), *Solder Mechanics: A State of the Art Assessment*, Minerals, Metals and Materials Society, Washington, DC, 1990.
104. Manko, H. H., *Soldering Handbook for Printed Circuits and Surface Mounting*, 2nd ed., Van Nostrand Reinhold, New York, 1994.
105. Judd, M., and K. Brindley, *Soldering in Electronics Assembly*, BH Newness, East Kilbride, Scotland, 1992.
106. Pecht, M. (ed.), *Soldering Processes and Equipment*, Wiley, New York, 1993.
107. Manko, H. H., *Solders and Soldering: Materials, Design, Production, and Analysis for Reliable Bonding*, McGraw-Hill, New York, 1992.
108. Christian, J., and G.-A. Nazri, *Solid State Batteries: Materials Design and Optimization*, Kluwer Academic Publishers, Boston, MA 1994.
109. Suhir, E., *Structural Analysis in Microelectronic and Fiber-Optic Systems*: Vol. I, Van Nostrand Reinhold, New York, 1991.
110. Engel, P. A., *Structural Analysis of Printed Circuit Board Systems*, Mechanical Engineering Series, Springer-Verlag, New York, 1993.
111. Caswell, G. (organizer), *Surface Mount Technology*, International Society for Hybrid Microelectronics, Silver Spring, MD, 1984,.
112. Classon, F., *Surface Mount Technology for Concurrent Engineering and Manufacturing*, McGraw-Hill, New York, 1993.
113. Vardaman, J., *Surface Mount Technology: Recent Japanese Developments*, IEEE Press, New York, 1993.
114. Furkay, S. S., R. F. Kilburn, G. Monti, Jr. (eds.), *Thermal Management Concepts in Microelectronic Packaging: from Component to System*, International Society for Hybrid Microelectrinics, Silver Spring, MD, 1984.
115. Messner, G., I. Turlik, J. W. Balde, and P. E. Garrou, *Thin Film: Multichip Modules*, International Society of Hybrid Microelectronics, Reston, VA, 1992.
116. Perry, D. L., *VHDL*, 2nd ed., McGraw-Hill, New York, 1994.
117. Pick, J., *VHDL Techniques, Experiments, and Caveats*, McGraw-Hill, New York, 1996.
118. Steinberg, D. S., *Vibration Analysis for Electronic Equipment*, 2nd ed., Wiley, New York, 1988.
119. Bar-Cohen, A., and A. D. Kraus (eds.), *Advances in Thermal Modeling of Electronic Components and Systems*, Vol. 3, ASME Press, New York, 1993.
120. Leidheiser, H., Jr., *Corrosion Control by Coatings*, Science Press, Princeton, NJ, 1979.
121. Koch, W. E., *Engineering Applications of Lasers and Holography*, Plenum Press, New York, 1975.
122. Hinch, S. W., *Handbook of Surface Mount Technology*, Longman Scientific & Technical, Essex, England, 1988.
123. Mandelkern, L., *An Introduction to Macromolecules*, Springer-Verlag, New York, 1972.
124. Brisky, M., *Mastering SMT Manufacturing*, SMT Plus, Inc., New York, 1992.
125. Chung, D. D. L., *Materials for Electronic Packaging*, Butterworth-Heinemann, Boston, MA 1995.
126. Lea, C., *A Scientific Guide to Surface Mount Technology*, Electrochemical Publications, Ayr, Scotland, 1988.
127. Prasad, Ray P., *Surface Mount Technology: Principles and Practice*, Van Nostrand Reinhold, New York, 1989.
128. Prasad, R. P., *Surface Mount Technology: Principles and Practice*, 2nd ed., Chapman & Hall, New York, 1997.
129. Lall, P., M. G. Pecht, and E. B. Hakim, *Influence of Temperature on Microelectronics and System Reliability*, CRC Press, Boca Raton, FL, 1997.
130. Harper, C. A., *Electronic Packaging and Interconnection Handbook*, 2nd ed., McGraw-Hill, New York, 1997.
131. Woodgate, R. W., *The Handbook of Machine Soldering: SMT and TH*, 3rd ed., Wiley, New York, 1996.
132. Hwang, J. S., *Modern Solder Technology for Competitive Eelctronics Manufacturing*,

McGraw-Hill, New York, 1996.
133. Licari, J. J., *Multichip Module Design, Fabrication, & Testing*, McGraw-Hill, New York, 1995.
134. Alvino, W. M., *Plastics for Electronics: Materials, Properties, and Design Applications*, McGraw-Hill, New York, 1995.
135. Pecht, M., A. Dasgupta, J. W. Evans, and J. Y. Evans (eds.), *Quality Conformance and Qualification of Microelectronic Packages and Interconnects*, Wiley, New York, 1994.
136. Harman, G., *Wire Bonding in Microelectronics: Materials, Processes, Reliability, and Yield*, McGraw-Hill, New York, 1997.
137. Intel Corp., *Packaging*, Intel Publications, Mt. Prospect, IL, 1997.
138. Fjelstad, J., *An Engineer's Guide to Flexible Circuit Technology*, Electrochemical Publications, 1997.
139. Barr, A., and W. J. Barr (eds.), *Smart Cards: Seizing Strategic Business Opportunities*, Irwin Professional Publishing, Chicago, 1997.
140. Elshabini-Riad, A., and F. D. Barlow, III, *Thin Film Technology Handbook*, McGraw-Hill, New York, 1997.
141. Lee, Y. C., and W. T. Chen (eds.), *Manufacturing Challenges in Electronic Packaging*, Chapman & Hall, London, 1998.
142. Konsowski, S. G., and A. R. Helland, *Electronic Packaging of High Speed Circuitry*, McGraw-Hill, New York, 1997.
143. DiGiacomo, G., *Reliability of Electronic Packages and Semiconductor Devices*, McGraw-Hill, New York, 1996.
144. Jawitz, M. W., *Printed Circuit Board Materials Handbook*, McGraw-Hill, New York, 1997.
145. Mahidhara, R. K., D. R. Frear, S. M. L. Sastry, K. L. Murty, P. K. Liaw, and W. L. Winterbottom (eds.), *Design and Reliability of Solders and Solder Interconnections*, TMS Minerals Metals Materials, Warrendale, PA, 1997.
146. Srihari, K., C. R. Emerson, S. Krishnan, Y. Hwang, C. H. Wu, and C. W. Yeh, *A "Design for Manufacturing" Environment for Surface Mount PCB Assembly: A Concise User's Guide*, Department of Mechanical and Industrial Engineering., Watson School, SUNY Binghamton, NY, 2012.
147. Ginsberg, G. L. (ed.), *Connectors and Interconnections Handbook*, Vol. 1: *Basic Technology*, Electronic Connector Study Group,. Fort Washington, PA, 1977, Lib. Congress No. 77-088086.
148. Ginsberg, G. L. (ed.), *Connectors and Interconnections Handbook*, Vol. 4: *Materials*, Electronic Connector Study Group, Fort Washington, PA, 1983.
149. Bardes, B. P., *Metals Handbook*, 9th ed., Vol. 1: *Properties and Selection: Irons and Steels*, American Society of Metals, New York, 1978.
150. Bardes, B. P. (ed.), *Metals Handbook*, 9th ed., Vol. 2: *Properties and Selection: Nonferrous Alloys and Pure Metals*, American Society of Metals, New York, 1979.
151. Bardes, B. P. (ed.), *Metals Handbook*, 9th ed., Vol. 3: *Properties and Selection: Stainless Steels, Tool Materials and Special-Purpose Metals*, American Society of Metals, New York, 1980.
152. Dallas, D. B. (ed.), *Tool and Manufacturing Engineers Handbook*, 3rd ed., McGraw-Hill, New York, 1976.
153. Gabriel, B. L., *SEM: A User's Manual for Materials Science*, American Society for Metals, New York, 1985.
154. Oechsner, H. (ed.), *Topics in Current Physics: Thin Film and Depth Profile Analysis*, Springer-Verlag, New York, 1984.
155. Sterling, D. J., Jr., *Technician's Guide to Fiber Optics*, Delmar Publishers, New York, 1987.
156. Senior, J. M., *Optical Fiber Communications: Principles and Practice*, Prentice-Hall, Englewood Cliffs, NJ, 1985.
157. Pecht, M. G., et al., *Electronic Packaging Materials and their Properties*, CRC Press, Boca Raton, FL, 1999.
158. McKeown, S. A., *Mechanical Analysis of Electronic Packaging Systems*, Marcel Dekker, New York, 1999.
159. Liu, J., *Conductive Adhesives for Electronics Packaging*, Electrochemical Publications, Isle of Man, UK, 1999.
160. Woishnis, W. A. (ed.), *Engineering Plastics and Composites*, 2nd ed., ASM International, Materials Park, OH, 1993.
161. Ross, R. J., C. Boit, and D. Staab (eds.), *Microelectronic Failure Analysis: Desk*

Reference, 4th ed., ASM International, Materials Park, OH, 1999.
162. Blackwell, G. R. (ed.), *The Electronic Packaging Handbook*, CRC Press, Boca Raton, FL, 1999.
163. Brown, W. (ed.), *Advanced Electronic Packaging: With Emphasis on Multichip Modules*, IEEE Press, Piscataway, NJ, 1999.
164. Avallone, E. A., and T. Baumeister, III (eds.), *Marks' Standard Handbook for Mechanical Engineers*, 10th ed., McGraw-Hill, New York, 1996.
165. Shewhart, W. A., *Statistical Method from the Viewpoint of Quality Control*, Dover Publications, New York, 1986.
166. Mickelson, A. R., N. R. Basavanhally, Y.-C. Lee (eds.), *Optoelectronic Packaging*, Wiley, New York, 1997.
167. Wagner, L. C. (ed.), *Failure Analysis of Integrated Circuits: Tools and Techniques*, Kluwer Academic Publishers, Boston, MA 1999.
168. Harper, C. A. (ed.), *Electronic Packaging and Interconnection Handbook*, 3rd ed., McGraw-Hill, New York, 2000.
169. McCluskey, F. P., R. Grzybowski, and T. Podlesak (eds.), *High Temperature Electronics*, CRC Press, Boca Raton, FL, 1997.
170. Lyshevski, S. E., *Nano- and Microelectromechanical Systems: Fundamentals of Nano- and Microengineering*, CRC Press, Boca Raton, FL, 2001.
171. Remsburg, R., *Thermal Design of Electronic Equipment*, CRC Press, Boca Raton, FL, 2001.
172. Harris, M. C., and G. E. Moore (eds.), *Prospering in a Global Economy: Linking Trade and Technology Policies*, National Academy Press, Washington, DC, 1992.
173. Whitaker, J. C., *Microelectronics*, CRC Press, Boca Raton, FL, 2000.
174. Madou, M., *Fundamentals of Microfabrication*, CRC Press, Boca Raton, FL, 1997.
175. Gibson, D. D., and R. W. Smilor (eds.), *Technology Transfer in Consortia and Strategic Alliances*, Rowman & Littlefield, Lanham, MD, 1992.
176. Brett, A., D. V. Gibson, and R. W. Smilor (eds.), *University Spin-Off Companies*, Rowman & Littlefield, Lanham, MD, 1991.
177. Pecht, M., *Integrated Circuit, Hybrid, and Multichip Module Package Design Guidelines: A Focus on Reliability*, Wiley, New York, 1994.
178. Chigrinov, V. G., *Liquid Crystal Devices: Physics and Applications*, Artech House, Boston, MA 1999.
179. Mohamed Gad-el-Hak (ed.), *The MEMS Handbook*, CRC Press, Boca Raton, FL, 2002.
180. Pecht, M. (ed.), *Product Reliability, Maintainability, and Supportability Handbook*, CRC Press, Boca Raton, FL, 1995.
181. Azar, K. (ed.), *Thermal Measurements in Electronic Cooling*, CRC Press, Boca Raton, FL, 1997.